Springer Proceedings in Complexity

Springer Proceedings in Complexity publishes proceedings from scholarly meetings on all topics relating to the interdisciplinary studies of complex systems science. Springer welcomes book ideas from authors. The series is indexed in Scopus.

Proposals must include the following:

- name, place and date of the scientific meeting
- a link to the committees (local organization, international advisors etc.)
- scientific description of the meeting
- list of invited/plenary speakers
- an estimate of the planned proceedings book parameters (number of pages/articles, requested number of bulk copies, submission deadline)

Submit your proposals to: Hisako.Niko@springer.com

Anna Visvizi · Orlando Troisi ·
Enric Serradell-López
Editors

Research and Innovation Forum 2025 Volume 1

Competitiveness, Growth and Well-Being Through Technology, Innovation and Education

Springer

Editors
Anna Visvizi
SGH Warsaw School of Economics
and Effat University
Warsaw, Poland

Orlando Troisi
University of Salerno
Fisciano, Italy

Enric Serradell-López
Open University of Catalonia
Barcelona, Spain

ISSN 2213-8684 ISSN 2213-8692 (electronic)
Springer Proceedings in Complexity
ISBN 978-3-032-09065-2 ISBN 978-3-032-09066-9 (eBook)
https://doi.org/10.1007/978-3-032-09066-9

This Springer imprint is published by the registered company Springer Nature Switzerland AG
The registered company address is: Gewerbestrasse 11, 6330 Cham, Switzerland

Organization

Research & Innovation Forum 2025: The Team

The Chairs

Anna Visvizi, Ph.D. (Dr. hab.), Professor, SGH Warsaw School of Economics, Poland and Effat University, Jeddah, KSA
Orlando Troisi, Ph.D., Professor, University of Salerno, Italy
Enric Serradell-López, Ph.D., Professor, Universitat Oberta de Catalunya, Spain

The Steering Committee

Xavier Baraza-Sanchez, Ph.D., Professor, Dean of the Faculty of Business and Economics, Universitat Oberta de Catalunya, Spain
Vincenzo Corvello, Ph.D., Professor, University of Calabria, Italy
Mara Grimaldi, Ph.D., Research Fellow, University of Salerno, Fisciano, Italy
Giuseppe Ioppolo, Ph.D., Professor, University of Messina, Italy
Jolanta Klemens, M.D., PSYCHOMedical, Poland
Higinio Mora, Ph.D., Professor, University of Alicante, Spain
Maria Jesús Martínez-Argüelles, Ph.D., Vice-Rector of Governance and Academic Policy, Universitat Oberta de Catalunya, Spain
Francesco Polese, Ph.D., Professor, Director of SIMAS [Centre for Innovation Systems and Health Management], University of Salerno, Fisciano, Italy

The Academic Committee

Andrew Buks, MBA Program Director, School of Business at Black Hills State University in Spearfish, South Dakota, USA
Antonio Botti, Ph.D., Professor, University of Salerno, Fisciano, Italy
Igor Calzada, Ph.D., Professor, University of the Basque Country/Basque Science Foundation, Spain and Cardiff University/WISERD, UK
Nicola Cobelli, Ph.D., Associate Professor, University of Verona, Italy
Linda Daniela, Ph.D., Professor, Dean of the Faculty of Education, Psychology and Art, University of Latvia
Amal Elasri-Ejjaberi, Ph.D., Associate Dean of the Faculty of Business and Economics, Universitat Oberta de Catalunya, Spain
Raquel Ferreras-Garcia, Ph.D., Associate Professor, Universitat Oberta de Catalunya
Pablo Lara, Ph.D., Professor, Universitat Oberta de Catalunya, Spain
Roberto de Luca, Ph.D., Professor, University of Salerno, Fisciano, Italy
Alberto Ferraris, Ph.D., Associate Professor, University of Torino, Italy
Akinul I. Jony, Ph.D., Associate Professor, Universitat Oberta de Catalunya, Spain
Jari Jussila, Ph.D., Director of HAMK Design Factory, Häme University of Applied Sciences, Finland
Krzysztof Kozłowski, Ph.D., Professor, Vice-Rector for Students and Teaching, SGH Warsaw School of Economics, Warsaw, Poland
Mady Mohamed, Ph.D., Effat University, Jeddah, KSA
Anna Moni, Instructional Designer, The American College of Greece, Athens, Greece
Cintia Pla-Garcia, Ph.D., Associate Professor, Academic Director of Permanent Marketing Training, Universitat Oberta de Catalunya, Spain
Placido Pinheiro, Ph.D., Professor, University of Fortaleza, Brazil
Jose Carlos Ramos, Ph.D., Associate Professor, Universitat Oberta de Catalunya, Spain
Lucie Rohlikova, Ph.D., Professor, University of Western Bohemia, Plzen, Czechia
Annika Steiber, Ph.D., Stanford University, California, USA
Katarzyna Szopik-Depczyńska, Ph.D., Professor, University of Szczecin, Poland
Lloyd George Waller, Ph.D., Professor, University of the West Indies (Mona Campus), Kingston, Jamaica; Chairman of the Broadcasting Commission, Jamaica

Preface

This volume is the outcome of the vibrant intellectual exchange that took place during the 7th Research and Innovation Forum (Rii Forum 2025), held in a hybrid format on April 24–25, 2025, in Barcelona, Spain. Convened under the overarching theme "Competitiveness, Growth and Well-Being Through Technology, Innovation and Education", the Rii Forum reaffirmed its standing as a high-impact platform fostering interdisciplinary collaboration among academia, industry, and civil society. Hosted in partnership with the Universitat Oberta de Catalunya (UOC), a global leader in digital education and applied research, also this edition of Rii Forum brought together scholars and practitioners committed to advancing rigorous, evidence-based dialogue on the most pressing technological, economic, and societal challenges of our time. The opening remarks of Prof. Àngels Fitó Bertran, President of the Universitat Oberta de Catalunya (UOC), confirmed the value of Rii Forum and the role and impact research excellence and collaboration have in advancing science, education, and innovation.

The Rii Forum 2025 Chairs

Rii Forum 2025 was co-chaired by Prof. Anna Visvizi (SGH Warsaw School of Economics, Poland), Prof. Orlando Troisi (University of Salerno, Italy), and Prof. Enric Serradell-López (Universitat Oberta de Catalunya, Spain), all well-respected researchers, experienced academics, and leaders. The program of the conference was organized around six core thematic tracks, alongside a dedicated Ph.D. workshop:

1. The individual, the state, and technology—exploring cities, urbanism, democracy, and resilience
2. Innovation, quality, and technology management exploring their local, regional, and global contexts
3. Education in the GenAI era addressing pedagogical, content-related, and managerial challenges

4. Design thinking and agile methodologies highlighting foresight, future studies, and strategic planning
5. Regulatory frameworks and policy-making dwelling on issues pertaining to competitiveness, growth, and well-being
6. Ph.D. Track focused on research impact, publication strategy, and supervisory collaboration, served also as an opportunity to empower early career researchers.

Key Events and Highlights

Keynote Address

The Rii Forum 2025 opened with a spot-on keynote by Professor Johan Magnusson, Director of the Swedish Center for Digital Innovation (www.scdi.se) at the University of Gothenburg. Professor Magnusson's research covers issues pertaining to how governance needs to be designed to facilitate transformation. In his roles as researcher, lecturer, educator, and thought-leader, Prof. Magnusson is highly visible in the national, Swedish, and international debates concerning digital and green transformation. He is ranked as one of the most influential individuals in tech. His keynote address offered strategic insights into digital governance, public sector transformation, and the interplay between regulatory frameworks and innovation ecosystems.

Round Table Debate

The high-profile Round Table Debate, titled "*The Great (?) Leap (?): Tech, Talent, and Transformation in the 21st Century*", was moderated by Prof. Anna Visvizi and featured:

- Prof. Johan Magnusson (University of Gothenburg, Sweden)
- Prof. Enric Serradell-López (UOC, Spain)
- Prof. Igor Calzada (University of the Basque Country and Cardiff University)
- Prof. Giuseppe Ioppolo (University of Messina, Italy)

The debate addressed the strategic risks and opportunities posed by accelerated technological change, talent mobility, and geopolitical fragmentation.

Online Panel Debate

The Rii Forum online panel debate "*From Community Engagement to Labor Market Participation: Rethinking Talent in the Web 3.0 Era*", moderated by Prof. Radosław Malik (SGH Warsaw School of Economics), explored the issue of how digital inclusion, migration, and mobility are reshaping the very notion of "talent" today. Regional case studies from Greece, Brazil, and Lithuania provided practical, grounded perspectives. The speakers included:

- Anis Barnat (CAN Network and El Sistema Greece)
- Prof. Regina Negri-Pagani (Federal University of Technology Paraná, Brazil)
- Prof. Mangirdas Morkūnas (Vilnius University, Lithuania).

The Rii Forum Industry-Academia Dialogue

A highlight of the conference the Rii Forum Industry-Academia Dialogue explored how cross-sector collaboration can catalyze innovation and social impact. Moderated by Prof. Lloyd Waller (University of the West Indies, Jamaica), the session brought together voices from business, design, entrepreneurship, and academia:

- Petros Georgiakakis, MBA (Senior Clinical Operations Manager, ICON Pharma Development Solutions)
- Dr. Francesc Font-Cot (Co-Founder and Co-CEO, Incapto Coffee)
- Dr. Alessandro Manetti (Vice President Europe, LCI Education; Barcelona Design Week Curator)
- Prof. Katarzyna Szopik-Depczyńska (University of Szczecin, Poland)

The panel debate focused on the question of how academia can meaningfully contribute to industry through talent development, strategic foresight, knowledge co-creation, and technology transfer, emphasizing the value of long-term, sustainable partnerships.

Academic Diplomacy

A special moment was marked by the remarks of Her Excellency Ilona Kałdońska, Consul General of the Republic of Poland in Barcelona, who emphasized the role of research and innovation in shaping EU-wide policy, particularly in the context of Poland's Presidency of the Council of the European Union (January–June 2025). The Rii Forum 2025 delegates had the opportunity to meet also His Excellency Łukasz Knurowski, Consul of the Republic of Poland in Barcelona. During the networking meeting that followed the Rii Forum Industry-Academia Dialogue, the scholars, academics, administrators and industry representatives had the opportunity to explore the prospects of collaboration. In this way, the Rii Forum 2025 aligned itself with goals and objectives of the Polish Presidency of the Council of the European Union.

Ph.D. Workshop

Convened by Dr. Mara Grimaldi (University of Salerno), the Ph.D. Workshop focused on the thesis-by-compendium model, offering doctoral candidates actionable strategies for writing, publishing, engaging with supervisors, and achieving research impact.

Social Impact: The Power of Music, Inclusion, and Education

At the heart of the Rii Forum lies a deep commitment to social impact. This year, the Rii Forum proudly reaffirmed its collaboration with El Sistema Greece, a pioneering initiative inspired by El Sistema Venezuela, by Anis Barnat. Since 2016, El Sistema Greece, led by Anis Barnat, experienced manager, leader, and art manager, has provided music education to over 3000 children and youth from more than 40 countries, including refugee and migrant communities. Through structured instrumental, choral, and orchestral training, El Sistema Greece promotes self-esteem, teamwork, emotional resilience, and cultural inclusion. Programs culminate in public performances, such as the Annual Christmas Gala Concert, bringing together families and communities through the unifying power of music. El Sistema Greece exemplifies how education, creativity, and compassion intersect to create tangible social change. The Rii Forum community is proud to support its mission.

Acknowledgments and Gratitude

We extend our deepest thanks to all authors and contributors to this volume. Your work continues to advance the conversation at the intersection of technology, innovation, education and their social impact, policy-making considerations including. We thank our peer reviewers for their diligence and commitment to academic excellence.

Special appreciation goes to the Rii Forum 2025 organizing team, including:

- Jolanta Klemens, M.D., for her unconditional support, and for enabling and facilitating collaboration with Rii Forum strategic partners.
- Anna Moni, Krzysztof Kozłowski, and Bilal Afzal, for their leadership in managing the on-site component of the conference. A simple “thank you” is not enough to express the value of your input to Rii Forum 2025.

- The UOC local team, whose hospitality and professionalism made the event in Barcelona a truly memorable experience.

We would also like to thank Anna Guerrero i Olivella (UOC) who generously offered her time and expertise to enable the Rii Forum team to run the conference.

Strategic Partners

We are grateful to our strategic partners, who continue to support the Rii Forum's values of research excellence, innovation, and social impact:

- Universitat Oberta de Catalunya (UOC)—Host institution
- EkaMedica (Poland)—advocates for natural health and wellness
- Polskie Piekarnie (Poland)—traditional and innovative bakery products
- PSYCHOMedical (Poland)—holistic healthcare and mental wellness
- INTER-TECH (Poland)—advanced solutions for industrial automation and machine relocation
- Polmotors (Poland)—leader in lightweight, eco-friendly automotive components for global OEMs

We also express our sincere gratitude to Springer Nature, our publishing partner, for their trust, professionalism, and continuous support.

The Rii Forum Community

The Research and Innovation Forum is more than just a conference. Over the years it has evolved into a community of researchers and thought leaders committed to research excellence and ethical integrity; interdisciplinary, thought-provoking dialogue; constructive collaboration across academia, industry, and policy-making; supporting early-stage researchers; promoting research that is rigorous, relevant, and socially responsible. By participating in the Rii Forum, scholars gain exposure to critical debates, receive constructive feedback, and connect with a global network of like-minded professionals. This book is a testament to that ethos, and an invitation to carry the conversation forward.

We look forward to reading your work, collaborating throughout the year, and welcoming you to the 8th edition of the Research and Innovation Forum, scheduled to take place in Athens in Spring 2026.

Warsaw, Poland — Anna Visvizi
Fisciano, Italy — Orlando Troisi
Barcelona, Spain — Enric Serradell-López

Contents

Contributors

Bilal Afzal Department of Business Management, Universitat Rovira i Virgili, Reus, Tarragona, Spain

Cristina Alenda Department of Pathology, Dr. Balmis General University Hospital, Alicante Institute for Health and Biomedical Research (ISABIAL), Alicante, Spain

Asmaa Hussain Alganawi King Abdulaziz University, Al Jamiaa District, Jeddah, Saudi Arabia

Rawan Almesbahi Effat College of Architecture and Design, Effat University, Jeddah, Saudi Arabia

Adriana Apuzzo Department of Political Science and Communication, University of Salerno, Fisciano, Italy

Hossny Azizalrahman King Abdulaziz University, Al Jamiaa District, Jeddah, Saudi Arabia

Emese Belenyesi Ludovika University of Public Service, Budapest, Hungary

Andrew G. Buks Black Hills State University, Spearfish, SD, USA

Grazia Calabrò Department of Economics, University of Messina, Messina, Italy

Igor Calzada University of the Basque Country, Public Policy & Economic History Department, Donostia-San Sebastian, Spain;
Basque Foundation for Science, Ikerbasque, Bilbao, Spain;
Cardiff University, WISERD, Cardiff, UK;
Decentralization Research Centre, Toronto, Canada;
US-UK Fulbright Commission, London, UK

Giuseppe Caristi Department of Economics, University of Messina, Messina, Italy

Cristina Ciliberto Department of Economics, University of Messina, Messina, Italy

Pablo de-Gracia-Soriano Department of Sociology I, University of Alicante, San Vicente del Raspeig, Spain

Carlotta D'Alessandro Department of Economics, University of Messina, Messina, Italy

João Batista Furlan Duarte University of Fortaleza (UNIFOR), Graduate Program in Applied Informatics, Fortaleza, CE, Brazil

Mert Mecit Erim Department of Economics, University of Messina, Messina, Italy

Aseel Fakieh Effat College of Architecture and Design, Effat University, Jeddah, Saudi Arabia

David Bernabeu Ferrer University Institute for Computer Research, University of Alicante, San Vicente del Raspeig, Spain

Raquel Ferreras-Garcia Universitat Oberta de Catalunya, Barcelona, Spain

Carmen García-Barceló University Institute for Computer Research, University of Alicante, San Vicente del Raspeig, Spain;
Department of Computer Science Technology and Computation, University of Alicante, Alicante, Spain

Ana Martínez García Department of Computer Science Technology and Computation, University of Alicante, Alicante, Spain

Swapan Ghosh California State University, Northridge, CA, USA

David Gil University Institute for Computer Research, University of Alicante, San Vicente del Raspeig, Spain;
Department of Computer Science Technology and Computation, University of Alicante, Alicante, Spain

Coral Grainger Manchester Metropolitan University, Manchester, UK

Giovanni Guzmán Centro de Investigacón en Computación, Instituto Politécnico Nacional, CDMX, Mexico

Pablo Herrero Gómez University Institute for Computer Research, University of Alicante, San Vicente del Raspeig, Spain

Ana Beatriz Hernández-Lara Department of Business Management, Universitat Rovira i Virgili, Reus, Tarragona, Spain

Yaimara Granados Hondares KULeuven, Leuven, Belgium

Mario Pezzillo Iacono Department of Economics, University of Campania Luigi Vanvitelli, Capua, CE, Italy

Asmaa Ibrahim Urban Design Lab, College of Architecture and Design, Effat University, Jeddah, Saudi Arabia

Giuseppe Ioppolo Department of Economics, University of Messina, Messina, Italy

Sebastien Ivanovici KULeuven, Leuven, Belgium

Levy Galas Jacob University of Fortaleza (UNIFOR), Graduate Program in Applied Informatics, Fortaleza, CE, Brazil

Francisco Javier Jiménez-Loaisa Department of Sociology I, University of Alicante, San Vicente del Raspeig, Spain

Xiaoni Li Department of Business Management, Universitat Rovira i Virgili, Reus, Tarragona, Spain

Claudia Carratalá Lillo Department of Computer Science Technology and Computation, University of Alicante, Alicante, Spain

Francesca Loia Department of Economics, University of Campania Luigi Vanvitelli, Capua, CE, Italy

Talha Mahmood KULeuven, Leuven, Belgium

Hajar Makry Polydisciplinary Faculty Sidi Bennour, Chouaib Doukkali University, El Jadida, Morocco

Radosław Malik SGH Warsaw School of Economics, Warsaw, Poland

Cecilia Maltempo Department of Economics, University of Campania Luigi Vanvitelli, Capua, CE, Italy

Kateryna Mamchur Department of Computer Languages and Systems, University of Alicante, San Vicente del Raspeig, Spain

Stefania Manetti Department for the Promotion of Human Sciences and Quality of Life, San Raffaele University of Rome, Rome, Italy

Rosario Marrapodi University of Naples Parthenope, Naples, Italy

Marcello Martinez Department of Economics, University of Campania Luigi Vanvitelli, Capua, CE, Italy

Stefania Mele Department of Economics, University of Campania Luigi Vanvitelli, Capua, CE, Italy

Higinio Mora University Institute for Computer Research, University of Alicante, San Vicente del Raspeig, Spain

Samanta Ortuño-Miquel Alicante Institute for Health and Biomedical Research (ISABIAL), Alicante, Spain;
Department of Computer Science Technology and Computation, University of Alicante, Alicante, Spain

Mandy Parkinson University of Salford, Salford, UK

Vincenzo Piluso University of Salerno, Fisciano, Italy

Plácido Rogério Pinheiro University of Fortaleza (UNIFOR), Graduate Program in Applied Informatics, Fortaleza, CE, Brazil

Cintia Pla-Garcia Universitat Oberta de Catalunya, Barcelona, Spain

Rolando Quintero Centro de Investigacón en Computación, Instituto Politécnico Nacional, CDMX, Mexico

Emanuel Saldana-Perez Centro de Investigacón en Computación, Instituto Politécnico Nacional, CDMX, Mexico

Jordi Sales-Zaguirre IQS, Universitat Ramon Llull, Barcelona, Spain

Mohammad Salman Faculty of Architecture, Planning & Design, Integral University, Lucknow, India

Peter Sasvari Ludovika University of Public Service, Budapest, Hungary; University of Miskolc, Miskolc, Hungary

Giustina Secundo Department of Engineering, LUM Giuseppe Degennaro University, Casamassima, Italy

Enric Serradell-López Universitat Oberta de Catalunya, Barcelona, Spain

Khodor Shatila Universitat Rovira i Virgili, Reus (Tarragona), Spain

Monique Snoeck KULeuven, Leuven, Belgium

Claudia Spilotro Department of Management, Finance and Technology, LUM Giuseppe Degennaro University, Casamassima, Italy

Annika Steiber Management Insights, Menlo Park, CA, USA

Katarzyna Szopik-Depczyńska Department of Corporate Management, Institute of Management, University of Szczecin, Szczecin, Poland

Mario Testa Department of Political Science and Communication, University of Salerno, Fisciano, Italy

Orlando Troisi Department of Political Science and Communication, University of Salerno, Fisciano, Italy

Anna Visvizi SGH Warsaw School of Economics, Warsaw, Poland; Effat University, Jeddah, Saudi Arabia

Felice Vitulano QuestIT, Siena, Italy

Maddalena Della Volpe University of Salerno, Fisciano, Italy

Yves Wautelet KULeuven, Leuven, Belgium

Kate Webb The KWP Ltd, Compton, UK

Dominika Weglarz Universitat Oberta de Catalunya, Barcelona, Spain

Roman Wosiek SGH Warsaw School of Economics, Warsaw, Poland

Ana Isabel Jiménez Zarco Universitat Oberta de Catalunya, Barcelona, Spain

Innovation in Context of the Industry 5.0 Paradigm

Industry 5.0's Innovation Dynamics: Modelling and Explaining its Potential

Bilal Afzal, Anna Visvizi, Ana Beatriz Hernández-Lara, and Xiaoni Li

Abstract While Industry 4.0 has introduced transformative advances in automation and efficiency, it often falls short in addressing the broader goals of sustainability, resilience, and human centricity. To address this gap, this paper develops a unified conceptual framework to explain the innovation dynamics of Industry 5.0 by integrating insights from research on platform ecosystems, the interplay between sustainability and resilience, the Sextuple Helix Model, and the emerging paradigm of Innovation 5.0. The proposed framework positions Industry 5.0 as a redefinition of innovation ecosystems-one that shifts focus from automation-led efficiency to inclusive, sustainable, and human-centric value creation. Core components include participatory AI, circular economy principles, and socially embedded ecosystems, which together can generate synergistic effects and enhance innovation potential. The framework also identifies key technological, organizational, and environmental enablers that support this transformation. Central to the model are AI-human collaboration, ecosystem integration, and dynamic adaptability, which emerge as critical drivers of innovation in the I50 context.

Keywords Innovation 5.0 · Industry 5.0 · Innovation performance · Enterprise resilience · Innovation emergence · SDG-9

B. Afzal (✉) · A. B. Hernández-Lara
Department of Business Management, Universitat Rovira i Virgili, Reus, Tarragona, Spain
e-mail: bilal.afzal@outlook.com

A. Visvizi
SGH Warsaw School of Economics, Warsaw, Poland

X. Li
Effat University, Jeddah, Saudi Arabia

A. Visvizi et al. (eds.), *Research and Innovation Forum 2025 Volume 1*,
Springer Proceedings in Complexity, https://doi.org/10.1007/978-3-032-09066-9_1

1 Introduction

Although Industry 4.0 (I40), primarily driven by automation, Internet of Things (IoT), and Artificial Intelligence (AI), brought significant technological advances, it has also been criticized for prioritizing efficiency at the expense of human agency, sustainability, and resilience [10, 14]. In response, Industry 5.0 (I50) addresses these gaps by emphasizing human-centricity, resilience, and sustainability through collaborative innovation and advanced technologies like AI and robots [5, 8, 24] (Visvizi et al., 2025), aligning with growing calls for inclusive, value-driven industrial practices [8]. Whereas I40's linear value chains and stakeholder inclusion (though limited) do promote sustainability, a great number of companies continues to struggle with the integration of technology, human needs, and ecological priorities [14, 16]. However, even if I50 shifts the focus of the industry paradigm toward human-AI collaboration, open innovation, and circularity, also here, the question of its effective implementation in dynamic, turbulent environments, which constrain the realization of sustainable innovation and enterprise resilience [13], remains underdiscussed.

Existing research on I40I40 tends to emphasize productivity and technological efficiency, often overlooking critical dimensions, such as human-centricity, sustainability, and adaptability [7, 10, 14]. Interestingly, although literature on I50 explores questions related to human-centric AI and sustainability-oriented practices, it explores these two factors in isolation one from the other. This divide limits our understanding of how organizations can effectively align technological advancement with human, ecological, and strategic resilience goals in volatile environments. As firms transition toward more complex innovation ecosystems, integrative frameworks are needed to capture the co-evolution of capabilities, ecosystem roles, and platform-based collaboration in shaping this new innovation ecosystem. Given the vital connection that exists among these variables, it is imperative to explore ways in which these variables may be brought together in one framework. In particular, the roles of dynamic capabilities, innovation ecosystems, and multi-stakeholder governance, such as through the sextuple helix, are under-theorized and empirically unvalidated [13].

Accordingly, the objective of this paper is to propose a unified I50 framework that combines human-centric, resilient, and sustainable innovation. It draws insights from literature on platform ecosystems, open innovation, dynamic capabilities, and the sextuple helix. The following research questions are thus explored:

RQ1: How does I50 redefine innovation compared to I40, particularly regarding human centricity, sustainability, and resilience?

RQ2. How to conceptualize the mechanisms and processes that boost innovation emergence under I50 paradigm?

To address these questions, the remainder of the paper is structured as follows. In Sect. 2, the methodological approach underpinning the discussion in the paper is outlined. In Sect. 3, the relationship between the precepts of I50 and Innovation is conceptualized. In Sect. 4, the conceptual framework is operationalized, i.e. applied in the context of I50. Discussion and conclusions follow.

2 Methods and Materials

This conceptual article offers a foundational contribution to understanding enterprise resilience and innovation performance in I50 contexts. It draws upon and extends insights from four prior studies by the authors, spanning the bibliometric mapping of platform ecosystems [1], a systematic review of sustainability-resilience-human-centeredness intersections [3], a conceptual study on platform business models and Innovation 5.0 [2] and an application of the sextuple helix model to emerging innovation systems [4]. These studies were purposefully designed and conducted by the authors as part of a coherent research trajectory and were systematically reviewed and synthesized to build the conceptual foundation of this framework. These studies informed the theoretical base through iterative synthesis, drawing on deductive reasoning and thematic convergence across ecosystemic, organizational, and behavioral dimensions.

The framework was developed through a theory-driven integration of themes and constructs repeatedly emphasized in these prior studies. Core dimensions were identified through conceptual triangulation across findings, aligning thematic codes from prior systematic reviews with theoretical propositions from earlier conceptual analyses. This approach ensured internal coherence and theoretical saturation without departing from the established literature base.

The article draws upon foundational theoretical perspectives that address innovation emergence across ecosystemic, organizational, and behavioral levels. The Sextuple Helix Model [3, 4, 6] offers a structural basis for understanding innovation ecosystems, expanded to include AI, digital platforms, and civil society. Open Innovation Theory [28] provides the foundation for cross-boundary collaboration and knowledge flows, while the Technology-Organization-Environment (TOE) framework [9] adds insights into organizational adoption drivers under technological and market turbulence. Finally, the Unified Theory of Acceptance and Use of Technology (UTAUT) [27] introduces a behavioral layer focused on perceived utility, social influence, and technology adoption.

By integrating these perspectives, the study constructs a cohesive framework for analyzing how the precepts of the I50paradigm may be meaningfully transposed from theory to (business) practice, through human-centric, adaptive, and sustainable processes. The latter unfold in a dynamic, platform-enabled industrial/business contexts. This framework lays the conceptual groundwork for subsequent empirical investigations.

3 Modelling the I50-Innovation Nexus: Toward the Conceptual Framework

The transition from the efficiency-driven paradigm of I40 to the human-centric, resilient, and sustainable vision of I50 opens new avenues for rethinking innovation in turbulent and policy-driven environments. At the heart of this transformation lies Innovation 5.0, a framework that emphasizes human–machine collaboration, societal value creation, and adaptive innovation ecosystems. It emerges not only as a technological response but also as a systemic necessity shaped by ecological, social, and economic imperatives. The dynamics of Innovation 5.0 are driven by a network of stakeholders, including firms, governments, civil society, and AI platforms, whose interactions are mediated through mechanisms of co-creation, digital collaboration, and open innovation. These innovation networks increasingly rely on data-driven ecosystems, where the activation of a data culture, diverse digital skills, and user participation serve as key enablers of sustainability and innovation [23].

To develop a comprehensive framework for Innovation 5.0 in volatile, policy-driven environments, we integrate the Sextuple Helix Model, the Technology–Organization–Environment (TOE) framework, and the Unified Theory of Acceptance and Use of Technology (UTAUT), based on both theoretical complementarity and practical relevance. These frameworks were selected deliberately for their layered and mutually reinforcing perspectives. The Sextuple Helix provides a macro-level systems view, TOE addresses meso-level organizational responses to innovation drivers, and UTAUT contributes a micro-level behavioral lens. This triangulated approach aligns with the multi-actor, multi-level, and multi-directional character of Innovation 5.0, allowing us to model not only ecosystem dynamics but also the internal and behavioral processes through which innovation is realized under external pressures and policy shifts.

The Sextuple Helix Model was chosen for its systemic capacity to represent innovation ecosystems that include not only traditional actors-government, industry, and academia-but also civil society, the environment, and digital media/AI [6]. This is particularly relevant for Innovation 5.0, which demands a broader, value-driven governance structure. The Technology-Organization-Environment (TOE) framework was included to capture how firms' internal capabilities and external pressures (such as policy, market dynamics, and technological shifts) influence their ability to adapt and innovate in the face of uncertainty [9]. TOE anchors the macro-systemic view into the meso level strategic organizational responses. Meanwhile, the Unified Theory of Acceptance and Use of Technology (UTAUT) introduces a crucial micro-level behavioral lens, highlighting how individual perceptions of usefulness, social norms, and institutional support affect technology adoption [27]. This behavioral layer ensures that the framework accounts not only for structural and institutional factors but also for the human-centric focus at the heart of I50.

By combining these three frameworks, we bridge macro (ecosystem), meso (organizational), and micro (user behavior) dimensions of innovation. This integrated synthesis presents Innovation 5.0 as a dynamic input–throughput–output process in which environmental turbulence and stakeholder expectations (input) shape organizational responses and inter-actor collaboration (throughput), ultimately generating adaptive, human-centric, and sustainable innovation outcomes (output) [12, 22]. Beyond these structural and behavioral aspects, data-driven business models (DDBMs) within innovative ecosystems stress the strategic centrality of data in fostering innovation, highlighting the enabling dimensions of such models in different contexts [22]. Each framework thereby adds a distinct but complementary layer to the conceptual model, aligning with the adaptive, human-centric, and sustainability-driven nature of Innovation 5.0.

To extend this foundation and reduce the risk of analytical narrowness, the framework is further enriched with Dynamic Capabilities Theory, Resilience Theory, Complexity Theory, and Socio-Technical Systems (STS) Theory. These perspectives emphasize firms' ability to sense, seize, and reconfigure resources in uncertain environments [20], systemic robustness through feedback loops and non-linear dynamics [11, 26], and the harmonization of human, organizational, and technological subsystems [21]. Together, these theories complement the Sextuple Helix, TOE, and UTAUT by deepening the account of adaptability, resilience, and socio-technical balance within evolving innovation ecosystems. The multi-dimensional nature of innovation in this model reflects diverse shades, including technological, processual, data-driven, social, and eco-sustainable forms of innovation, that can be observed within innovative and interconnected industrial systems [25].

4 Operationalizing the I50-Innovation Framework

4.1 Innovation in I50 (RQ1)

I50 reconceptualizes the innovation ecosystem by shifting from I40I40's automation-centric model towards a socio-technical ecosystem where humans and AI collaborate to deliver sustainability, resilience, and personalized value creation. While I40 focused on optimizing efficiency through technologies like IoT, AI, and automation, I50 positions workers as co-creators with cobots and intelligent systems, freeing humans from rote tasks and focusing their effort on creative problem-solving and decision-making (Table 1).

Cobots manage repetitive work, enabling humans to apply contextual understanding and creativity [17]. Participatory AI design tailors algorithms to worker preferences, enhancing innovation through real-time data analytics for customized production. Sustainability becomes integral: factories adopt renewable energy, closed-loop processes, and IoT-driven resource optimization [15, 18].

Furthermore, I50 embraces open innovation and dynamic capabilities to build resilience against market disruptions. Cross-sector collaboration among industry, academia, suppliers, and civil society cultivates agile, transparent supply chains [24].

Table 1 Comparison of I40 and 5.0

Aspect	I40	I50
Focus	Automation, efficiency, and digitization	Human–machine collaboration and sustainability
Human role	Humans oversee automated systems	Humans co-create with AI/cobots
Sustainability	Energy efficiency through technology	Circular economy, eco-design, and waste reduction
Production Model	Mass customization	Hyper-personalization and on-demand manufacturing
Resilience	Optimized supply chains	Adaptive ecosystems and dynamic capabilities

Source Authors

AI-driven predictive analytics enhance resilience, equipping companies to foresee market changes and operational risks.

Technological advances such as augmented reality for training, digital twins for rapid prototyping, and AI-driven process orchestration augment human skills rather than replace them. This shift marks a transition from purely technical innovation to holistic socio-technical systems that balance productivity with ethical and ecological imperatives, unlocking new pathways for sustainable growth (Fig. 1).

4.2 A Unified Framework (RQ2)

The literature synthesis reveals a diverse yet converging body of work that frames Innovation 5.0 as a networked, multi-actor ecosystem. We integrate six theoretical lenses into an interdisciplinary framework:

1. Stakeholder Integration (Sextuple Helix Model): Maps the multi-actor architecture of Innovation 5.0, emphasizing collaborative governance, civil society inclusion, and the convergence of human and artificial intelligence. It underpins human centricity by enabling participatory innovation models.
2. Knowledge Flow and Collaboration (Open Innovation Theory): Describes the flow of knowledge across boundaries, both within and beyond firms, facilitating collaborative problem-solving and technological co-development.
3. Contextual Fit (TOE Framework): Identifies technological capabilities (AI, IoT), organizational readiness (absorptive capacity, leadership), and environmental turbulence (policy shifts, market volatility) as drivers or constraints of innovation adoption.
4. Behavioral Adoption (UTAUT Model): Examines perceptions of usefulness, ease of use, and social influence to explain micro-level acceptance of human-AI collaborative technologies.

Fig. 1 I50

5. Adaptive Capacity (Dynamic Capabilities & Resilience): Focuses on sensing, seizing, and reconfiguring assets, framing resilience as forward-looking transformation rather than mere recovery.
6. Emergent System Dynamics (Complexity and STS): Interprets Innovation 5.0 ecosystems as non-linear and emergent, where feedback loops, decentralized control, and path dependency play central roles.

These layers interact dynamically: stakeholder integration influences knowledge flows and collaboration; organizational and technological readiness mediates technological adoption; behavioral dimension shapes how users interact with emergent technologies, feeding back into the broader socio-technical system (Fig. 2).

By integrating these perspectives into a cohesive diagnostic and design tool, the framework guides both empirical measurement (e.g., maturity assessments, performance metrics) and strategic action, helping firms and policymakers cultivate Innovation 5.0 ecosystems that thrive amid complexity.

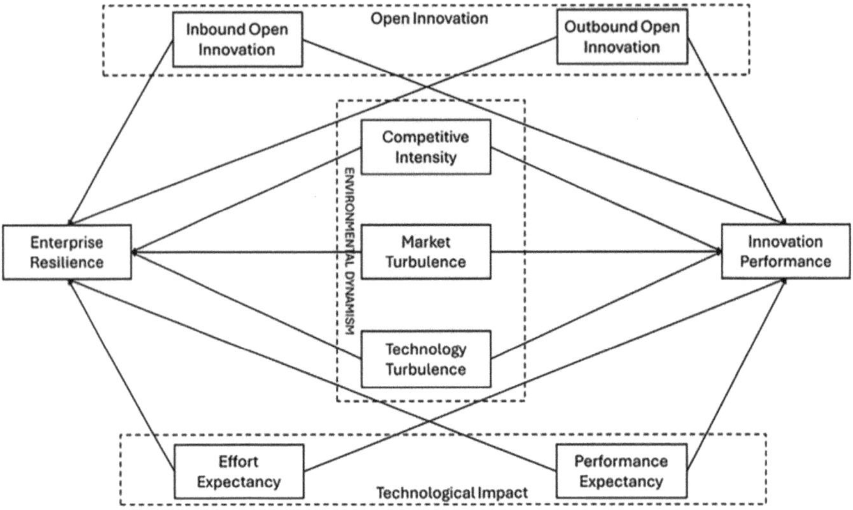

Fig. 2 I50 Innovation emergence framework: from paradigm to ecosystem. *Source* Authors

5 Discussion and Conclusions

The unified framework developed in this paper underscores that Innovation 5.0 is not merely an extension of I40 technologies but a fundamental reorientation toward socio-technical co-creation. This has three key implications:

1. Strategic Alignment: Firms must realign innovation strategies to integrate human creativity and ecological considerations alongside technological investments. This entails cultivating leadership that champions participatory design and environmental stewardship.
2. Capability Development: Beyond deploying IoT and AI, organizations need to develop dynamic capabilities, such as sensing weak signals, rapidly reconfiguring resources, and fostering cross-functional teams, to navigate uncertainty and leverage emergent opportunities.
3. Policy Enablement: Policymakers should design regulatory sandboxes and incentive mechanisms that encourage cross-sector partnerships, ethical AI development, and circular-economy practices, thereby accelerating responsible I50 transitions.

For managers responsible for promoting innovation in their organizations, the framework provides a structured lens to diagnose organizational readiness, including mapping stakeholder networks, assessing behavioral adoption barriers, and evaluating resilience potential. Scholars can operationalize the framework in empirical studies by developing survey instruments aligned with each theoretical layer and evaluating their interactions in diverse industrial contexts.

This paper lays the conceptual groundwork for understanding Innovation 5.0 within the evolving context of I50 era, synthesizing six theoretical perspectives into a multi-layered framework that prioritizes human-AI collaboration, resilience, and sustainability. Our analysis reveals how I50 transitions from automation-centric paradigms to collaborative, ethically grounded innovation ecosystems. The proposed framework captures the multi-actor, adaptive, and emergent nature of innovation in this new era through a layered integration of theoretical perspectives, including TOE, UTAUT, Dynamic Capabilities, STS, and Complexity Theory. This foundation enables future empirical studies to explore the dynamic interactions among technologies, stakeholders, and environmental conditions that drive enterprise resilience and innovation performance. In practical terms, the proposed Innovation 5.0 framework offers strategic guidance for firms navigating the transition to I50 by emphasizing the integration of human-centricity, resilience, and sustainability into core business operations. For business leaders and managers, it underscores the urgency of developing adaptive capabilities, fostering stakeholder co-creation, and embedding socio-technical alignment to remain competitive in dynamic environments.

While this study offers a novel and comprehensive framework for understanding Innovation 5.0 in the context of I50, several limitations must be acknowledged. One key limitation is its purely conceptual nature, which, while offering a robust theoretical foundation, lacks empirical validation to evaluate the proposed Innovation 5.0 framework in real-world industrial settings. Additionally, the integration of multiple theoretical lenses, such as the Sextuple Helix, TOE, UTAUT, and others, though comprehensive, may introduce conceptual overlap and complexity, making practical application and operationalization challenging in diverse, policy-driven, and turbulent environments.

Future research should focus empirically on validating the proposed Innovation 5.0 framework by developing and testing robust measurement scales for each of its theoretical dimensions across diverse industrial sectors and geographical contexts. This would help assess the framework's applicability and relevance in different environments and facilitate cross-sectoral benchmarking. Longitudinal studies are also essential for capturing the dynamic evolution of Innovation 5.0 maturity over time, particularly in response to varying degrees of environmental turbulence, market disruptions, or policy changes. Such temporal analyses will provide valuable insights into how firms adapt, reconfigure, or transform their innovation capabilities to maintain resilience and competitiveness. Additionally, examining the impact of institutional mechanisms, such as regulatory incentives, policy interventions, and public–private partnerships, on promoting inclusive, human-centric, and sustainable innovation ecosystems will offer guidance for evidence-based policymaking. These lines of inquiry can collectively enhance our understanding of Innovation 5.0's operational mechanisms, bridging the gap between conceptual models and real-world practices, and accelerating the transition toward more resilient and ethically grounded industrial futures.

Funding This research was funded by the Martí Franquès Fellowship at Universitat Rovira i Virgili (URV), with partial support from Diputació de Tarragona.

References

1. Afzal, B., Hernández-Lara, A.B., Li, X.: Platform enterprises and SMEs: a bibliometrics analysis. In: Research and Innovation Forum 2023, pp. C1–C1. SPCOM (2024). https://doi.org/10.1007/978-3-031-44721-1_61
2. Afzal, B., Hernández-Lara, A.B., Li, X.: Platform business model: an innovation 5.0 perspective. In: Research and Innovation Forum 2024, pp. 659–667 (2025). https://doi.org/10.1007/978-3-031-78623-5_52
3. Afzal, B., Li, X., Hernández-Lara, A.B.: The innovation journey and crossroads of sustainability, resilience and human-centeredness: a systematic literature review. Transf. Gov. People Process Policy **18**(3), 368–383 (2024). https://doi.org/10.1108/TG-10-2023-0151
4. Afzal, B., Li, X., Hernández-Lara, A.B.: Industrial revolution and innovation evolution: analyzing the past to envision the future. In: Digital Transformation and Innovation in Organizations, pp. 21–44. Springer Nature Switzerland (2025). https://doi.org/10.1007/978-3-031-90794-4_2
5. Breque, M., De Nul, L., & Petridis, A.: Industry 5.0: towards a sustainable, human-centric and resilient European industry. Policy Brief (2021)
6. Carayannis, E.G., Barth, T.D., Campbell, D.F.: The Quintuple Helix innovation model: global warming as a challenge and driver for innovation. J. Innov. Entrep. **1**(1), 2 (2012). https://doi.org/10.1186/2192-5372-1-2
7. Carvalho, A.V., Chouchene, A., Lima, T.M., Charrua-Santos, F.: Cognitive manufacturing in industry 4.0 toward cognitive load reduction: a conceptual framework. Appl. Syst. Innov. **3**(4), 55 (2020). https://doi.org/10.3390/asi3040055
8. Directorate-General for Research and Innovation: Industry 5.0 community of practice discuss future directions to achieve its goals. European Commission (2024)
9. Eveland, J., Tornatzky, L.G.: Technological innovation as a process, pp. 27–50 (1990)
10. Grabowska, S., Saniuk, S., Gajdzik, B.: Industry 5.0: improving humanization and sustainability of industry 4.0. Scientometrics **127**(6), 3117–3144 (2022). https://doi.org/10.1007/s11192-022-04370-1
11. Holling, C.S.: Resilience and stability of ecological systems. Annu. Rev. Ecol. Syst. **4**, 1–23 (1973). http://www.jstor.org/stable/2096802
12. Karwowski, K., Visvizi, A., Troisi, O.: Explaining China's Pivots and Priorities Through the Aerospace Industry Development Strategy, pp. 543–558 (2024). https://doi.org/10.1007/978-3-031-44721-1_41
13. Martini, B., Bellisario, D., Coletti, P.: Human-centered and sustainable artificial intelligence in industry 5.0: challenges and perspectives. Sustainability **16**(13), 5448 (2024). https://doi.org/10.3390/su16135448
14. Mayer, C.-H., Oosthuizen, R.M.: Sustainability in industry 4.0 business practice: insights from a multinational technology company. Front. Sustain. **3** (2022). https://doi.org/10.3389/frsus.2022.886986
15. Mouazen, A.M., Hernández-Lara, A.B., Chahine, J., Halawi, A.: Triple bottom line sustainability and Innovation 5.0 management through the lens of Industry 5.0, society 5.0 and digitized value chain 5.0. Eur. J. Innov. Manag. (2025). https://doi.org/10.1108/EJIM-04-2024-0339
16. Papa, A., Dezi, L., Gregori, G.L., Mueller, J., Miglietta, N.: Improving innovation performance through knowledge acquisition: the moderating role of employee retention and human resource management practices. J. Knowl. Manag. **24**(3), 589–605 (2020). https://doi.org/10.1108/JKM-09-2017-0391
17. Ramírez, T., Mora, H., Pujol, F.A., Maciá-Lillo, A., Jimeno-Morenilla, A.: Management of heterogeneous AI-based industrial environments by means of federated adaptive-robot learning. Eur. J. Innov. Manag. **28**(1), 50–64 (2025). https://doi.org/10.1108/EJIM-09-2023-0831
18. Shen, C., Koziel, A., Cheng, Y.: Topic discovery on sustainable smart city development, pp. 133–150 (2024). https://doi.org/10.1007/978-3-031-44721-1_11

19. Taratori, R., Rodriguez-Fiscal, P., Pacho, M.A., Koutra, S., Pareja-Eastaway, M., Thomas, D.: Unveiling the evolution of innovation ecosystems: an analysis of triple, quadruple, and quintuple helix model innovation systems in European case studies. Sustainability **13**(14), 7582 (2021). https://doi.org/10.3390/su13147582
20. Teece, D. J., Pisano, G., Shuen, A.: Dynamic capabilities and strategic management. Strat. Manag. J. **18**(7), 509–533 (1997). http://www.jstor.org/stable/3088148
21. Trist, E.L., Bamforth, K.W.: Some social and psychological consequences of the longwall method of coal-getting. Hum. Relat. **4**(1), 3–38 (1951). https://doi.org/10.1177/001872675100400101
22. Troisi, O., Grimaldi, M., Visvizi, A.: Digital Transformation in Tourism Ecosystems: What Impact on Sustainability and Innovation? pp. 73–81 (2023). https://doi.org/10.1007/978-3-031-19560-0_5
23. Troisi, O., Visvizi, A., Grimaldi, M.: Digitalizing business models in hospitality ecosystems: toward data-driven innovation. Eur. J. Innov. Manag. **26**(7), 242–277 (2023). https://doi.org/10.1108/EJIM-09-2022-0540
24. Troisi, O., Visvizi, A., Grimaldi, M.: Rethinking innovation through industry and society 5.0 paradigms: a multileveled approach for management and policy-making. Eur. J. Innov. Manag. **27**(9), 22–51 (2024). https://doi.org/10.1108/EJIM-08-2023-0659
25. Troisi, O., Visvizi, A., Grimaldi, M.: The different shades of innovation emergence in smart service systems: the case of Italian cluster for aerospace technology. J. Bus. Ind. Mark. **39**(6), 1105–1129 (2024). https://doi.org/10.1108/JBIM-02-2020-0091
26. Uhl-Bien, M., Marion, R.: Complexity leadership in bureaucratic forms of organizing: a meso model. Leadersh. Q. **20**(4), 631–650 (2009). https://doi.org/10.1016/j.leaqua.2009.04.007
27. Venkatesh, V., Morris, M.G., Davis, G.B., Davis, F.D.: User acceptance of information technology: toward a unified view. MIS Q. **27**(3), 425–478 (2003)
28. William Chesbrough, H., Vanhaverbeke, W., West, J.: Open Innovation: Researching a New Paradigm (2006). https://www.researchgate.net/publication/232957368

Velocity Inequalities in Digital Public Service Ecosystem: David Against Goliath in the Agri-Food Value Chain?

Stefania Manetti, Orlando Troisi, Mario Testa, and Adriana Apuzzo

Abstract Digital transformation is fundamentally reshaping public service ecosystems (PSEs), generating velocity asymmetries that influence governance structures, service interactions, and power dynamics, contributing to a paradigm shift from traditionally industrially oriented ecosystems to service logic. Addressing key gaps in previous PSE studies, this research examines how digital transformation shapes velocity inequalities among different actors. The agri-food sector was selected as the empirical context due to the profound digital and structural transformations currently reshaping its ecosystem and governance mechanisms. This study adopts a qualitative, multi-method approach, combining a systematic desk review and participatory observation within a stable smart agri-food network with strong European linkages. Findings indicate that tech giants and startups operate at high velocity, leveraging data-driven service innovation, whereas small farmers and public institutions struggle to adapt due to resource constraints, bureaucratic inertia, and fragmented digital governance. The study highlights the emergence of data-centric service models, underscoring the need for strategic public–private coordination to ensure equitable participation in the evolving PSE. By advancing theoretical insights on velocity inequalities in PSEs, this research contributes to a deeper understanding of how governance mechanisms and institutional adaptation shape service-driven digital transitions in complex, multi-actor ecosystems.

S. Manetti (✉)
Department for the Promotion of Human Sciences and Quality of Life, San Raffaele University of Rome, Rome, Italy
e-mail: stefania.manetti@uniroma5.it

O. Troisi · M. Testa · A. Apuzzo
Department of Political Science and Communication, University of Salerno, Fisciano, Italy
e-mail: otroisi@unisa.it

M. Testa
e-mail: mtesta@unisa.it

A. Apuzzo
e-mail: adapuzzo@unisa.it

A. Visvizi et al. (eds.), *Research and Innovation Forum 2025 Volume 1*,
Springer Proceedings in Complexity, https://doi.org/10.1007/978-3-032-09066-9_2

Keywords Digital transformation · Public service ecosystem · Agri-food · Inequalities · SDG 9 (industry · innovation and infrastructure) · SDG-10 (reduced inequalities)

1 Introduction

The concept of public service ecosystems (PSEs) has increasingly attracted attention in public management and service research as a comprehensive framework that connects institutional, service, and individual levels of public service delivery [1–4]. Rooted in service-dominant logic [5], PSE extends beyond traditional public service perspectives by emphasizing multi-actor integration, dynamic stakeholder roles, and institutional arrangements in value co-creation [1]. Key elements of PSE include its co-creative, multi-level approach, encompassing micro, meso, and macro interactions rather than limiting analysis to dyadic exchanges [3, 6].

Despite these advancements, two major theoretical gaps persist. First, PSE studies predominantly conceptualize ecosystems statically, overlooking their dynamic evolution in response to digital transformation and technological advancements [2, 6]. This gap limits PSE's applicability in fast-changing environments where actors must continuously adapt to shifting service logics, regulatory changes, and digital infrastructures [7, 8]. Second, service logic remains underexplored in PSE studies [6], with existing literature prioritizing industrial logic, institutional roles, and efficiency mechanisms rather than the co-creative nature of service-based interactions [9, 10].

The urgency for theoretical advancement becomes more urgent with the digital transformation and the rise of Artificial Intelligence, which is redistributing power, transforming regulatory landscapes, and accelerating service-driven interactions from end-users [11, 12]. Government roles are also shifting: historically reliant on location-based incentives and sanctions, traditional policy levers now struggle to address the rapid service logic transformation [13].

Velocity disparities extend beyond public sector organisations, reflecting broader differences among ecosystem actors. In this paper, "velocity" is defined as the relative pace at which different actors within a PSE adapt to exogenous transformations, such as digitalisation, through changes in governance, service delivery models, and institutional configurations. While technology giants operate globally with high adaptability and rapid innovation cycles, other players exhibit contrasting dynamics. Startups and highly innovative firms often face constraints on velocity, scalability, and market penetration. Conversely, traditional local providers embedded in legacy supply chains tend to adapt more slowly, largely due to their historical reliance on conventional distribution channels and incremental innovation models.

While previous studies provide valuable insights into PSE's structural and institutional dimensions, they often remain highly abstract, overlooking service logic dynamics and digital shocks [1, 9, 10]. Globalization and digitalization are reshaping economic and institutional ecosystems, accelerating the transition from industrial ecosystems to service-driven models [6].

One illustrative example is the agri-food sector, which is historically structured around industrial production and distribution chains. The entry of digital platforms, multinational corporations, and large cooperatives has accelerated hyper-personalization, transforming agricultural products into service-based, data-driven offerings similar to sophisticated, digitally-based and personalized para-medical products [14, 15]. This shift is not unique to agri-food but reflects broader trends across multiple industries [12, 16, 17].

Understanding how different actors adapt at varying paces is essential to detect asymmetries in responsiveness that can reshape power relations and value co-creation across the ecosystem. These divergences in adaptation velocity highlight the need to explore how different PSE actors navigate digital transformation and how public sector organizations can address emerging inequalities. Building on this premise, the study investigates the impact of digital transformation on velocity inequalities within agri-food ecosystems. Specifically, it addresses the following research question: *How does digital transformation influence velocity inequalities among different public service ecosystem (PSE) actors?*

To ground the analysis empirically, the agri-food sector has been selected as a representative and theoretically significant case of a traditionally industrial ecosystem undergoing a rapid transition toward service logic. Characterized by high actor heterogeneity and historically fragmented governance structures, this sector exemplifies how digitalization can reconfigure value co-creation dynamics, accelerate service-based interactions, and redefine institutional roles within the ecosystem.

2 Methodology

This study adopts a public service ecosystem (PSE) perspective to investigate velocity inequalities among ecosystem actors in response to digital transformation, which is reshaping traditional industrial ecosystems and driving their transition toward a service-dominant logic. The agri-food sector, with a specific focus on Italy and Europe, has been theoretically and empirically chosen as a reference case due to the profound and ongoing transformations that are fundamentally altering its structure, governance, and market dynamics. This sector is particularly relevant because of the increasing integration of digital technologies, the rising role of non-traditional actors such as tech giants, and the reconfiguration of value co-creation mechanisms, making it an ideal setting for analysing PSE dynamics under accelerated change. The research follows a qualitative, multi-method approach, integrating a systematic desk review, document analysis, and participatory observation of a stable Italian network in smart agri-food with strong international linkages. To operationalize the PSE framework, we derived a set of analytical variables, such as 'velocity', 'influencing factors for velocity', and "digital innovation trajectories", that were mapped across key ecosystem categories identified during the desk review and observation phases. These constructs guided the development of data collection instruments and informed the coding scheme used during thematic analysis. Creating a comparative matrix (Table 1) involved systematically mapping these constructs against empirical observations, ensuring a coherent translation of theoretical concepts into measurable elements within the study.

The desk review was conducted to examine policy documents, industry reports, and regulatory frameworks, providing a theoretical foundation for understanding the evolving role of digital transformation within agri-food PSEs. This was complemented by an analysis of institutional and market reports, which allowed the authors to identify key actors, structural dynamics, and digital innovation trajectories shaping the sector.

A core methodological emphasis was placed on participatory observation, as the research team had a privileged position as direct observers and facilitators of a stable and internationally connected multi-stakeholder network comprising public and private actors, including policymakers, regulatory bodies, technology providers,

Table 1 Comparative matrix of empirical findings: velocity, influencing factors, and digital innovation trajectories in agri-food PSEs

PSE actor	Velocity	Influencing factors for velocity	Digital innovation trajectories
Tech giants	Rapid across the supply chain	Resources, advanced competencies, client data access, AI-driven decision-making, predictive analytics, global reach	Advanced software integration, AI for analytics and risk management, big data platforms
Startups	Rapid in innovation, slow in scaling	Creativity, IT expertise, limited management and communication skills, AI adoption	AI-driven risk management and monitoring solutions
Farmers' cooperatives	Mixed: quick to low	Market reorientation, structured training, investment access, co-innovation, size and price volatility constraints	Farm management systems, satellite imagery integration
Small farmers	Low to stagnant	Limited skills, low training, cultural barriers, minimal innovation access, price volatility, individualistic actions	Low digital adoption, minimal Agriculture 4.0 tools
Regulators	Quick in policy, low in implementation	Bureaucracy, political negotiation, fragmented oversight, regulatory complexity	Blockchain-based compliance, AI-powered policy monitoring
Public administration	Low investment and responsiveness	Resource constraints, slow policy cycles, focus on open-data platforms	Public data repositories, macro-data management tools

Source Authors' elaboration based on seminal works [1–6], and empirical evidence from desk review, document analysis, and participatory observation of a multi-stakeholder agri-food network
Acronyms: public sector ecosystem (PSE); public sector organizations (PSOs)

farmers' cooperatives, and industry associations. This allowed us to capture first-hand insights into the interactions, narratives, and strategic priorities of key stakeholders. Immersion in discussions and industry-led workshops provided deep insights into how digital transformation is reshaping service logic and governance models. Additionally, our role as embedded observers enabled us to critically reflect on the negotiation processes, institutional tensions, and power asymmetries faced by different PSE actors in adapting to service-driven, data-intensive business models. However, as with other participatory research approaches [18], this study is not presented as an objective or purely rational account of events, but rather as a theoretically informed narrative shaped by evolving perspectives throughout the research process.

3 Results

In this section, the results of the analysis are presented, based on the conceptual view previously outlined. Findings are now illustrated through two complementary tables that categorize public service ecosystem (PSE) actors and related dynamics across distinct analytical dimensions.

Table 1 categorizes PSE actors in the rows and presents three key variables influencing ecosystem dynamics across the columns. The first column, *Velocity*, classifies actors based on their speed of adaptability, distinguishing between highly dynamic entities (e.g., tech giants and startups) and slower or stagnant actors (e.g., small farmers and public sector institutions). The second column, *Influencing Factors for Velocity*, outlines the primary enablers and constraints affecting each actor's capacity to adapt and innovate. The third column, *Digital Innovation Trajectories*, identifies the technological pathways pursued by different actors. High-velocity actors, such as tech giants, demonstrate advanced digital adoption, leveraging AI-driven automation, big data analytics, and predictive modelling. Conversely, small farmers and public sector entities exhibit limited technological uptake, often constrained by financial, infrastructural, and organizational barriers.

Table 2 complements this analysis by focusing on interaction mechanisms and competitiveness. The first column, *PSE Mechanisms of Interactions*, captures the relational dynamics among ecosystem actors. Tech giants and farmers' cooperatives exhibit strong engagement with networks, governance structures, and digital service providers, fostering co-innovation and technological diffusion. In contrast, small farmers and public administration actors display deficits in multi-actor interactions, limiting their ability to integrate into the broader agri-food innovation landscape. The second column, *Competitiveness*, assesses the relative market positioning of each actor. Tech giants maintain a dominant position due to their technological leadership and global market integration, while startups struggle with scalability despite their high innovation potential. Farmers' cooperatives demonstrate moderate competitiveness, leveraging collective innovation strategies, whereas small farmers and public institutions exhibit the lowest levels of competitiveness, often due to structural inefficiencies and digital marginalization.

Table 2 Comparative matrix of empirical findings: interaction mechanisms and competitiveness in agri-food PSEs

PSE actor	PSE mechanisms of interactions	Competitiveness
Tech Giants	Strong engagement with networks, PSOs, governance structures, consumer interaction via analytics	High
Startups	High interaction with digital innovators and firms; low with PSOs and regulators	Medium to high
Farmers' cooperatives	High collaboration with innovation brokers and digital providers; medium to high interaction with regulators	Medium to low
Small farmers	Deficit across supply chains and weak engagement with providers, regulators, and markets	Low to none
Regulators	Active in policy dialogue; low interaction with ecosystem actors; struggle with agile governance	Low
Public administration	Limited engagement with innovators; primarily interacts with regulators and policymakers; slow digitization	Low

Source Authors' elaboration based on seminal works [1–6], and empirical evidence from desk review, document analysis, and participatory observation of a multi-stakeholder agri-food network
Acronyms: public sector ecosystem (PSE); public sector organizations (PSOs)

The categories of actors that emerged from the empirical analysis are tech giants, start-ups, farmers' cooperatives, small farmers, regulators, and public administration. The analysis reveals substantial velocity inequalities among the key PSE actors, shaped by a range of influencing factors.

Tech giants demonstrate high velocity across the supply chain, leveraging global market positioning, advanced digital infrastructure, and AI-driven decision-making. Their PSE interactions are extensive, engaging with public sector organizations, governance structures, and consumers through structured data analytics and service-driven profiling. These firms orchestrate co-innovation processes, reinforcing a service-dominant logic where value derives not only from products but also from data-driven personalization, efficiency optimization, and strategic decision-making.

Startups exhibit high innovation velocity but face scalability constraints, hindered by limited managerial competencies despite strong IT expertise in predictive data analytics and AI-driven risk management. Their PSE role is dynamic, collaborating with agri-food firms and digital innovators, yet their low engagement with regulators and public administration presents challenges in navigating policy cycles and regulatory compliance.

Farmers' cooperatives display heterogeneous velocity, with digitally integrated networks adapting faster than traditional product-driven producers. Their digital innovation trajectory is shaped by farm management systems, satellite imagery, and precision agriculture tools. While strong interactions with innovation brokers facilitate technology adoption, barriers such as size constraints, price volatility, and generational shifts persist. Large consortia, such as Granarolo, illustrate how digital governance strategies, training programs, and consumer assistance models enhance competitiveness.

Small farmers operate at low to stagnant velocity, hindered by limited digital competencies, cultural resistance, and restricted access to innovation. Their digital adoption remains minimal, reinforcing marginalization within data-driven value chains. Although awareness of data analytics is increasing, actual adoption rates remain low, highlighting the need for targeted public investment in skill development and digital infrastructure.

Regulators engage actively in policy formulation but exhibit low implementation velocity, constrained by bureaucratic complexity and fragmented regulatory frameworks. Their digital innovation strategy focuses on blockchain-based compliance systems and AI-powered policy monitoring, yet limited interaction with PSE actors hinders the scalability of emerging technologies.

Public administration demonstrates low investment velocity and limited responsiveness to digital innovation needs, constrained by resource allocation issues and slow policy adaptation. Macro-data platforms and public repositories signify a shift towards data centralization for evidence-based policymaking, yet slow digital governance and fragmented EU coordination hinder widespread innovation adoption.

Overall, PSE velocity inequalities are shaped by human capital, knowledge, access to innovation, market positioning, bureaucratic inefficiencies, and ecosystem-level multi-actor interaction mechanisms.

4 Discussion

This paper advances a more nuanced understanding of public service ecosystem (PSE) dynamics in the context of digital transformation, with a particular focus on velocity inequalities among PSE actors. It is argued that the evolution of PSEs should be conceptualized as a dynamic and velocity-sensitive process, in which the integration of service logic acts as a catalyst for change while also amplifying disparities. Through a multi-method approach, this study explores how different PSE actors respond to necessary digital transformation, paying particular attention to public sector organizations, which face the tension between regulatory stability and digital acceleration. The findings contribute to the broader debate on public–private interactions, emphasizing the need for governance models that accommodate service-driven velocity while ensuring equity and sustainability in emerging digital ecosystems.

The study identifies four critical insights relevant to public sector organizations, managers, and practitioners seeking to navigate digital transformation.

Firstly, velocity disparities within PSEs impact not only digital adoption but also power balances. As traditional sectoral boundaries blur, non-traditional digital actors increasingly shape service co-creation processes, reinforcing service logic as a competitive differentiator. While this shift fosters co-creation opportunities, it also challenges PSE equilibrium, raising concerns over public sector organizations' ability to ensure equitable participation. The growing asymmetry in velocity risks marginalizing slower-adapting actors, particularly public institutions, whose bureaucratic structures struggle to match the rapid evolution of digital service models. This fundamental paradigm shift, driven by digital transformation, highlights the need for strategic public–private coordination, balancing digital investment, regulatory modernization, and infrastructure expansion.

Secondly, decision-making horizons within PSEs are increasingly influenced by data-driven service delivery. Tech giants, as pivotal ecosystem orchestrators, shape digital transformation trajectories, leveraging data analytics, automation, and personalization tools to engage both customers and other ecosystem actors. This acceleration, however, fosters a short-term market-driven orientation, where decisions prioritize immediate responsiveness over long-term public value creation. This raises concerns about the role of public sector organizations in balancing technological acceleration with sustainable governance models.

Thirdly, the logics guiding PSE actors' actions reveal divergent strategic orientations. Tech giants align with service-dominant logic, leveraging AI-driven service co-creation to address sustainability and digital inclusion challenges. Startups act as innovation enablers, integrating product- and service-oriented models. Collaborative entities, such as cooperatives, position themselves between product- and service-driven transitions, emphasizing market integration and adaptability. Meanwhile, regulators and public sector organizations operate within product-dominant governance structures, focusing on policy standardization and compliance, yet struggling with digital velocity asymmetries, requiring new governance frameworks to sustain public value in digital ecosystems.

Lastly, the agri-food sector exemplifies these dynamics, with tech giants fully embracing service logic, transforming products into data-driven, traceable, and sustainability-certified offerings. *Farmers' cooperatives* exhibit heterogeneity, with some embracing digital farm management and sustainability monitoring while others remain anchored to product-dominant logic due to size limitations. *Small and medium agri-food firms*, however, continue to operate within product-centric models, facing low digital adoption and weak ecosystem integration, exacerbating their marginalization in data-driven markets. Public sector organizations, particularly public administration and regulatory bodies, remain slow in adapting governance frameworks, struggling with harmonizing digital regulations and investing in open-data platforms, which, despite their potential, remain underutilized due to fragmented governance structures. The findings highlight an urgent need to strengthen the role of universities, research centres, and knowledge-sharing platforms in fostering competency-building and collaborative innovation within the PSE. The lack of public agencies for macro-data governance in agriculture, both in Italy and the EU, limits the scalability of digital innovation and AI-based sustainability initiatives.

Importantly, the long-term social and territorial consequences of such inequalities deserve further attention. Uneven digital adaptation may exacerbate regional disparities, especially in rural or underserved areas, limiting citizens' access to smart services and perpetuating socio-economic exclusion. The still limited integration of the citizen or consumer perspective in digital PSE transformation undermines the full potential of service logic, which inherently requires user engagement, transparency, and value-in-use. Including these perspectives is essential to ensure that digital transition pathways are not only efficient but also inclusive and territorially balanced.

The digital transformation of PSE is currently in progress, exhibiting varying rates of readiness, implementation, and levels of technological maturity. Velocity inequalities in the pace of digital transition and persistent structural barriers remain, particularly affecting small to medium-sized independent farmers and cooperative agricultural enterprises. These disparities are further compounded by the challenges faced by start-ups in achieving scalability, as well as the difficulties encountered by public sector organizations in keeping pace with rapid digital innovation and responsiveness. Focusing on the agri-food sector, the increasing convergence of the ecosystem with digital technology sectors reinforces the need for coordinated policy frameworks, stronger public–private partnerships, and inclusive digital infrastructure investments to ensure equitable participation in the evolving service-driven agri-food ecosystem. The PSE framework has been instrumental in capturing the complex interdependencies between actors, institutional arrangements, and digital innovation trajectories. By applying this theoretical lens, the study moves beyond static analyses to reveal how service logic, when unevenly integrated, generates velocity gaps that reshape both governance and equity in the ecosystem. This provides new conceptual clarity on how public value is co-created—or inhibited—under digital transformation. Future research should explore velocity disparities within PSE actor categories, particularly among collaborative entities such as cooperatives, to understand organizational structures, governance mechanisms, and access to digital networks. Additionally, examining how velocity inequalities shape power dynamics and service co-creation across different industrial ecosystems would provide broader insights into public sector organizations' role in sustaining equilibrium in service-driven economies.

5 Conclusion

Applying the public service ecosystem (PSE) perspective, this study reveals how service logic and velocity inequalities reshape governance, value co-creation, and actor interactions. The agri-food case enables a detailed analysis of uneven digital adaptation across sectors. Theoretical insights highlight how multi-actor dynamics and institutional constraints influence service co-creation, offering guidance for managing digital complexity and sustaining inclusive public value. From a managerial standpoint, for-profit organizations should invest in flexible governance structures and digital capacity-building initiatives, particularly for smaller actors (e.g., SMEs), to mitigate velocity inequalities and foster sustained collaboration. Politically, policymakers should develop regulatory frameworks that balance rapid innovation with equitable access, leveraging public–private partnerships to harmonize data governance and digital infrastructure development. Moreover, transparent multi-actor engagement across the supply chain can enhance legitimacy and trust in digital transformation, while targeted funding for under-resourced groups will help safeguard social cohesion and territorial equity.

References

1. Osborne, S.P., Powell, M., Cui, T., Strokosch, K.: Value creation in the public service ecosystem: an integrative framework. Public Adm. Rev. **82**(4), 634–645 (2022)
2. Trischler, J., Røhnebæk, M., Edvardsson, B., Tronvoll, B.: Advancing public service logic: Moving towards an ecosystemic framework for value creation in the public service context. Public Manag. Rev. 1–29 (2023)
3. Petrescu, M.: From marketing to public value: towards a theory of public service ecosystems. Public Manag. Rev. **21**(11), 1733–1752 (2019)
4. Polese, F., Troisi, O., Carrubbo, L., Grimaldi, M.: An integrated framework toward public system governance: insights from viable systems approach. In: Anthopoulos, L.G. (ed.) Cross-Sectoral Relations in the Delivery of Public Services, vol. 6, pp. 23–51. Emerald Publishing Limited (2018)
5. Vargo, S.L., Lusch, R.F.: Institutions and axioms: an extension and update of service-dominant logic. J. Acad. Mark. Sci. **44**, 5–23 (2016)
6. Trischler, J., Charles, M.: The application of a service ecosystems lens to public policy analysis and design: exploring the frontiers. J. Public Policy Mark. **38**(1), 19–35 (2019)
7. Elia, G., Margherita, A., Passiante, G.: Digital entrepreneurship ecosystem: how digital technologies and collective intelligence are reshaping the entrepreneurial process. Technol. Forecast. Soc. Chang. **150**, 119791 (2020)
8. Gupta, R., Mejia, C., Kajikawa, Y.: Business, innovation and digital ecosystems landscape survey and knowledge cross sharing. Technol. Forecast. Soc. Chang. **147**, 100–109 (2019)
9. Grönroos, C.: Adopting a service logic for marketing. Mark. Theory **6**(3), 317–333 (2006)
10. Grönroos, C.: Value co-creation in service logic: a critical analysis. Mark. Theory **11**(3), 279–301 (2011)
11. Troisi, O., Visvizi, A., Grimaldi, M.: Digitalizing business models in hospitality ecosystems: toward data-driven innovation. Eur. J. Innov. Manag. **26**(7), 242–277 (2023)
12. Visvizi, A., Troisi, O., Corvello, V.: Research and Innovation Forum 2023: Navigating Shocks and Crises in Uncertain Times—Technology, Business, Society. Springer Nature, Cham (2024)
13. Troisi, O., Visvizi, A., Grimaldi, M.: Rethinking innovation through Industry and Society 5.0 paradigms: a multileveled approach for management and policy-making. Eur. J. Innov. Manag. **27**(9), 22–51 (2023)
14. Schnebelin, É., Labarthe, P., Touzard, J.M.: How digitalisation interacts with ecologisation? Perspectives from actors of the French Agricultural Innovation System. J. Rural. Stud. **86**, 599–610 (2021)
15. Manetti, S., Cumetti, M., De Benedictis, A., Lettieri, E.: Adoption of novel biomarker test parameters with machine learning–based algorithms for the early detection of sepsis in hospital practice. J. Nurs. Manag. **30**(8), 3754–3764 (2022)
16. Manetti, S., Lettieri, E., Ni, M.Z.: Development and validation of Medical Device Key Evidence Tool ('MeDKET'): an evidence-based framework to explain success in selected European and US companies. PLoS ONE **18**(7), e0288126 (2023)
17. Visvizi, A., Troisi, O., Grimaldi, M.: Mapping and conceptualizing big data and its value across issues and domains. In: Big Data and Decision-Making: Applications and Uses in the Public and Private Sector, pp. 15–25. Emerald Publishing Limited, Bingley (2023)
18. Chua, W.F.: Experts, networks and inscriptions in the fabrication of accounting images: a story of the representation of three public hospitals. Acc. Organ. Soc. **20**(2–3), 111–145 (1995)

Digital Literacy and Entrepreneurial Mindset in the Middle East: A Structural Equation Approach

Khodor Shatila and Ana Beatriz Hernandez-Lara

Abstract This study investigates the interplay between digital accessibility, digital skills, and digital literacy in shaping the entrepreneurial mindset and ecosystem in the Middle East. Despite growing digital entrepreneurship initiatives in the region, there remains limited empirical understanding of how digital readiness translates into entrepreneurial capacity and innovation. Grounded in the Resource-Based View (RBV) and Technology Acceptance Model (TAM), the research employs Structural Equation Modeling (SEM) to analyze survey data from 250 entrepreneurs across diverse industries. Findings reveal that digital accessibility, skills, and literacy significantly contribute to the development of an entrepreneurial mindset, which, in turn, mediates their impact on the broader entrepreneurial ecosystem. Specifically, digital literacy enhances critical thinking, risk evaluation, and cybersecurity awareness, while digital skills foster innovation and adaptability. Digital accessibility reduces geographic and financial barriers, enabling broader participation in entrepreneurial activity. The results emphasize the critical role of a digitally empowered mindset in fostering resilient and sustainable entrepreneurship. The study offers practical insights for policymakers and educators aiming to strengthen digital inclusion and entrepreneurial capacity in the MENA region. By addressing the interdependencies between digital readiness and mindset, this research contributes a comprehensive framework to the digital entrepreneurship literature and supports more inclusive innovation-driven development in emerging economies.

Keywords Digital literacy · Entrepreneurial mindset · Structural equation modeling · Middle East · SDG-8

K. Shatila (✉)
Universitat Rovira i Virgili, Reus (Tarragona), Spain
e-mail: khodor.shatila@estudiants.urv.cat

A. B. Hernandez-Lara
Department of Business Management, Universitat Rovira i Virgili, Reus (Tarragona), Spain
e-mail: anabeatriz.hernandez@urv.cat

A. Visvizi et al. (eds.), *Research and Innovation Forum 2025 Volume 1*,
Springer Proceedings in Complexity, https://doi.org/10.1007/978-3-032-09066-9_3

1 Introduction

The rapid pace of digitalization in society has profoundly affected the entrepreneurship landscape, shaping the emergence and operation of businesses. The development of digital technologies is making an increasing contribution to entrepreneurship [1]. Digital accessibility, digital skills, and digital literacy are essential components of any person's striving to step up in entrepreneurial activity and become a cog in the vast machine of entrepreneurship. Nevertheless, these digital components cannot be considered in isolation as they are closely related to the levels of an entrepreneurial mindset [22]. The entrepreneurial ecosystem is a primary driver of economic growth, innovation, and sustainable development in the turmoil of today's business setting. In this respect, a comprehensive analysis of the interplay between levels of digital accessibility, digital skills, and digital literacy helps us understand under what conditions these digital enablers can become hindrances and vice versa [9]. The past years have seen an upsurge in entrepreneurial activities in the Middle East, driven by government-sector enterprises, efforts to diversify the economy, and increased attention to digital evolution. Countries like the United Arab Emirates, Saudi Arabia, and Qatar have invested heavily in a digital base to promote innovation and entrepreneurs [22]. At the same time, the population is still mostly risk-averse, which might be one of the reasons why not many people want to become entrepreneurs [9]. Despite the increasing interest in digital entrepreneurship from scholars, few studies have analyzed the impact of digital accessibility, digital skills, and digital literacy on the ecosystem of entrepreneurship, specifically in interaction with the entrepreneurial mindset, and never regarding the Middle East [9]. Most other studies have investigated digital adoption on the organizational level, largely neglecting the ways people acquire and use digital competencies to promote entrepreneurship. This paper aims to summarize the existing body of knowledge while identifying current research gaps, thereby contributing to the ongoing scholarly discourse. Despite the increasing scholarly interest in digital entrepreneurship, there remains a notable gap in the literature concerning the interconnected effects of digital accessibility, digital skills, and digital literacy on the development of an entrepreneurial mindset and, subsequently, the broader entrepreneurial ecosystem, especially within the context of the Middle East. While existing research focuses on these digital factors in isolation or within organizational contexts, few studies have holistically examined how these dimensions collectively shape entrepreneurial attitudes, behaviors, and outcomes in emerging economies undergoing rapid digital transformation. Hence, this research aims to answer the question: *How do digital accessibility, digital skills, and digital literacy collectively influence the entrepreneurial mindset and contribute to developing the entrepreneurial ecosystem in the Middle East?*

2 Hypothesis Development

Digital accessibility entails the ability of individuals to access and use digital technologies, such as the internet, digital platforms, and online resources. It is a powerful enabler of entrepreneurship because it provides individuals with the tools and platforms to do business, communicate with clients, and obtain information on various markets [8]. Conversely, the lack of digital access or its limited access can close opportunities and hinder access to digital and online markets. In such cases, the enhanced accessibility of these platforms can create a more level playing field, thus preventing disparities [29]. Entrepreneurs in settings with enhanced accessibility to digital platforms are likely to experience reduced operational costs, greater market reach, and more efficient business processes, among other operational and marketing efficiencies [20]. ***As such, the hypothesis (H1a) suggests that digital accessibility positively influences the entrepreneurial ecosystem.***

Digital skills refer to using digital tools, applications, and platforms effectively. The digital skills may be basic, medium, or high and encompass such sophisticated competencies as digital marketing, cybersecurity, or data analytics [13]. Professionally digitally literate entrepreneurs are better placed to use technology for business growth and development and to increase their firms' competitive advantage. Secondly, digital skills are the key to business innovation in entrepreneurship [21]. Entrepreneurs proficient in digital fields can develop new, innovative business models that are revolutionary and dynamic. Additionally, such entrepreneurs can adapt to changes in technology or innovation, explore new markets, and partake in e-commerce and digital finance [16]. Acquisition of digital skills also enables the entrepreneur to be a better analytical person. ***Consequently, the hypothesis (H1b) posits that digital skills positively impact the entrepreneurial ecosystem.***

Digital literacy is an interconnected set of cognitive, technical, and socio-emotional factors that define an individual's capacity to use the internet effectively [31]. Central to digital literacy is a person's ability to critically understand and evaluate digital information. Digital literacy is an individual's ability to critically understand, evaluate, and use digital information from various digital sources to make informed decisions and act on them [5]. One's good use of digital information uses a person's technical expertise in applying various digital devices and encoders, an understanding of the sectors of digital protection, online ethics, and the competence to think critically about the products that come from the Internet. Such a concept also works similarly in entrepreneurship [21]. Whereas some critics argue that digital literacy is a skill that a person could learn if one has the resources, it is also a key aspect of flourishing in a modern environment. ***The hypothesis (H1c) asserts that digital literacy significantly contributes to the development of the entrepreneurial ecosystem.***

Internet availability, digital infrastructure's affordability, and digital platforms' inclusiveness refer to digital accessibility [33]. They found that digital accessibility has a highly positive effect on entrepreneurial intentions and behaviors. This is because digital tools open up possibilities to obtain information, establish networks, and access broader markets that can increase an entrepreneur's innovative potential [10]. Additionally, as real-time information, market analytics, and e-commerce platforms are obtainable for potential entrepreneurs, they can make more evidence-based decisions. Another key factor is that digital accessibility reduces geographical and economic challenges [17]. Even a start-up in a remote village can find and attract investors or suppliers worldwide. However, there are several issues [10]. ***Thus, H2a posits that digital accessibility positively influences the development of an entrepreneurial mindset.***

Furthermore, research reveals that individuals with digital skills acquire high levels of self-efficacy and adaptability, essential for an entrepreneurial mindset [21]. Thus, digital entrepreneurs can optimize technology for competitive advantage as more of them engage in e-commerce, digital marketing, and platform-based business models. AI-driven analytics technology can help entrepreneurs improve customer engagement and personalize services using consumer behavior trends [32]. In addition, the rise of remote work and the gig economy has made digital skills even more vital. Evidence suggests that freelancers and digital entrepreneurs with digital skills are likelier to display highly innovative behavior and resilience amidst economic shocks [14]. ***Hence, H2 b suggests that digital skills positively contribute to developing an entrepreneurial mindset.***

Thus, digital literacy allows entrepreneurs to understand market trends better, check business information, and ensure no misinformation is obtained to make well-informed decisions [5]. Cybersecurity and risks, including data breaches and cyber fraud, are avoided thanks to digital entrepreneurs' knowledge. Moreover, digital literacy helps promote collaborative innovation, and digital literacy entrepreneurs can participate in online communities, crowdsourcing platforms, and knowledge-sharing forums [14]. These findings support a more dynamic innovator mindset in which digital literacy entrepreneurs continually adapt and learn due to digital transformation. However, despite these benefits, previous research also shows that the lack of digital literacy is a critical barrier to all these business practices [6]. ***Based on this, H2c hypothesizes that digital literacy positively influences the entrepreneurial mindset.***

The entrepreneurial mindset encompasses innovation, opportunities, risk-taking, and resiliency. It has been grounded as a vital cue between the digital considerations – accessibility, skills, and literacy- and the more comprehensive entrepreneurial framework [18]. In this sense, an ecosystem refers to the intertwined environment of startups, investors, incubators, policymakers, and support structures that underpin environments conducive to entrepreneurship. Entrepreneurial mindset feeds the ecosystem by promoting collaboration, knowledge sharing, and innovating on the business model [12]. This mindset benefits the system by creating high-impact entrepreneurship and attracting venture funding, with job opportunities. ***Thus, H3 posits that an entrepreneurial mindset positively impacts the entrepreneurial ecosystem.***

The research model is visualized in the *Fig. 1.*

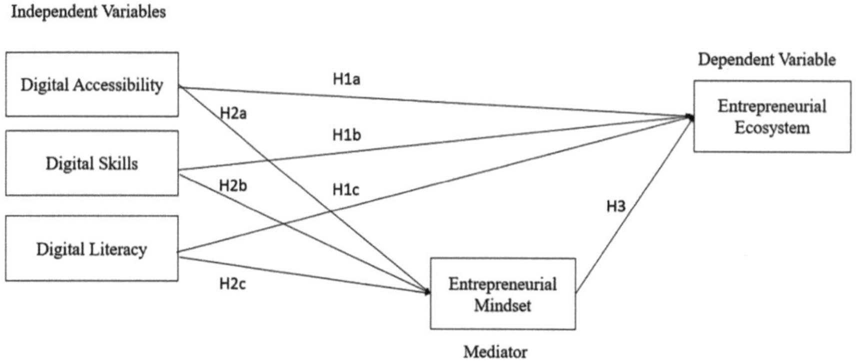

Fig. 1 Research model. *Source* Author work

3 Methods and Materials

Grounded in the Resource-Based View (RBV) theory, a firm's sustained competitive advantage arises from effectively deploying both tangible and intangible resources. In the entrepreneurial context, digital accessibility, digital skills, and digital literacy are strategic resources that enable entrepreneurs to identify opportunities, innovate, and create value through new business ventures (Zhang et al. 2023). These digital resources are foundational for building the entrepreneurial mindset, a non-material asset that enhances an entrepreneur's capacity to leverage digital tools for competitive differentiation. Complementing this, the Technology Acceptance Model (TAM) provides a behavioral framework to understand how entrepreneurs adopt and utilize digital technologies based on perceived usefulness and ease of use [8]. Digital accessibility and skills directly influence perceptions of ease of use, facilitating smoother technology adoption. In contrast, digital literacy enhances the perceived usefulness of digital tools by enabling entrepreneurs to interpret, evaluate, and apply digital information effectively [29]. Together, these constructs shape the entrepreneurial mindset, which mediates the impact of digital readiness on entrepreneurial success.

This study employs a quantitative research approach to investigate the relationships between digital accessibility, digital skills, digital literacy, the entrepreneurial mindset, and the entrepreneurial ecosystem. Data were collected using a structured survey targeting entrepreneurs and small business owners across various industries in the Middle East, ensuring diverse representation in terms of business size and sector. A sample of 250 respondents participated in the study. Participants were recruited based on their active engagement in entrepreneurial activities across the Middle East. The sample consisted of entrepreneurs and small business owners operating in diverse industries such as information technology, manufacturing, education, healthcare, retail, and finance. The inclusion criteria emphasized individuals with varying levels of digital experience and exposure to entrepreneurial decision-making in digitally evolving environments.

The questionnaire was designed to measure each variable using established and validated scales from prior research on digital entrepreneurship and technology adoption. Specifically, the constructs in the study were measured using items adapted from well-established scales in the literature, enhancing the validity and reliability of the research instrument. Digital literacy, accessibility, and skills were measured using a scale focusing on competencies related to solving technological problems, learning new digital tools, cybersecurity awareness, and digital collaboration [2]. Entrepreneurial mindset was assessed through items adapted from studies capturing the extent of entrepreneurs' relationships across diverse industries [15, 30]. Finally, entrepreneurial success was measured using a combination of subjective and objective indicators based on frameworks evaluating dimensions such as personal satisfaction, business growth, and organizational performance. All items were rated on a 5-point Likert scale, ranging from 1 (Strongly Disagree) to 5 (Strongly Agree), to capture the intensity of respondents' perceptions and experiences [4, 28]. To analyze the relationships among these variables, the study applied SEM, which allows for examining both direct and indirect effects. This analytical technique is particularly suitable for testing the mediating role of the entrepreneurial mindset in shaping the broader entrepreneurial ecosystem. The survey was conducted online, ensuring accessibility for respondents across multiple Middle Eastern countries.

4 Findings

Table 1 presents the demographic distribution of the 250 respondents is presented in Table 1, which provides an overview of key characteristics such as age, gender, years of experience, company age, and industry distribution. The largest age group among respondents falls within the 25–29 years range (22.00%), followed by those aged 30–34 years (19.20%) and 35–39 years (18.00%), indicating a relatively young entrepreneurial demographic. Regarding gender, most respondents are male (54.00%), while females constitute 46.00%, highlighting a near-balanced representation of male and female entrepreneurs. Regarding professional experience, 28.00% of the respondents have less than five years of experience, while 24.00% have between 5 and 10 years, showing a substantial presence of early-career entrepreneurs. The distribution of company age reveals that 32.00% of businesses have been operating between 6 and 10 years, while 24.00% fall within the 1–5 year category, suggesting that many companies are in their growth phase. Industry-wise, the largest sector represented is Information Technology (24.00%), followed by Manufacturing (18.00%) and Education (16.00%), reflecting the digital transformation and industrial diversity of the sampled entrepreneurs.

Table 2 addresses the validity and reliability of the research model; several fit indices were evaluated, as summarized in Table 2. The Chi-Square ($\chi 2$) value is 512.67, with 230 degrees of freedom, which, when divided, results in a Chi-Square/df ratio of 2.23, well within the acceptable range of less than 3.00. The Comparative Fit Index (CFI) is 0.94, and the Tucker-Lewis Index (TLI) is 0.92, exceeding the

Table 1 Sample profiling

Variable	Category	Frequency (n)	Percentage (%)
Age	25–29 years	55	22.00
	30–34 years	48	19.20
	35–39 years	45	18.00
	40–44 years	40	16.00
	45–49 years	37	14.80
	50 years and above	25	10.00
Gender	Male	135	54.00
	Female	115	46.00
Years of experience	Less than 5 years	70	28.00
	5–10 years	60	24.00
	11–15 years	50	20.00
	16–20 years	40	16.00
	More than 20 years	30	12.00
Company age	Less than 1 year	20	8.00
	1–5 years	60	24.00
	6–10 years	80	32.00
	11–15 years	50	20.00
	More than 15 years	40	16.00
Field/Industry	Information technology	60	24.00
	Manufacturing	45	18.00
	Education	40	16.00
	Healthcare	35	14.00
	Retail	35	14.00
	Finance	35	14.00

Source Author work

recommended threshold of 0.90, indicating a well-fitting model. The Root Mean Square Error of Approximation (RMSEA) is 0.05, and the Standardized Root Mean Square Residual (SRMR) is 0.04 below the 0.08 cutoff, confirming an adequate model fit. Furthermore, the Goodness of Fit Index (GFI) is 0.91, and the Adjusted Goodness of Fit Index (AGFI) is 0.85, exceeding their respective minimum levels and reinforcing model appropriateness. Lastly, the Normed Fit Index (NFI) of 0.93 further supports the strong validity of the model.

Table 3 addresses the reliability and validity of the constructs, which were assessed through Cronbach's Alpha, Composite Reliability (CR), and Average Variance Extracted (AVE), as illustrated in Table 3. Cronbach's Alpha values for all variables exceed 0.80, with Digital Skills (0.87) and Entrepreneurial Mindset (0.89) demonstrating high internal consistency, surpassing the acceptable threshold of 0.70 [7]. The Composite Reliability (CR) scores range from 0.85 to 0.91, indicating strong

Table 2 Model fit

Test	Value	Minimum required level
Chi-square (χ^2)	512.67	N/A
Degrees of freedom (df)	230	N/A
Chi-square/df ratio	2.23	<3.00
Comparative Fit Index (CFI)	0.94	>0.90
Tucker-Lewis Index (TLI)	0.92	>0.90
Root Mean Square Error of Approximation (RMSEA)	0.05	<0.08
Standardized Root Mean Square Residual (SRMR)	0.04	<0.08
Goodness of Fit Index (GFI)	0.91	>0.90
Adjusted Goodness of Fit Index (AGFI)	0.85	>0.80
Normed Fit Index (NFI)	0.93	>0.90

Source Author work

construct reliability. The Average Variance Extracted (AVE) values are above 0.50, with Entrepreneurial Mindset having the highest AVE at 0.71, signifying a high level of convergent validity. The AVE values confirm that the majority of variance in the indicators is captured by the respective constructs rather than by error.

Figure 2 illustrates the relationships between digital accessibility, digital skills, and digital literacy as independent variables, entrepreneurial mindset as a mediator, and the entrepreneurial ecosystem as the dependent variable. The model presents significant direct and indirect relationships, indicated by standardized path coefficients. Notably, *digital accessibility (H1a: 0.369)***, *digital skills (H1b: 0.423), and digital literacy (H1c: 0.387)* exert direct positive effects on the entrepreneurial ecosystem, signifying that improved digital access, skill development, and literacy contribute to a stronger business environment. Additionally, digital accessibility (H2a: 0.317*), digital skills (H2b: 0.301*), and digital literacy (H2c: 0.139*) positively influence the entrepreneurial mindset, confirming its mediating role. The most substantial impact in the model is observed in the relationship between

Table 3 Robustness checks

Variable	Cronbach's alpha	Composite Reliability (CR)	Average Variance Extracted (AVE)
Digital accessibility	0.85	0.88	0.62
Digital skills	0.87	0.9	0.68
Digital literacy	0.83	0.85	0.59
Entrepreneurial mindset	0.89	0.91	0.71
Entrepreneurial ecosystem	0.86	0.89	0.66

Source Author work

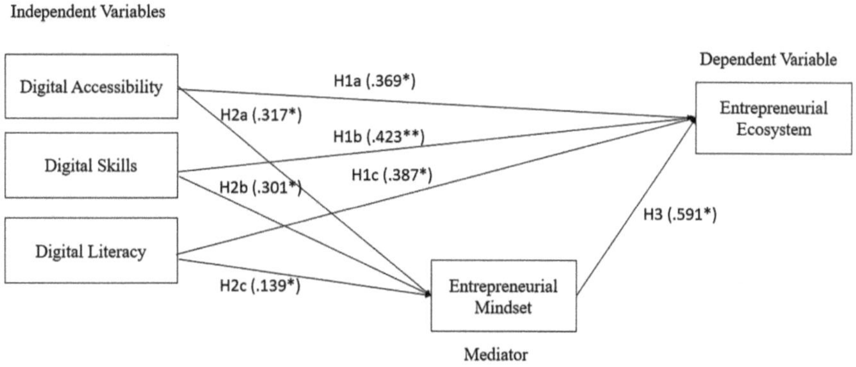

Fig. 2 Mediation-moderation model. *Source* Author work

entrepreneurial mindset and the entrepreneurial ecosystem (H3: 0.591), underscoring the critical role of mindset in fostering a robust entrepreneurial landscape.

5 Discussion

These results support H1a, which states that digital accessibility significantly impacts the entrepreneurial ecosystem. It was expected that digital accessibility would positively correlate with such economic factors because of the importance of technological infrastructure, internet penetration, and the use of digital platforms for business development. In the modern globalized world, entrepreneurs engage in online marketplaces, utilize global supply chains, and offer their products and services to foreign consumers. The positive association between digital accessibility and the entrepreneurial ecosystem found in this study aligns with the extensive body of literature emphasizing the foundational role of digital infrastructure in entrepreneurial development. Access to digital platforms, affordable internet, and ICT infrastructure enables entrepreneurs to harness business intelligence, reach global markets, and participate in innovation networks, factors essential to ecosystem dynamism [3, 33].

The analysis results support H1b, indicating that digital skills positively influence the entrepreneurial ecosystem. The implications of this result emphasize the importance of digital literacy in shaping successful entrepreneurship. Thus, entrepreneurs with the right skills can use digital marketing, online commerce, data analytics, and automation to drive business performance and long-term viability. Digital competencies support business operations and enhance strategic thinking and innovation, allowing entrepreneurs to navigate increasingly complex digital markets [26, 27]. Similarly, entrepreneurs with robust digital skills exhibit higher levels of innovation, resilience, and market responsiveness—traits that feed into a more vibrant and sustainable entrepreneurial ecosystem [23, 25].

Thus, the results confirmed H1c since digital literacy significantly positively affects the entrepreneur's ecosystem. In this case, digital literacy is more complex than knowing how to use technology. This factor includes the ability to evaluate the quality of digital content and the awareness of cybersecurity and ethics in digital issues. As a result, digital literacy ensures that entrepreneurs can access and evaluate high-quality information, protect themselves from cyber fraud, and make data-based decisions. This finding complements other researchers who argue that digital literacy involves cognitive, emotional, and ethical competencies essential for making informed decisions in a digital business environment [31]. The role of digital literacy in enhancing entrepreneurs' ability to critically assess market signals, evaluate risks, and innovate responsibly [5].

The results provide empirical support for H2a, as digital accessibility significantly and positively affects the entrepreneurial mindset. This aligns with the existing literature, underscoring the importance of internet availability, digital platforms, and the state-of-the-art digital infrastructure in forming an entrepreneur's behavior. Entrepreneurs with broad access to digital resources are more likely to discover business opportunities, develop innovative solutions, and participate in digital markets. Digital accessibility increases real-time access to business intelligence, market analyses, and online networking – all important for developing the entrepreneurial mindset. However, access to digital resources remains uneven. Zhou and Cen [33] argue that digital accessibility enhances opportunity recognition and decision-making by increasing exposure to real-time data and entrepreneurial networks. Rutherford et al. [17] further demonstrate that entrepreneurs in rural or underserved regions can overcome geographic and economic constraints through digital tools, supporting the claim that accessibility shapes entrepreneurial intent.

The results also support H2b; thus, possessing digital skills positively affects the entrepreneur's mindset. Digital marketing, data analytics, and e-commerce are examples of why this component's high level of development allows people to promote their business ideas more successfully. Digital skills help to develop self-confidence in their abilities, improve the ability to view problems, and prepare business ideas, which are essential to developing an entrepreneur's mindset. The positive relationship between digital skills and the entrepreneurial mindset reflects a critical synergy between competence and cognition. Digitally skilled individuals exhibit stronger self-efficacy and creative problem-solving, two key drivers of entrepreneurial initiative [14]. Likewise, AI-driven digital skills enhance entrepreneurs' ability to personalize services, engage customers, and iterate business strategies, reinforcing a future-oriented mindset [32].

The findings confirm the test H2c since a statistically significant positive relationship exists between digital literacy and the entrepreneurial mindset. While conventional literacy is associated with the ability to read and write, digital literacy encompasses critical thinking, digital ethics, cyber wellness, and the judgment of sound information. Therefore, entrepreneurs with higher levels of digital literacy will likely be more proficient at knowing market trends, evaluating business information, and playing down online risks. Digital marketing literacy enhances self-efficacy and informed risk-taking, enabling entrepreneurs to conceptualize and execute business strategies effectively [6]. Digital literacy fosters higher-order thinking, ethical judgment, and cyber-resilience—qualities that directly influence entrepreneurial orientation [24]. Educational interventions targeting digital literacy boost technical aptitude and reshape entrepreneurial attitudes and motivations, especially among students and early-stage founders [11].

H3 is notably strongly supported, as the entrepreneurial mindset is a major contributor to entrepreneurial ecosystems and their development. Thus, initiative-taking individuals who take risks in novel ways are a driving force behind what makes an ecosystem an entrepreneurial one in the first place. By facilitating collaboration and exchanging knowledge to innovate business models, the entrepreneurial mindset enables a productive landscape for entrepreneurial activities. Mindset-driven entrepreneurs are instrumental in shaping innovation ecosystems through collaborative behavior, risk-taking, and sustained engagement with change [19].

6 Contributions, Recommendations, and Conclusion

The implications of this study for digital entrepreneurship are also considerable. Using the actual background based on established relationships among such digital accessibilities as availability, accessibility skills, and literacy, along with the entrepreneur mechanism and ecosystem, has provided an essential underpinning. While the existing research has established the relationships of digital entrepreneurs with their sources, this study enhances the knowledge of how the entrepreneurs' mindset influences the use of digital opportunities. The research also developed existing theories by combining RBV and TAM. It also has essential practical implications because the MENA region is highly prospective but underexplored in the context of digital entrepreneurship. Improving digital inclusion and eliminating barriers to access play a primary role in fostering a more inclusive and competitive entrepreneurial ecosystem. Governments and industry stakeholders can facilitate internet coverage extension to underprivileged areas and reduce the costs of digital tools and platforms to make the digital economy more accessible. As a result, with minimized digital accessibility gaps, governments can eliminate geographical and financial limitations, enabling startups and small businesses to access the same digital tools and use them for innovation and development.

References

1. Alzamel, S.: Exploring the role of E-entrepreneurship in fostering future green economy and environmental policies: a study on Saudi women entrepreneurs. JEAPM **26**(01), 2450002 (2024)
2. Avinç, E., Doğan, F.: Digital literacy scale: Validity and reliability study with the Rasch model. Educ. Inf. Technol. 1–47 (2024)

3. Barra, C., Grimaldi, M., Muazzam, A., Troisi, O., Visvizi, A.: Digital divide, gender gap, and entrepreneurial orientation: how to foster technology adoption among Pakistani higher education students? Socioecon. Plan. Sci. **93**, 101904 (2024). https://doi.org/10.1016/j.seps.2024.101904
4. Buttner, E.H., Moore, D.P.: Women's organizational exodus to entrepreneurship: self-reported motivations and correlates with success. J. Small Bus. Manag. **35**, 34–46 (1997)
5. Ganefri, G., Kamdi, W., Makky, M., Hidayat, H., Rahmawati, Y.: Entrepreneurship education and entrepreneurial intention among university students: the roles of entrepreneurial mindset, digital literacy, and self-efficacy. J. Soc. Stud. Educ. Res. **15**(4), 85–134 (2024)
6. Guo, T., Kiratikarnkul, S.: The influence of digital marketing literacy and self-efficacy on the intention to be an e-commerce entrepreneur (2024)
7. Hair, J.F., Risher, J.J., Sarstedt, M., Ringle, C.M.: When to use and how to report the results of PLS-SEM. Europ. Bus. Rev. **31**(1), 2–24 (2019)
8. Hajli, N., Baydarova, I., Nisar, T.: Digital entrepreneurial ecosystem: the role of the sharing economy in driving innovation. Entrep. Reg. Dev. 1–31 (2025)
9. Kakeesh, D.F.: Female entrepreneurship and entrepreneurial ecosystems. J. Res. Mark. Entrep. **26**(3), 485–526 (2024)
10. Kamutuezu, E.U., Winschiers-Theophilus, H., Peters, A.: An exploration of factors influencing the adoption of ICT-enabled entrepreneurship applications in Namibian rural communities (2021). arXiv:2108.09789
11. Khodor, S., Aránega, A.Y., Ramadani, V.: Impact of digitalization and innovation on women's entrepreneurial orientation on sustainable start-up intention. Sustain. Technol. Entrep. **3**(3), 100078 (2024)
12. Kuratko, D.F., Fisher, G., Audretsch, D.B.: Unraveling the entrepreneurial mindset. Small Bus. Econ. **57**(4), 1681–1691 (2021)
13. Lopes, J.M., Gomes, S., Nogueira, E.: Educational insights into digital entrepreneurship: the influence of personality and innovation attitudes. J. Innov. Entrep. **14**(1), 16 (2025)
14. Mirhabibi, A., Shayan, A., Sahraei, S.: Improving digital entrepreneurship readiness of business students: the moderating roles of digital mindset and digital education. Int. J. Manag. Educ. **23**(2), 101151 (2025)
15. Möller, K., Halinen, A.: Relationship marketing theory: its roots and direction. J. Mark. Manag. **16**, 29–54 (2000). https://doi.org/10.1362/026725700785100460
16. Nigam, N., Shatila, K.: Entrepreneurial intention among women entrepreneurs and the mediating effect of dynamic capabilities: empirical evidence from Lebanon. Int. J. Entrep. Behav. Res. **30**(4), 916–937 (2024)
17. Rutherford, M.W., Whitacre, B.E., Captain, L., Ekin, S., Angle, J., Hensley, T., O'Hara, J.F.: Promoting rural entrepreneurship through technology: a case study using productivity enhancing technology experience kits (PETE-kits) (2023). arXiv:2311.04327
18. Samsami, M., Kolaly, H.E., González-Pernía, J.L., Boutaleb, F.: Gender roles shaping the entrepreneurial mindset: embedded in the entrepreneurial ecosystem and impacted by the pandemic. Eur. J. Int. Manag. **22**(4), 551–575 (2024)
19. Shatila, K., Aránega, A.Y., Soga, L.R., Hernández-Lara, A.B.: Digital literacy, digital accessibility, human capital, and entrepreneurial resilience: a case for dynamic business ecosystems. J. Innov. Knowl. **10**(3), 100709 (2025)
20. Shatila, K., Nigam, N., Mbarek, S.: Seeds of change: nurturing entrepreneurial ecosystems for sustainable enterprises in Lebanon and Jordan. J. Entrep. **33**(4), 897–924 (2024)
21. Soltanifar, M., Smailhodžić, E.: Developing a digital entrepreneurial mindset for data-driven, cloud-enabled, and platform-centric business activities: practical implications and the impact on society. Digit. Entrep. **3**, 3–21 (2021)
22. Syed, R.T., Alzahmi, R.A., Tariq, U.: Digital entrepreneurship education in universities through the lens of educators: evidence from the United Arab Emirates. Cogent Educ. **11**(1), 2409472 (2024)
23. Troisi, O., Visvizi, A., Grimaldi, M.: Rethinking innovation through industry and society 5.0 paradigms: a multileveled approach for management and policy-making. Eur. J. Innov. Manag. **27**(9), 22–51 (2024). https://doi.org/10.1108/EJIM-08-2023-0659

24. Ushakov, D.S., Ivanova, D.G., Rubinskaya, E.D., Shatila, K.: The mediating impact of innovation on green entrepreneurship practices and sustainability. In: Climate-Smart Innovation: Social Entrepreneurship and Sustainable Development in the Environmental Economy, pp. 3–18 (2023)
25. Visvizi, A., Troisi, O., Grimaldi, M.: Digital divide and entrepreneurial orientation in the global south: quantifying and explaining the nexus. In: Visvizi, A., Troisi, O., Grimaldi, M. (eds.) Research and Innovation Forum 2022. RIIFORUM 2022. Springer Proceedings in Complexity. Springer, Cham (2023). https://doi.org/10.1007/978-3-031-19560-0_62
26. Visvizi, A., Malik, R., Guazzo, G.M., Çekani, V.: The industry 5.0 (I50) paradigm, blockchain-based applications and the smart city. Eur. J. Innov. Manag. **28**(1), 5–26 (2025). https://doi.org/10.1108/EJIM-09-2023-0826
27. Visvizi, A., Calzada, I., Kozłowski, K., Troisi, O.: Road-mapping Generative AI (Gen-AI) and its impact: AI for whom and for what? In: Visvizi, A., Kozłowski, K., Calzada, I., Troisi, O. (eds.) Multidisciplinary Movements in AI and Generative AI: Society, Business, Education. Elgar (2025). 978 1 03535 865 6
28. Walker, E., Brown, A.: What success factors are important to small business owners? Int. Small Bus. J. **22**, 577–594 (2004). https://doi.org/10.1177/0126624260404711
29. Wibisono, E.: The digital entrepreneurial ecosystem in the European Union: evidence from the digital platform economy index. Eur. Plan. Stud. **31**(6), 1270–1292 (2023)
30. Xin, K.K., Pearce, J.L.: Guanxi: connections as substitutes for formal institutional support. Acad. Manag. J. **39**, 1641–1658 (1996). https://doi.org/10.2307/257072
31. Young, R., Wahlberg, L., Davis, E., Abhari, K.: Towards a theory of digital entrepreneurship mindset: The role of digital learning aptitude and digital literacy. In: 26th Americas Conference on Information Systems, AMCIS, vol. 1, no. 10 (2020)
32. Zaki, Y., Al Muwali, A., Mahdi, N.: The role of nurturing technopreneurship education and building university students' entrepreneurial mindsets and skill sets in fostering digital innovation and augmenting the tech start-up ecosystem in Bahrain. Int. J. Learn. Teach. Educ. Res. **20**(6), 152–173 (2021)
33. Zhou, J., Cen, W.: Digital entrepreneurial ecosystem embeddedness, knowledge dynamic capabilities, and user entrepreneurial opportunity development in China: the moderating role of entrepreneurial learning. Sustainability **16**(11), 4343 (2024)

The Dark Side of the Metaverse: Managing Conflict and Human Resource Management Challenges in Virtual Environments

Francesca Loia, Cecilia Maltempo, Rosario Marrapodi, Stefania Mele, Mario Pezzillo Iacono, and Marcello Martinez

Abstract Nowadays, the metaverse represents a disruptive technology that companies are increasingly integrating into their business models, particularly in human resource management. One of the most promising aspects of its adoption concerns knowledge management, as virtual environments can enhance collaboration, knowledge sharing, and training. However, despite its recognized potential, the metaverse can also generate conflicts among employees, which may negatively impact organizational dynamics. Virtual interactions differ significantly from traditional face-to-face communication, potentially leading to misunderstandings, power imbalances, and increased tensions among team members. This paper explores this "dark side" of the metaverse by focusing on conflict management within virtual teams. To investigate this phenomenon, a case study methodology was adopted, which is particularly suitable for analyzing contemporary and emerging topics. Specifically, the study examines relevant cases of innovative companies that have implemented metaverse-based solutions in their operations. The findings highlight key challenges in managing team conflicts within virtual environments and provide valuable insights into mitigating such tensions. Moreover, the study offers practical implications for HR managers and IT specialists, outlining strategies to foster a more effective and conflict-resilient

F. Loia (✉) · C. Maltempo · S. Mele · M. P. Iacono · M. Martinez
Department of Economics, University of Campania Luigi Vanvitelli, Capua, CE, Italy
e-mail: francesca.loia@unicampania.it

C. Maltempo
e-mail: cecilia.maltempo@unicampania.it

S. Mele
e-mail: stefania.mele@unicampania.it

M. P. Iacono
e-mail: mario.pezzilloiacono@unicampania.it

M. Martinez
e-mail: marcello.artinez@unicampania.it

R. Marrapodi
University of Naples Parthenope, Naples, Italy
e-mail: rosario.marrapodi@studenti.uniparthenope.it

A. Visvizi et al. (eds.), *Research and Innovation Forum 2025 Volume 1*,
Springer Proceedings in Complexity, https://doi.org/10.1007/978-3-032-09066-9_4

virtual workspace. By addressing these challenges, organizations can better harness the full potential of the metaverse while minimizing its risks.

Keywords HRM · Metaverse · Conflict · Virtual team · Decent work · Case study

1 Introduction

The rise of the metaverse has sparked considerable interest across various sectors, offering new opportunities for innovation and transforming how we interact in digital spaces. As this immersive virtual world continues to evolve, its potential to reshape the landscape of work and collaboration becomes increasingly apparent. The application of metaverse technologies in the workplace promises enhanced interactivity, more engaging environments, and novel approaches to team collaboration. Companies like Walmart and Samsung have leveraged virtual environments to improve employee learning, transcending geographical boundaries [4, 26]. KFC's 2017 VR training program, "The Hard Way," demonstrated how virtual reality can reduce training time and enhance efficiency. This highlights the potential for metaverse-based HR training to evolve from task-oriented learning to fostering social collaboration and teamwork. Similarly, Deloitte's use of virtual onboarding highlights the potential of the metaverse to foster integration in remote work settings [16]. However, these innovations are not without unintended consequences. While the metaverse opens new frontiers for innovation and workplace engagement, it also unveils a less explored yet critical dimension of risk. Concerns around privacy, behavioral regulation, social detachment, and digital inequality are becoming increasingly pressing [9]. The use of avatars and the lack of physical cues can obscure emotional nuance and complicate communication, especially in conflict-prone scenarios [22]. At the same time, the rise of immersive performance-tracking tools introduces complex ethical questions about surveillance and the potential for excessive monitoring [6]. Beyond individual interactions, the transition from task-based activities to more socially intricate forms of engagement—such as informal hierarchies, alliances, and peer competition—may unintentionally amplify organizational tensions [37]. Without appropriate safeguards, problems like exclusion, harassment, or implicit bias—already documented on public VR platforms—could easily permeate corporate settings [32], calling for urgent reflection on how to manage these emerging dynamics responsibly. For instance, a Texas mother reported that her 12-year-old daughter received inappropriate messages from an adult posing as a teenager on the metaverse app Zepeto, highlighting the risks of grooming and cyberbullying in virtual environments [23]. Additionally, China's proposal to extend its social credit system into the metaverse raises significant privacy and surveillance concerns, as it would involve extensive data collection and monitoring of users' virtual behaviors [3].

Thus, with these possibilities come a set of challenges that require careful consideration, particularly in the realm of human resources management (HRM). Despite its promise of enhanced interactivity and novel ways to collaborate, the metaverse introduces complexities that may complicate traditional organizational dynamics. These challenges raise important questions about the impact of these virtual spaces on interpersonal relationships, team cohesion, and the fundamental principles of resource management in organizations. Accordingly, this paper aims to examine the potential negative implications of metaverse applications, with a focus on the challenges they pose in managing conflict and human resources effectively. To conceptually frame the investigation of the unintended consequences associated with metaverse adoption in HRM, this study draws on the Dark Side of Information Technology (DSIT) framework [28, 29]. This perspective critically examines how digital technologies, while designed to improve efficiency, communication, and engagement, may simultaneously produce negative outcomes that impact individual well-being, organizational performance, and ethical standards.

Leveraging these considerations, this paper aims to explore the impact of the metaverse on HRM, focusing on both the opportunities it presents and the complexities it introduces. The research question guiding this study is: "How does the metaverse influence the emergence and management of conflict in virtual teams, and what are the key implications for human resource management?" To address this question, a qualitative research approach is employed, utilizing the Case Study model. The study analyzes the experiences of several international companies utilizing metaverse technologies in their operations, focusing on insights from HR professionals and organizational leaders regarding the effects of virtual environments on human resources management.

Despite these challenges, the study highlights the transformative potential of the metaverse for human resource management. By carefully addressing the obstacles it presents—such as ensuring equitable access to technology, supporting digital literacy, and fostering inclusive work practices—organizations can harness the benefits of the metaverse while mitigating its risks.

This paper is structured as follows: Sect. 2 explores the opportunities and challenges of digital technologies in HRM; Sect. 3 outlines the methodological approach; Sect. 4 presents a discussion of the findings; Sect. 5 concludes with key insights and recommendations for future research.

2 Literature Review. Between Innovation and Disruption: Rethinking HRM and Conflict in the Age of the Metaverse

Initially conceived within the realm of digital gaming, the metaverse has rapidly evolved into one of the most compelling and strategically significant frontiers for contemporary enterprises, reshaping the paradigms of interaction, branding, and virtual engagement. Specifically, the adoption of the metaverse in organizations is redefining human resource management (HRM) practices, presenting both new opportunities and challenges for training, recruitment, and collaboration among distributed teams (Table 1). By fostering active participation and engagement, the metaverse has been shown to enhance learners' motivation and comprehension, promoting deeper and more meaningful experiences [37]. In particular, digital and virtual simulations allow users to apply abstract concepts in practical scenarios, encouraging exploration, experimentation, and critical thinking [5, 13]. The integration of the Metaverse into professional and learning settings is further supported by its ability to simulate collaborative experiences, enabling communication, problem-solving, and peer learning within realistic, controlled environments [14]. These platforms also allow learners to navigate and reflect on complex interactions, enhancing metacognitive skills and decision-making abilities [7]. According to [37], engagement in the metaverse is not primarily driven by social dynamics, as one might commonly think, but rather by game tasks and reward mechanisms in the initial stages. This approach, observed in the game Nova Empire, suggests that user engagement is initially driven by performance-based incentives and the completion of specific objectives. In later stages, as players engage with alliances and more complex social interactions, the social aspect plays a crucial role in maintaining engagement [37]. These findings have significant implications for human resource management in the metaverse, particularly concerning training and recruiting.

However, the debate on the "dark side" of the metaverse is becoming increasingly relevant, as the immersive nature of these environments can exacerbate challenges related to anonymity, behavior moderation, social isolation, and data security [9]. Despite its promise, other concerns persist regarding ethical dilemmas, data security vulnerabilities, and the lack of universal standards for regulation and behavior [24]. The anonymity and reduced nonverbal cues typical of virtual environments can intensify misunderstandings and miscommunications, creating tensions within the team. Therefore, HR must not only encourage effective social interactions but also integrate conflict resolution mechanisms and moderation into virtual experiences to ensure that social dynamics do not become fertile ground for conflict or exclusion but instead support the creation of an inclusive and productive environment [22, 37]. Yang et al. [34] explored the adoption of virtual reality in medical staff training, highlighting gains in efficiency and cost-effectiveness while also cautioning against technostress, social isolation, and the erosion of professional identity—concerns equally pertinent to corporate environments. Also, one of the primary risks associated with the metaverse is its potential to amplify conflicts within virtual teams. The anonymous nature of avatars and the absence of nonverbal cues make it more challenging to interpret emotions and intentions, thereby increasing the likelihood of misunderstandings and tensions among colleagues [22]. Moreover, the phenomenon of "diminished reality"—the gradual detachment from the physical world in favor of virtual immersion—can lead to feelings of alienation and social isolation, ultimately compromising team cohesion and the psychological well-being of employees [32]. The immersive nature of the metaverse may also facilitate cyberbullying, harassment, and discrimination, issues that have already been documented on platforms such as VRChat and Horizon Worlds [9].

Table 1 Impacts of the metaverse in corporate environments in HRM dynamics

Aspect	Description	Literature source
HRM innovation	Enhances recruitment, onboarding, and training by offering immersive, engaging, and accessible experiences for distributed or remote teams	[6, 20, 37]
Immersive learning and engagement	Immersive environments foster deeper learning through gamification, exploration, and practical application of abstract concepts	[5, 13, 37]
Skill development and collaboration	Simulated environments improve collaboration, critical thinking, problem-solving, metacognitive reflection, and team interaction	[14]
Inclusivity and flexibility	Virtual platforms can increase inclusiveness and flexibility (e.g., through remote hiring, virtual hours), supporting work-life balance	[6, 25]
Emotional disconnection and isolation	Excessive immersion can lead to alienation, detachment from reality, and weakened team cohesion and psychological well-being	[6, 32, 34]
Communication barriers and conflict risk	Avatar anonymity and lack of nonverbal cues increase the risk of misunderstanding, unresolved tensions, and interpersonal conflicts	[22, 37]
Cyber risks and ethical issues	Data collection for surveillance, profiling, or biometric analysis raises ethical concerns around privacy, discrimination, and trust	[6, 9, 15]
Cultural and human interaction loss	Virtual environments may weaken the transmission of corporate culture and reduce the richness of interpersonal relationships and soft skill cues	[6, 20]

(continued)

Table 1 (continued)

Aspect	Description	Literature source
Moderation and governance challenges	AI-based moderation can assist conflict management but raises concerns about transparency and the absence of clear, universal standards	[2, 6, 21, 24]
Online misconduct and abuse	Risks of cyberbullying, harassment, and discrimination are prevalent in some immersive platforms if unregulated	[9]

Source Authors' elaboration

Another critical concern regarding virtual teams in the metaverse is security and privacy. Metaverse environments collect vast amounts of sensitive data, including eye movements, facial expressions, and social interactions, which can be leveraged for purposes that are not always ethical, such as advanced employee profiling or predictive performance evaluations [9]. This digital surveillance could introduce new forms of workplace discrimination, as career decisions may increasingly be based on biometric and behavioral data extracted from virtual worlds without proper regulation. Persistent digital immersion, where colleague interactions are mediated by avatars, may lead to a sense of disconnection and exacerbate preexisting tensions, making it more difficult to build trust and maintain organizational cohesion [6]. One of the most critical aspects of this scenario is the need to rethink traditional HR strategies to align with metaverse dynamics. Del Barone et al. [6] highlight that startups leveraging immersive virtual spaces for recruitment and onboarding can enhance accessibility and inclusivity but simultaneously reduce the quality of human interaction and complicate the assessment of soft skills. While effective in attracting digitally skilled talent, this approach carries the risk of excessive detachment from physical reality, leading to difficulties in managing corporate culture and organizational values [20]. On the other hand, performance management and employee evaluation must be redefined in virtual work environments. Some organizations are already experimenting with biometric tracking and behavioral analytics within immersive metaverse spaces, raising ethical concerns about employee privacy and workplace surveillance [6]. The risk of "hyper-monitoring" could contribute to increased stress and conflict between employees and management, ultimately undermining organizational well-being [15]. Another critical challenge concerns work-life balance, as the boundaries between professional and personal life become increasingly blurred. Research by [6] shows that some startups are introducing "virtual working hours" and tools designed to indicate the transition between work and leisure, aiming to mitigate hyper-connectivity and burnout risks. However, the effectiveness of these initiatives depends on whether companies truly prioritize employee well-being rather than merely implementing superficial policies [25].

Regarding conflict management in virtual teams, the metaverse introduces new challenges that organizations must address to ensure inclusive and secure work environments. The anonymity of avatars and the reduction of nonverbal communication make it more difficult to accurately interpret intentions and emotions, thereby increasing the likelihood of misunderstandings and unresolved tensions [22]. Recent studies suggest that AI-driven virtual facilitators capable of monitoring and mediating team interactions could serve as an effective solution, yet concerns persist regarding transparency and trust in automated moderation systems [6].

These critical dynamics reflect broader concerns raised in the emerging literature on the dark side of information technology (DSIT), which explores how digital innovations—while enabling productivity and engagement—can also lead to unintended and often detrimental outcomes such as technostress, privacy erosion, social isolation, and workplace surveillance [28, 29]. This perspective provides a useful lens for understanding how immersive technologies like the metaverse may exacerbate organizational tensions and challenge fundamental HRM principles, especially in relation to employee well-being, autonomy, and psychological safety.

In conclusion, the integration of the metaverse into HRM, therefore, presents both opportunities and risks. While this technology offers new modes of engagement and collaboration, it also introduces unprecedented challenges that necessitate a fundamental reassessment of human resource management practices. Organizations must develop clear governance strategies, invest in workforce training, and implement moderation and support mechanisms to prevent the deterioration of workplace relationships in virtual teams [15]. Only by adopting a balanced approach between innovation and organizational well-being can companies fully harness the metaverse's potential without amplifying its "dark side" in workplace dynamics. To unlock its full potential, the Metaverse requires strategic planning, technological infrastructure, and clear ethical frameworks to ensure safe, equitable, and meaningful adoption across contexts [2, 21].

3 Research Design: A Multiple Case Study Approach

According to [35], a case study is an empirical investigation that places real phenomena at the center of analysis and seeks to observe them in their uniqueness. Case studies have many strengths represented by the depth of analysis, high conceptual validity, understanding of context and process, and finally, the possibility of promoting new hypotheses and new research questions [36].

While the Single Case Study typology is generally used in the observation and analysis of particular and unique events, we use the Multiple Case Study to examine recurring processes and/or phenomena through the analysis of a number of similar cases. Such an approach allows for a more direct comparison of the similarities and differences in implementation practices in the various contexts considered [27] and allows for more general conclusions [10].

Following a qualitative and exploratory research approach, the cases under analysis involve five international companies active in the application of the technologies metaverse-based. Adopting a qualitative and exploratory research design, this study relied on secondary data—namely, documents and official information related to the five selected case studies. Secondary data present several advantages, including access to extensive datasets readily available online through systematic search strategies, as well as efficient data retrieval [12]. The study relies on secondary qualitative data obtained from a variety of publicly accessible and organizational sources, including company reports, official websites, white papers, press releases, HR blogs, and third-party industry analyses. For each of the five selected companies, we gathered documents that specifically describe their metaverse-related HR initiatives. The selection criteria prioritized relevance to HRM functions (e.g., onboarding, training, recruitment), recency (published after 2020), and source credibility (authentic corporate documents or peer-reviewed sources). To ensure data validity, we employed source triangulation by cross-referencing information from multiple documents for each case. Where possible, corporate sources were validated against independent analyses or academic literature. We adopted criteria of credibility, authenticity, and representativeness [19, 35] to select and assess the quality of the data used.

The cases examined involve five prominent multinational corporations from distinct sectors (as shown in Table 2): Accenture, Microsoft, BMW, Johnson and Johnson, and H&M. The selection of Accenture, Microsoft, BMW, Johnson and Johnson, and H&M as case studies is grounded in both their pioneering roles in integrating metaverse technologies into human resource management and the accessibility of comprehensive, verifiable documentation regarding their initiatives. These organizations, operating across diverse sectors—professional services, technology, automotive, healthcare, and retail—offer a heterogeneous yet complementary perspective on the application of immersive technologies in onboarding, training, and employee engagement. Their initiatives represent some of the most mature and well-documented examples of metaverse-driven HRM strategies currently available, thus providing fertile ground for comparative analysis and the extrapolation of broader trends. The availability of official reports, corporate publications, and third-party analyses further substantiates the relevance and analytical value of these cases within an exploratory research framework. Case selection, thus, was guided by both data accessibility and the potential to extrapolate recurring themes across diverse contexts, thereby enabling broader generalizations.

The data were analyzed using a qualitative thematic analysis approach, guided by the theoretical framework of the Dark Side of Information Technology (DSIT). Following [19] model, we organized the data into matrices capturing the opportunities and risks associated with metaverse adoption. Recurring themes across the cases (e.g., technostress, surveillance, training innovation, inclusion) were identified through inductive coding. Comparative analysis was conducted to identify patterns and divergences between companies.

The analysis yielded meaningful insights into the strategic adoption of metaverse-based solutions and their associated organizational benefits and challenges.

Table 2 Case studies selection on metaverse integration in HRM

Company	Sector	Metaverse HR initiative	Implications for HRM
Accenture	Professional Services/ Consulting	Development of a proprietary metaverse platform ("Nth floor") for onboarding, learning, and virtual collaboration among its global workforce	Demonstrates scalability of immersive onboarding; supports hybrid workforce models and fosters global cohesion
Microsoft	Technology/IT	Development of microsoft mesh, enabling immersive collaboration, virtual meetings, and training for remote teams	Facilitates remote team collaboration and engagement; enhances productivity through immersive tools
BMW	Automotive/ Mobility	Implementation of virtual training environments for maintenance, machinery handling, and product design simulations	Provides risk-free, scalable training for technical staff; accelerates skill development in a controlled virtual setting
Johnson and Johnson	Healthcare/ Biotechnology	Virtual simulations for medical staff training, including surgical procedures and healthcare-related skills development	Reduces training costs, facilitates global access to medical education, and enhances surgical training
H&M	Retail/Fashion	Virtual fashion collections, immersive experiences on platforms, and virtual showrooms aimed at enhancing customer interaction and promoting sustainability	H&M's metaverse initiatives, though primarily marketing-focused, indirectly benefit HRM by strengthening employer branding and attracting digitally native talent. These efforts foster an innovation-oriented culture and stimulate cross-functional skill development, yet the strategic use of immersive technologies for core HR functions remains largely unexplored

Source Authors' elaboration from online sources

4 Findings and Discussion

The cross-case analysis of Accenture, H&M, Microsoft, BMW, and Johnson and Johnson offers valuable insights into the opportunities and challenges associated with the integration of metaverse technologies in human resource management (HRM). The cases highlight a variety of emerging themes, which can be grouped into both advantageous and challenging dimensions (Table 3). The critical dimensions emerging from the case study analysis can be meaningfully interpreted through the lens of the Dark Side of Information Technology (DSIT) framework [30, 31], which highlights the unintended and often detrimental consequences of digital innovation. While metaverse technologies hold transformative potential for human resource management, their implementation may also introduce forms of technostress, erosion of privacy, emotional detachment, and perceived loss of control—all of which fall within the scope of DSIT. In the following discussion, we explore how metaverse adoption can both facilitate and undermine HRM goals, with a focus on how tensions and conflicts manifest across the five cases.

Opportunities in Metaverse-Enhanced HRM

1. ***Immersive Learning and Skill Development***

One of the most striking benefits of the metaverse is its capacity to revolutionize experiential learning and accelerate skill acquisition. Companies have successfully implemented immersive, scenario-based training environments that replicate real-world operational contexts. This approach allows employees to acquire competencies more engagingly and efficiently compared to traditional learning methods. By simulating complex scenarios that would otherwise be difficult or costly to replicate in the physical world, the metaverse enables employees to practice critical skills without the associated risks or limitations. Moreover, the interactive nature of these environments increases employee engagement, as it immerses learners in dynamic and challenging situations that encourage active problem-solving. This leads to improved retention of knowledge and quicker mastery of new skills, making the metaverse a powerful tool for learning and development in HRM—as in the case of Johnson and Johnson [11].

2. ***Employer Branding and Talent Engagement***

The metaverse also provides significant advantages in terms of employer branding and talent engagement. Companies have leveraged virtual platforms to craft innovative recruitment experiences, allowing them to showcase their corporate cultures and engage with potential candidates in immersive environments. Other companies indirectly benefits HRM by strengthening employer branding and attracting digitally native talent. Thus, these virtual experiences are particularly effective in attracting digitally native talent, who are more likely to resonate with forward-thinking, technology-driven organizations. By offering virtual tours, interactive interviews, gamified recruitment processes, and digital experiences, these companies are able to differentiate themselves in a competitive job market. Additionally, the metaverse provides a platform for creating personalized, engaging experiences that allow prospective employees to better understand the company's values, work culture, and career opportunities. This not only attracts top talent but also strengthens employer branding, positioning these organizations as pioneers in digital transformation. For instance, H&M leveraged the metaverse to attract young audiences through immersive experiences like "Loooptopia" on Roblox, where users can explore virtual environments, participate in interactive games, and create digital clothing for their avatars.

Table 3 Opportunities and challenges of metaverse integration in HRM through the lens of the Dark Side of IT (DSIT) framework

Dimension	Description	Link to DSIT framework	Implications
Opportunities			
Immersive learning	Virtual training environments accelerate skill acquisition and engagement	**Intended Positive Consequences**: Enhances learning efficacy and employee development	Improved knowledge retention and faster mastery of skills
Employer branding	Virtual recruitment experiences attract digitally native talent	**Intended Positive Consequences**: Strengthens employer brand and engagement	Increased talent attraction and differentiation in competitive markets
Inclusivity and well-being	Customizable, accessible virtual environments promote psychological safety and equity	**Intended Positive Consequences**: Fosters inclusivity and employee well-being	Healthier work-life balance and reduced bias
Global collaboration	Real-time immersive spaces enable geographically dispersed teams to collaborate effectively	**Intended Positive Consequences**: Enhances communication and cross-cultural collaboration	Better knowledge exchange and stronger organizational cohesion
Challenges			
Technological inequality	Unequal access to infrastructure and digital skills limits participation	**Unintended Negative Consequences**: Digital divide causes exclusion and inequity	Risk of alienation and disparities in development opportunities
Privacy and surveillance	Collection of biometric and behavioral data raises ethical and legal concerns	**Unintended Negative Consequences**: Invasion of privacy, loss of control, increased technostress	Need for data governance policies and robust security to maintain trust
Cultural resistance	Organizational inertia and skepticism hinder adoption of virtual tools	**Unintended Negative Consequences**: Resistance limits technology acceptance and use	Necessity for change management and authentic virtual design

(continued)

Table 3 (continued)

Dimension	Description	Link to DSIT framework	Implications
Behavioral risks	Harassment and misconduct possible due to anonymity and immersive nature of virtual spaces	**Unintended Negative Consequences**: Increased risk of unethical behavior and conflict	Implementation of codes of conduct and monitoring tools
Psychosocial challenges	Prolonged virtual interaction leads to isolation, fatigue, and reduced social connection	**Unintended Negative Consequences**: Emotional detachment and mental fatigue	Importance of psychological support and balancing virtual and real interaction

Source Authors' elaboration

Additionally, H&M hosted a virtual concert with the Filipino music group BGYO, drawing over 11,000 visitors and connecting with youth through music and pop culture [8].

3. ***Inclusivity and Employee Well-Being***

A key advantage of the metaverse is its potential to foster inclusivity and enhance employee well-being [1]. Companies have utilized virtual environments to create customizable and accessible experiences that cater to a wide range of user needs. These environments offer employees a sense of psychological safety by enabling them to participate in work-related activities in a more inclusive and comfortable setting. For instance, virtual platforms can accommodate employees with disabilities by offering customizable avatars and interfaces that make participation easier for individuals with physical or cognitive impairments. Furthermore, the anonymity provided by virtual spaces can reduce biases related to physical appearance, gender, or ethnicity, promoting a more inclusive and equitable work environment. The metaverse also facilitates employee well-being by providing virtual spaces for relaxation, team bonding, and stress relief, contributing to a healthier work-life balance.

4. ***Global Collaboration and Diversity***

The metaverse enables seamless global collaboration, particularly in organizations with geographically dispersed teams, as in the case of the Global Collaboration Village, powered by Microsoft Mesh, as a prime example of how metaverse technology can bring people and communities together in new ways [18]. Companies have taken advantage of this capability to create virtual spaces where employees from different parts of the world can collaborate in real-time, overcoming the limitations of time zones and physical distance. The immersive nature of these spaces enhances communication and fosters a sense of presence, making remote collaboration more natural and productive. This facilitates the exchange of ideas, knowledge, and expertise across diverse cultural contexts, helping organizations to harness the full potential of their global talent pool. The metaverse's ability to bring together employees from various backgrounds in a shared virtual environment can also promote diversity and inclusivity within teams, enhancing cross-cultural understanding and strengthening organizational cohesion.

Challenges and Conflicts in Metaverse Integration

1. ***Technological Inequality and the Digital Divide***

Despite its many advantages, the adoption of the metaverse is hindered by technological inequality and the digital divide [1]. Access to immersive tools and platforms remains uneven, both geographically and demographically. In some regions, employees may lack the necessary infrastructure—such as high-speed internet or advanced computing devices—to fully participate in metaverse-based activities. Additionally, the digital literacy required to navigate and engage with virtual environments may be lacking among certain segments of the workforce. Employees from older generations or those in lower-skilled positions may struggle to adapt to the metaverse, leading to a potential sense of exclusion or alienation. This divide risks creating disparities in access to career development opportunities, training, and engagement activities, making it essential for organizations to address this issue by providing digital literacy training and ensuring that the necessary infrastructure is in place to support all employees.

2. ***Data Privacy and Surveillance Concerns***

The integration of metaverse technologies into HRM also raises significant concerns related to data privacy and surveillance. As virtual platforms collect sensitive biometric, behavioral, and interactional data, organizations face increasing scrutiny over their data practices. The ability to track employees' movements, facial expressions, and emotional responses in virtual environments presents ethical and regulatory challenges, particularly with regard to user autonomy and consent. The collection and analysis of such data must be carefully managed to avoid infringements on employee privacy and to ensure compliance with data protection laws. Failure to address these concerns could lead to employee distrust, legal liabilities, and reputational damage. Companies must establish clear data governance policies and transparent practices to safeguard employee information and maintain ethical standards in their use of metaverse technologies. In such senses, Microsoft has acknowledged the privacy risks associated with the metaverse, particularly concerning biometric data collection. The company discusses the potential for fraud and phishing attacks in the metaverse, where avatars could impersonate real individuals, posing identity theft risks. Microsoft stresses the necessity for robust security measures to protect users' personal information in these immersive environments [17].

3. ***Cultural Resistance and Organizational Inertia***

The adoption of the metaverse within existing organizational structures is often met with cultural resistance and inertia. For instance, Accenture has acknowledged that cultural resistance can pose challenges to the adoption of metaverse technologies within organizations. In their article "Metaverse continuum set to redefine how the world operates," Accenture discusses how the metaverse stands to bring people-centric technology experiences to life, redefining how organizations work, collaborate, and interact. However, they emphasize the need for a new focus on interoperability and the importance of thinking ahead about platforms and partnerships to make these interfaces and interactions seamless. Many companies, particularly those with lower digital maturity or rigid hierarchical structures, may struggle to embrace the metaverse as a viable tool for HRM. Employees accustomed to traditional, face-to-face interactions may be skeptical of the artificiality of virtual environments, perceiving them as less authentic or effective than in-person communication. Additionally, organizations with strong, entrenched cultures may find it challenging to integrate the metaverse into their established ways of working. This resistance can manifest in reluctance to adopt new technologies, negative perceptions of the metaverse's relevance, and a general sense of discomfort with virtual workspaces. To overcome these barriers, organizations must invest in change management strategies that emphasize the value of the metaverse, provide training and support for employees, and ensure that virtual environments are designed in a way that feels authentic and relevant to their work culture.

4. ***Behavioral Risks and Misconduct***

Among the emerging criticalities associated with the integration of the metaverse into Human Resources practices is the potential for employee misconduct and inappropriate behavior within virtual environments. Instances of harassment and unethical conduct have already been documented on public metaverse platforms, raising concerns about the replicability of such issues in corporate virtual workspaces. The immersive and, at times, anonymous nature of these environments may inadvertently facilitate unprofessional behavior, thereby necessitating the establishment of clear codes of conduct and robust monitoring mechanisms.

This has been recognized by Microsft which regarding AltspaceVR have introduced personal space features in their virtual worlds to combat issues like groping and harassment. For instance, in their AltspaceVR platform, Microsoft implemented anti-harassment protections, including stopping avatars from getting uncomfortably close, closing public social hubs, automatically muting new attendees at events, increasing moderation, and requiring a Microsoft account to use VR social spaces [33].

5. ***Psychosocial Challenges***

Prolonged engagement in virtual interactions may contribute to psychosocial discomfort, particularly in the form of social isolation and diminished interpersonal connection. The absence of physical presence and organic social interaction can impair team cohesion and employee well-being, potentially leading to disengagement and mental fatigue. As such, organizations must proactively address these risks by fostering digital civility, providing psychological support structures, and balancing virtual collaboration with opportunities for authentic human connection. As reported in [1], 3ell-being is a critical human dimension in metaverse design, including addressing mental health risks from prolonged virtual immersion such as isolation and digital fatigue.

Thus, the integration of metaverse technologies into HRM practices presents a complex interplay of opportunities and challenges. While offering innovative avenues for learning, collaboration, and inclusivity, it also introduces conflicts related to technology access, privacy, cultural adaptation, and employee well-being. Organizations must navigate these dynamics thoughtfully, leveraging insights from industry leaders to harness the metaverse's potential responsibly.

5 Final Remarks and Conclusions: Unveiling the Dark Side of the Metaverse in Terms of Conflict Management

In conclusion, the integration of the metaverse into HRM practices presents both significant opportunities and considerable challenges. The positive emerging themes highlight the transformative potential of the metaverse in enhancing learning, engagement, inclusivity, and global collaboration. However, the challenges—ranging from technological inequality and data privacy concerns to cultural resistance —underscore the complexities organizations must navigate to fully capitalize on this technology. By addressing these challenges with careful planning, inclusive design, and ethical governance, companies can unlock the full potential of the metaverse while mitigating its risks.

The findings from this study underscore the metaverse's capacity to revolutionize HRM practices by offering technologically enhanced, immersive environments that support remote collaboration, experiential learning, and novel forms of employee engagement. These digitally mediated spaces can potentially redefine how organizations approach talent development, teamwork, and inclusion by transcending geographical boundaries and promoting flexible, user-centric work experiences. However, alongside these promising innovations, several critical challenges have emerged, particularly those about the "dark side" of metaverse adoption within the HRM domain. The study highlights significant concerns related to human and social sustainability, such as the widening digital skill gap, the emergence of technostress, and unequal access to virtual infrastructures. These issues reflect broader barriers to the democratization of technology and raise fundamental questions about inclusion and fairness in digitally transformed workplaces. The immersive nature of the metaverse, while fostering engagement, may also amplify risks of psychological fatigue and interpersonal disconnection. The absence of physical co-presence and non-verbal communication cues can exacerbate misunderstandings among employees, complicating conflict resolution and undermining team cohesion. Moreover, the persistent connectivity inherent in metaverse environments can blur boundaries between professional and personal life, intensifying stress and impairing work-life balance.

From an organizational perspective, these dynamics present unprecedented governance challenges. Human resource managers must not only navigate the technical complexities of deploying metaverse technologies but also address their socio-psychological implications. Without proper safeguards, these immersive platforms may reinforce existing inequalities, heighten employee vulnerability, and erode social capital within teams.

Given these challenges, companies must develop strategies to mitigate the risks associated with the metaverse and ensure inclusive and secure work environments. Establishing clear codes of conduct, implementing advanced moderation tools, and fostering a corporate culture that prioritizes collaboration in digital settings are essential steps. Combining virtual interactions with periodic physical meetings could help maintain team cohesion and reduce isolation. Moreover, businesses must invest in employee training on ethics and behavior in virtual worlds to promote responsible and sustainable metaverse integration. Although the metaverse presents a significant opportunity to redefine the future of work, without proper governance, it risks exacerbating rather than resolving existing organizational challenges. Human resource management plays a pivotal role in balancing the effects of this technology, ensuring that it enhances work quality rather than intensifying conflicts, isolation, and employee vulnerability.

This contribution, therefore, calls for a nuanced and ethically aware integration of metaverse technologies into HRM practices. It advocates for the implementation of inclusive digital literacy programs, the establishment of clear codes of conduct for virtual interactions, and the adoption of hybrid work models that balance immersion with physical connection. By proactively addressing the dark side of the metaverse, organizations can mitigate emerging risks and promote a more human-centric, resilient approach to digital transformation.

While this exploratory investigation yields valuable insights, it also acknowledges certain limitations. The analysis draws upon a limited pool of organizational cases and lacks systematic quantitative validation. Furthermore, the evolving nature of the metaverse renders any findings necessarily provisional. Future research should thus extend this preliminary inquiry by conducting longitudinal studies and engaging HR professionals. Such efforts will be crucial in refining our understanding of how immersive technologies can be responsibly embedded within human resource strategies to ensure not only innovation but also sustainability, equity, and employee well-being in the digital workplace.

References

1. Accenture: Building a responsible metaverse (2023)
2. Allahyegheh, A., et al.: Facilitating metaverse integration in education via extended reality and the Internet of Things. Int. Conf. Inf. Technol. **63**(8), 24–42 (2023)
3. BluSharkMedia: The metaverse and China's digital social credit system: a glimpse into a potential future (2023). https://blusharkmedia.medium.com/the-metaverse-and-chinas-digital-social-credit-system-a-glimpse-into-a-potential-future-6db7f89ed5b4

4. Bousquette, I.: Companies Finally Find a Use for Virtual Reality at Work. Wall Street J. (2024). https://www.wsj.com/articles/companies-finally-find-a-use-for-virtual-reality-at-work-9dddf8db?utm_source=chatgpt.com
5. Chang, C.Y., Hwang, G.J.: Trends in digital game-based learning in the mobile era: a systematic review of journal publications from 2007 to 2016. Int. J. Mob. Learn. Organ. **13**(1), 68–90 (2019)
6. Del Barone, A., De Gennaro, M., Buonocore, F.: HRM nel Metaverso: Sfide e Strategie per le Organizzazioni Digitali. Prospettive in Organizzazione (2024)
7. Díaz, J., Saldaña, C., Avila, C.: Virtual world as a resource for hybrid education. Int. J. Emerg. Technol. Learn. (iJET) **15**(15), 94–109 (2020)
8. Digital4biz:. Al Fashion Retail piace il Metaverso: nel mondo virtuale flagship, sfilate e collezioni NFT (2022). https://www.digital4.biz/marketing/brand-nel-metaverso-flagship-sfilate-collezioni-nft/?utm_source=chatgpt.com. Accessed 27 May 2025
9. Dwivedi, Y.K., Kshetri, N., Hughes, L., Rana, N.P., Baabdullah, A.M., Kar, A.K., Koohang, A., Ribeiro-Navarrete, S., Belei, N., Balakrishnan, J., Basu, S., Behl, A., Davies, G.H., Dutot, V., Dwivedi, R., Evans, L., Felix, R., Foster-Fletcher, R., Giannakis, M., Gupta, A., Hinsch, C., Jain, A., Patel, N.J., Jung, T., Juneja, S., Kamran, Q., Ab, S.M., Pandey, N., Papagiannidis, S., Raman, R., Rauschnabel, P.A., Tak, P., Taylor, A., Dieck, M.C.T., Viglia, G., Wang, Y., Yan, M.: Exploring the darkverse: a multi-perspective analysis of the negative societal impacts of the metaverse. Inf. Syst. Front. **25**, 2071–2114 (2023). https://doi.org/10.1007/s10796-023-10400-x
10. Eisenhardt, K.M., Graebner, M.E.: Theory building from cases: opportunities and challenges. Acad. Manag. J. **50**(1), 25–32 (2007)
11. Entrepreneur: Johnson & Johnson will train Mexican students and doctors in the metaverse (2022). https://www.entrepreneur.com/business-news/johnson-johnson-will-train-mexican-students-and-doctors/426426?utm_source=chatgpt.com. Accessed 27 May 2025
12. Hox, J.J., Boeije, H.R.: Data collection, primary vs. secondary. Encycl. Soc. Meas. **1**(1), 593–599 (2005)
13. Hsu, K., Levine, S., Finn, C.: Unsupervised learning via meta-learning (2018). arXiv:1810.02334
14. Koutromanos, G., Sofos, A., Avraamidou, L.: The use of augmented reality games in education: a review of the literature. Educ. Media Int. **52**(4), 253–271 (2015)
15. Lim, D.H., Lee, J.Y., Park, S.: The metaverse in the workplace: possibilities and implications for human resource development. Hum. Resour. Dev. Rev. 1–35 (2023)
16. Mauri, M.: Metaverso e gestione delle risorse umane: quali applicazioni in ambito HR? Osservatorio HR Innovation Practice (2022)
17. Microsoft: The metaverse is coming. Here are the cornerstones for securing it (2022). https://blogs.microsoft.com/blog/2022/03/28/the-metaverse-is-coming-here-are-the-cornerstones-for-securing-it/. Accessed 27 May 2025
18. Microsoft: Microsoft mesh: creating connections at the world economic forum 2023 (2023). https://www.microsoft.com/en-us/microsoft-365/blog/2023/01/16/microsoft-mesh-creating-connections-at-the-world-economic-forum-2023/. Accessed 27 May 2025
19. Miles, M.B., Huberman, A.M.: Drawing valid meaning from qualitative data: toward a shared craft. Educ. Res. **13**(5), 20–30 (1984)
20. Mystakidis, S.: Metaverse. Encyclopedia **2**(1), 486–497 (2022)
21. Ortiz, L.: Risks of the metaverse: a vrchat study case. J. Intell. Conflict Warfare **5**(2), 53–128 (2022)
22. Papagiannidis, S.: The impact of metaverses on social inclusion and wellbeing. Inf. Syst. Front. **25**, 2085–2100 (2023)
23. Parents: If your kid is using zepeto, you'll want to read this mom's warning (2025). https://www.parents.com/zepeto-app-dangers-for-kids-11700647
24. Roy, R., Babakerkhell, M.D., Mukherjee, S., Pal, D., Funilkul, S.: Development of a framework for metaverse in education: a systematic literature review approach. IEEE Access **11**, 57717–57734 (2023)

25. Rozak, H.A., Fachrunnisa, O., Sugiharti, T., Fitriati, I.R.: Metaverse and modification needs of human resources management practices and policies: An overview. In: International Conference on Intelligent Networking and Collaborative Systems, pp. 285–294. Springer, Cham (2023)
26. Samsung Biologics: Samsung biologics hosts its first metaverse job fair (2022). https://samsungbiologics.com/media/company-news-view?boardSeq=1642
27. Silverman, D., Marvasti, A.: Doing Qualitative Research: A Comprehensive Guide. Sage (2008)
28. Tarafdar, M., Gupta, A., Turel, O.: The dark side of information technology use. Inf. Syst. J. **23**(3), 269–275 (2013)
29. Tarafdar, M., D'arcy, J., Turel, O., Gupta, A.: The dark side of information technology. MIT Sloan Manag. Rev. (2014)
30. Tarafdar, M., Pullins, E.B., Ragu-Nathan, T.S.: Technostress: negative effect on performance and possible mitigations. Inform. Syst. J. **25**(2), 103–132 (2015)
31. Tarafdar, M., Cooper, C.L., Stich, J.F.: The technostress trifecta-techno eustress, techno distress and design: theoretical directions and an agenda for research. Inform. Syst. J. **29**(1), 6–42 (2019)
32. Taylor, A., Dieck, M.C.T., Jung, T.: Dark side of the metaverse: mental health and wellbeing perspective. Inf. Syst. Front. **25**, 2101–2114 (2023)
33. Techtarget: Microsoft, Meta tackle harassment in virtual worlds (2022). https://www.techtarget.com/searchunifiedcommunications/news/252513581/Microsoft-Meta-tackle-harassment-in-virtual-worlds. Accessed 27 May 2025
34. Yang, Y., Siau, K., Xie, W., Sun, Y.: Smart health: intelligent healthcare systems in the metaverse, artificial intelligence, and data science era. J. Organ. End User Comput. **34**(1) (2022). https://doi.org/10.4018/JOEUC.308814
35. Yin, R.K.: Designing case studies. Qual. Res. Methods **5**(14), 359–386 (2003)
36. Yin, R.K.: Case study methods (2012)
37. Zhang, G., Wu, J., Jeon, G., Chen, Y., Wang, Y., Tan, M.: Towards understanding metaverse engagement via social patterns and reward mechanism: a case study of Nova Empire. IEEE Trans. Comput. Soc. Syst. **10**(5), 2165–2176 (2023). https://doi.org/10.1109/TCSS.2022.3211679

Integrating Model-Driven Development Within Agile Methods: Insights from a Systematic Literature Review

Sebastien Ivanovici, Yaimara Granados Hondares, Yves Wautelet, Talha Mahmood, and Monique Snoeck

Abstract The integration of Model-Driven Development (MDD) within Agile Methods has emerged as a promising approach to combine the strengths of both approaches in software development. MDD furnishes abstraction and automation through model-based system specification and code generation, while agile development emphasizes flexibility, iterative cycles, and stakeholder collaboration. Despite their complementary advantages, their integration presents challenges such as steep learning curves, tool dependencies, and difficulties in generalization. This study conducts a systematic literature review to explore current implementations, benefits, challenges, and implications of using MDD within Agile methods. The findings highlight substantial benefits, including enhanced software quality, improved development efficiency, and strengthened stakeholder collaboration, alongside persistent barriers requiring tailored strategies. Recommendations for effective integration include targeted training, strategic tool selection, and project-specific customization of methodologies. This research offers insights for academics and practitioners aiming to optimize the use of MDD within Agile frameworks in evolving software development environments.

Keywords Model-driven development · Agile development · Agile methods · SCRUM

S. Ivanovici · Y. G. Hondares · Y. Wautelet (✉) · T. Mahmood · M. Snoeck
KULeuven, Leuven, Belgium
e-mail: yves.wautelet@kuleuven.be

S. Ivanovici
e-mail: sebastien.ivanovici@kuleuven.be

Y. G. Hondares
e-mail: yaimaragranados.hondares@kuleuven.be

T. Mahmood
e-mail: talha.mahmood@kuleuven.be

M. Snoeck
e-mail: monique.snoeck@kuleuven.be

A. Visvizi et al. (eds.), *Research and Innovation Forum 2025 Volume 1*,
Springer Proceedings in Complexity, https://doi.org/10.1007/978-3-032-09066-9_5

1 Introduction

In contemporary software engineering, the convergence of Model-Driven Development (MDD) and Agile Methods (AM) presents both an opportunity and a challenge. MDD emphasizes abstraction, automation, and rigor through model-based requirements elicitation and software design while AM prioritize flexibility, iterative development, and continuous stakeholder involvement. The potential synergy between these approaches lies in reducing the technical debt and enabling faster delivery of high-quality software that remains aligned with stakeholder expectations.

Integrating MDD into Agile practices is nevertheless far from trivial. The differences in process philosophy, tooling requirements, and team roles can lead to tensions, inconsistencies, or inefficiencies. Several academic and industrial efforts have explored this integration under various configurations but the comprehensive understanding remains limited.

This paper presents a systematic literature review (SLR) aimed at synthesizing current knowledge on the integration of MDD with AM. Following established SLR guidelines [11, 22], we investigate five core questions: (1) *What implementations exist?* (2) *How prevalent are they?* (3) *What motivations and benefits are associated with them?* (4) *What challenges and limitations emerge?* and (5) *What recommendations can be drawn for practitioners and researchers?* We analyze 45 peer-reviewed studies to eventually identify dominant patterns, evaluate empirical findings, and formulate practical guidance.

2 Methodology

The approach adheres to the well-established guidelines proposed by Kitchenham and Charters [22], and the ten-stage model from Brereton et al. [11]. These frameworks emphasize the value of SLRs in summarizing existing evidence, identifying research gaps, and providing a structured foundation for exploring complex research topics. Adhering to these standards ensures that the review process remains unbiased, systematic, and replicable.

Figure 1 illustrates the three primary phases of the SLR: planning (depicted in Sect. 2.1), conducting (depicted in Sect. 2.2), and documenting (depicted in Sect. 2.3).

2.1 Planning the Literature Review

2.1.1 Specifying the Research Questions (Stage 1)

Formulating the research questions is the first step of the SLR. These questions guide the search strategy and form the basis for data extraction and synthesis. The main research question (MRQ) is:

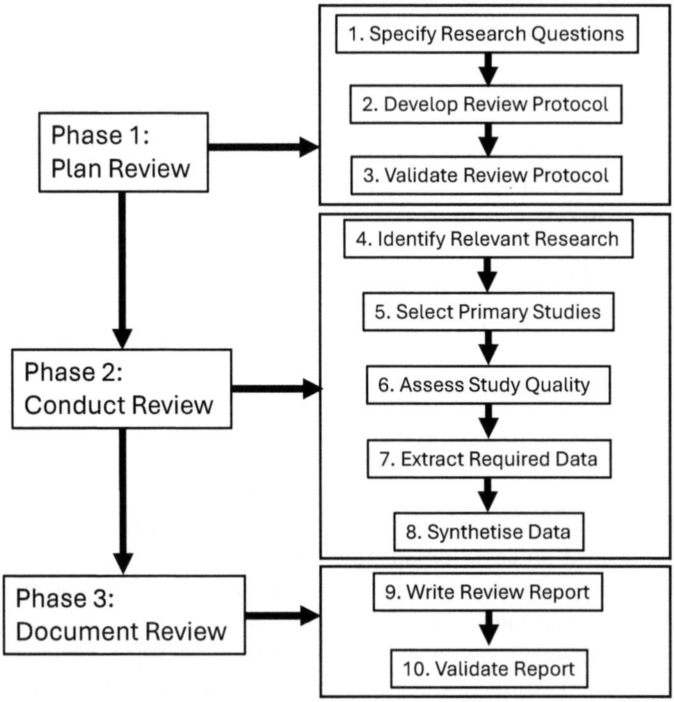

Fig. 1 SLR process. Adapted from [11]

MRQ: *What are the successes and flaws emerging from the implementation of MDD in AM, identified through an SLR?*

To explore this MRQ thoroughly, it is decomposed into five Specific Research Questions (SRQs):

- **SRQ1:** *What are the current implementations of MDD in AM, as reported in the literature?*
- **SRQ2:** *What is the relative importance of these implementations, based on frequency of occurrence in published studies?*
- **SRQ3:** *What are the key motivations and benefits associated with these implementations?*
- **SRQ4:** *What are the main challenges and limitations encountered?*
- **SRQ5:** *What implications and recommendations arise from the implementation of MDD in Agile contexts?*

2.1.2 Developing the Review Protocol (Stage 2)

The review protocol outlines the procedures and criteria for executing the review and consists of the following elements:

- **Background:** Justifies the need to investigate Agile MDD and its academic and practical implications.
- **Research Questions:** As outlined above, guiding the search, selection, and synthesis phases.
- **Search Strategy:** Includes keyword definitions, Boolean operators, and the target databases (LIMO, Google Scholar).
- **Selection Criteria:**
  - **Inclusion:** (1) Must address both MDD and AM; (2) Peer-reviewed; (3) Written in English.
  - **Exclusion:** (1) Not addressing both topics; (2) Not peer-reviewed; (3) Not in English.

2.1.3 Validating the Protocol (Stage 3)

The validity of the review protocol was ensured through a structured peer review and iterative refinement process. Initially, the protocol was developed based on established methodological frameworks [11, 22] to align with the research objectives and maintain rigorous standards. To further strengthen methodological rigor, the protocol underwent a review involving all of the authors of the paper. Feedback obtained from this validation process was considered and incorporated, resulting in targeted improvements such as clarification of inclusion and exclusion criteria, refinement of search strings, and enhancements to the strategy for data extraction and synthesis. These revisions reinforced the protocol's comprehensiveness, objectivity, and reproducibility, ensuring alignment with best practices for SLRs.

2.2 Conducting the Literature Review

2.2.1 Identifying Relevant Research (Stage 4)

A comprehensive search was conducted using LIMO (the integrated academic search interface of KULeuven Libraries) and Google Scholar. We argue that relying on KU Leuven's LIMO interface and Google Scholar is the most encompassing approach we can have in this SLR. LIMO is indeed not a standalone institutional catalogue but a sophisticated discovery platform powered by Ex Libris Primo, which integrates the **Central Discovery Index (CDI)**. CDI is a unified index of over 5 billion records from thousands of content providers. LIMO grants access not only to the university's physical and digital holdings, but also to "CDI, a central index with more than 4.5 billion publications from the most prominent publishers [25]. The CDI includes metadata from major databases such as **IEEE Xplore, ACM Digital Library, Scopus**, and **Web of Science**, ensuring substantial coverage of high-impact venues in computer science and engineering [15]. In parallel, Google Scholar adds an even more encompassing layer of access, indexing not only peer-reviewed literature but also grey literature, preprints, theses, and institutional repositories across disciplines and geographies. The combined use of LIMO and Google Scholar provides an overwhelming breadth of access where all the major scientific publishers in management and engineering are encompassed. The core search string was:

> ("model-driven development" OR "MDD" OR "model-driven" OR "Agile model-driven development" OR "AMDD") AND ("agile methods" OR "agile development" OR "agile software development" OR "agile model*") AND ("implementation" OR "integrat*" OR "applicat*")

Google Scholar was also queried using variations of these terms to identify additional relevant studies not indexed in LIMO.

2.2.2 Selecting and Filtering Studies (Stage 5)

The selection process involved four filtering steps for each database (see Tables 1and 2).

In total, 45 studies were included in the SLR. They are listed here [1–10, 12–14, 16–21, 23, 24, 26–49].

2.2.3 Assessing, Extracting, and Synthesizing Data (Stages 6–8)

All selected papers were reviewed in full and coded based on key themes: implementations, motivations and benefits, challenges and limitations, and implications and recommendations. The coding enabled systematic categorization to answer each SRQ. A qualitative synthesis followed, enabling cross-comparison of findings and trend identification.

Table 1 Filtering process for LIMO database

Step	Description	Papers
1	Initial search result	1145
2	Applied exclusion criteria 2 and 3	617
3	Filter by title, abstract, and conclusion	44
4	Full-text review	35

Table 2 Filtering process for Google Scholar

Step	Description	Papers
1	Initial search result	769
2	Applied exclusion criteria 2 and 3	523
3	Filter by title, abstract, and conclusion	14
4	Full-text review	10

2.3 Documenting the Literature Review

2.3.1 Writing the Review Report (Stage 9)

The findings were compiled in a structured document comprising: Introduction, Methodology, Results (per SRQ), Discussion, and Conclusion. Each section was iteratively refined to ensure clarity and coherence.

2.3.2 Validating the Report (Stage 10)

The validation process consisted of collaborative internal reviews among co-authors to ensure methodological consistency, traceability of evidence, and logical alignment of the synthesis with the defined RQs. Three of the authors independently reviewed a subset of the included studies during both the selection and data extraction phases in order to minimize potential biases and strengthen reliability. When disagreements appeared, they were discussed and resolved once unanimity was obtained. The coding categories have also been refined together by the three authors during regular discussions. The key findings were checked by all authors to ensure consistency. This collaboration throughout the systematic review process has helped to maintain transparency and strengthen the reliability of the results.

3 Results

3.1 Current Implementations of MDD in AM (SRQ1)

Table 3 summarizes the distribution of MDD usage across Agile development phases, addressing SRQ1 by identifying where MDD is most commonly applied. The table lists studies focusing on each phase, with 'Main Focus' indicating the primary phase discussed and other columns showing secondary mentions. In the table, each article is examined to determine in which phases of the Agile process the use of models was directly discussed. This is important for understanding how MDD integrates into the AM. The considered phases are:

- **Requirements Specification**: this phase involves capturing and refining requirements through the use of models;
- **Sprint Planning**: in this phase the focus is one ensuring that sprints are organized and aligned with the overarching project goals. Models are broken down in manageable tasks, which are allocated to different sprints;
- **Design and Development**: this phase involves the iterative development and refinement of models, which guide or generate the code and the construction of the system;

Table 3 Distribution of MDD usage across agile phases, showing main and secondary focus of studies

Agile phase	Papers reporting on the phase	Main focus phase of the paper
Requirements specification	[1, 5–7, 9, 10, 12–14, 16–19, 21, 23, 24, 26–28, 31, 34–43, 45, 48, 49]	[1, 10, 18, 27, 28, 31, 36–38, 45]
Sprint planning	[17, 40, 41]	[17, 40, 41]
Design and development	[5–7, 9, 12–14, 16–19, 21, 23, 24, 26–28, 31, 34–45, 48, 49]	[5–7, 9, 13, 14, 21, 23, 24, 26, 34, 35, 39, 42, 43, 49]
Testing and validation	[1, 5–7, 9, 12–14, 16–19, 23, 24, 26, 28, 31, 34, 37–45, 48, 49]	[19, 48]
Deployment	[5, 6, 9, 12, 21, 24, 42, 44]	None
Continuous feedback	[1, 5, 7, 9, 10, 12–14, 16, 17, 19, 23, 24, 26–28, 31, 34–38, 41–43, 45, 48]	[12, 16]
Documentation	[21, 39, 49]	None

- **Testing and Validation**: in this phase, the system is tested to ensure that it meets the specified requirements identified in requirement models. Models can enable automated testing;
- **Deployment**: this phase involves the delivery of the final product to the end-user;
- **Continuous Feedback**: during this phase, feedback from stakeholders is used to refine and adjust models continuously, ensuring constant alignment with requirements and user needs;
- **Documentation**: this phase ensures that the system design and its evolution are well-documented, providing a clear picture of how the system was developed.

Among the identified phases, the column "Main focus phase of the paper" was determined, based on the most emphasis placed by the article's author (so 1 main focus phase).

We can see that the integration of MDD is most prominent during *Requirements Specification* and *Design and Development* phases. In the requirements phase, MDD acts as a bridge between business needs and technical implementation. Collaborative modeling, for example, has been shown to facilitate near real-time discussion and iterative refinement with stakeholders [27]. Models like interface mockups and visual artifacts help to clarify requirements and reduce ambiguity, enhancing communication and ensuring alignment with business goals [37, 38]. In the design and development phase, MDD is often used to drive code generation and incremental refinement. Agile MDD promotes the use of lightweight, situational models that evolve iteratively to support rapid delivery without sacrificing structure [7]. More comprehensive integration frameworks also demonstrate how MDD can be adapted to support iterative, safe development even in large-scale contexts such as ERP systems [5, 21]. These findings underscore the versatility of MDD in Agile environments and its central role in bridging high-level conceptualization with executable software.

3.2 *Relative Importance of Agile MDD Implementations (SRQ2)*

Table 4 addresses SRQ2 by presenting the frequency of AM integrated with MDD across 45 studies. It lists the number of studies citing each AMDD combination, highlighting the most common integrations and their relative importance in the literature. Table 4 categorizes each article by identifying the specific AM used in combination with MDD. This categorization is performed to identify which integrations are most commonly explored in literature where the frequency of mentioned combinations is discussed in more detail. The relative importance of Agile MDD implementations can be assessed based on their frequency across the reviewed studies. The most prevalent integration is MDD with Scrum, mentioned in 12 articles. This underscores Scrum's dominant position in both industrial practice and research; its iterative structure, frequent feedback loops, and well-defined roles align naturally with the MDD approach. In this context, Scrum provides a project-level backbone that facilitates iterative and incremental refinement of models throughout the development process [4]. The integration of MDD at both the sprint and project levels enables ongoing validation of models, ensuring that development remains responsive to change while retaining structural rigor. Essebaa and Chantit [13] extend this approach by incorporating elements of Model-Driven Architecture and Model-Based Testing within the Scrum framework, emphasizing how such combinations can improve adaptability and reduce maintenance effort.

Table 4 Frequency of agile methods combined with MDD, indicating prevalence in literature

AMDD type	Number of studies	Cited papers
MDD with SCRUM	12	[5, 13, 14, 16–19, 26, 37, 39–41]
MDD with Agile Iterations	10	[6, 7, 9, 10, 23, 27, 34, 35, 43, 45]
MDD with XP	7	[5, 13, 17, 19, 24, 39, 48]
MDD with Agile Mockups	4	[28, 36–38]
MDD with Agile User Stories	2	[1, 31]
MDD with DevOps	2	[42, 44]
MDD with RUP	2	[13, 21]
MDD with SLAP	1	[49]
MDD with Rapid Development	1	[12]
Other	11	[2–5, 8, 20, 29, 32, 33, 46, 47]

The second most frequently observed combination is MDD with Agile Iterations, where MDD is applied in generic Agile settings without reference to a specific method such as Scrum or XP. These studies often emphasize Agile MDD practices and highlight iterative refinement cycles at the core of development. MDD with eXtreme Programming (XP) ranks third, with seven mentions. This confirms XP's relevance in contexts where practices such as Test-Driven Development (TDD) align well with modeling efforts and automation.

Other less common but noteworthy combinations include MDD with Agile Mockups and MDD with Agile User Stories both emphasizing the role of models in capturing and refining requirements early in the development lifecycle. MDD with RUP appears twice, reflecting RUP's hybrid positioning between traditional and AM. Finally, MDD with SLAP and MDD with Rapid Development are rare and tend to appear in niche or domain-specific applications. The "OTHER" category comprises 11 articles, mostly SLRs and theoretical analyses, which did not focus on one specific AMDD implementation but instead offered broader conceptual insights. These studies are primarily useful in addressing subsequent RQs.

3.3 Key Motivations and Benefits (SRQ3)

Tables 5 and 6 respectively outline the motivations and benefits of integrating Model-Driven Development (MDD) with Agile Methods (AM), as reported in 45 selected studies. Table 5 categorizes motivations into seven main themes, each with subcategories, listing studies that identify each motivation to address SRQ3. These tables present the results to answer SRQ3, we summarize the main findings in the rest of the section.

Motivations. The integration of MDD with AM is primarily motivated by the desire to leverage the strengths of both paradigms [6, 49]. MDD introduces automation and structure, while Agile provides flexibility and iterative development. This synergy aims to reduce development time and enhance code quality [38]. Responding to changing requirements is another major driver, supported by models that improve traceability and adaptation [10, 26].

Improved stakeholder collaboration is also a key motivator; models act as a shared language across technical and non-technical roles [27]. Further motivations include aligning IT with business goals, fulfilling project-specific needs, and—though less frequent—optimizing cost and scaling development.

Benefits. Literature confirms that Agile-MDD implementations lead to improved software quality through more robust, maintainable, and less error-prone code [3, 34]. Automation accelerates development and increases productivity [49], aligning well with Agile's iterative nature. Stakeholder involvement and satisfaction are enhanced via clearer communication and shared modeling artifacts [10, 35].

Collaboration across development teams also improves, aided by MDD's structured artifacts [42]. Flexibility and adaptability are strengthened through traceable, updatable requirements [20, 36]. Cost savings, efficient resource use, and reduced

Table 5 Key motivations for integrating MDD with AM, categorized by study

Leveraging strengths of both approaches	
Combine Agile and MDD for synergy	[2, 3, 5–8, 13, 17, 18, 20, 21, 24, 26–32, 38–41, 43, 46–49]
Focus on problem-solving	[6, 7, 18]
Reduce development time	[1–5, 7, 9, 12, 14, 29, 30, 33–35, 38, 40–42, 44–49]
Enhance code quality	[1, 9, 14, 16, 19, 20, 31, 33, 38, 41, 44, 46, 48, 49]
Increase productivity (through automation)	[1–5, 12–14, 20, 29, 32, 33, 38, 39, 41, 42, 44, 49]
Improve maintainability	[3, 42]
Responding to dynamic requirements	
Adapt to rapidly changing requirements	[1–5, 7, 13, 16, 18–20, 26, 30, 34, 35, 38–43, 46, 47]
Enable traceable requirements	[1, 10, 19, 28, 29, 34, 36–38, 40, 45]
Facilitating collaboration and communication	
Increase stakeholder involvement	[2–5, 7, 8, 10, 12, 17–20, 26–30, 33, 34, 36–43, 45–47]
Enhance cross-functional team collaboration	[2, 4, 5, 7, 8, 10, 13, 18, 19, 26, 28, 34, 39, 40, 42, 44]
Cost and resource optimization	
Reduce development costs	[9, 16, 23, 42]
Optimize use of resources	[45]
Minimize development waste	[39]
Scaling development	
Scale Agile and MDD	[33]
Aligning IT with business goals	
Support business objectives through IT	[8, 18, 23, 34, 43, 45]
Specific contextual motivations	
Fulfilling industry/project-specific requirements	[16–19, 21, 23, 27, 35, 37, 39]

waste are additional reported gains. Lastly, project management benefits include better progress tracking and integration testing supported by model-generated test cases [49].

3.4 *Main Challenges and Limitations (SRQ4)*

Table 7 addresses SRQ4 by categorizing challenges of integrating Model-Driven Development (MDD) with Agile Methods (AM) into five themes, each with subcategories, listing studies from 45 selected articles that report each challenge. Table 8 summarizes limitations of MDD-AM integration, organized by theme, identifying studies that highlight these constraints to further address SRQ4. These tables present the results to answer SRQ4, we summarize the main findings in the rest of the section.

Challenges. Despite their benefits, Agile MDD implementations face notable challenges. A key issue is the steep learning curve for teams unfamiliar with modeling practices [2, 44, 49]. Adoption is hindered without proper training or gradual onboarding.

Table 6 Key benefits of MDD and Agile integration, grouped by category

Improved software quality	
Enhanced code robustness/quality	[1–9, 13, 16, 19, 20, 24, 26–30, 32, 34, 37, 38, 40–43, 47–49]
Reduced defect rates	[3–9, 17, 24, 28, 32, 34, 36, 43, 48, 49]
Reduced complexity	[6, 8, 16, 21, 28]
Reduced size of specification (LOC)	[6]
Improved maintainability	[3, 6, 8, 14, 16, 19, 23, 34, 40, 42]
Increased development efficiency	
Accelerated development cycles	[1–9, 12–14, 16, 17, 20, 21, 23, 24, 26, 28–32, 34, 35, 37, 38, 40, 42, 43, 45, 47, 49]
Increased productivity (through automation)	[1–3, 5–9, 12–14, 17, 19–21, 23, 24, 26–29, 31, 32, 34, 36–41, 45, 47–49]
Enhanced stakeholder involvement	
Incorporated continuous stakeholder feedback	[3–8, 12, 18, 20, 24, 26–28, 31–35, 37–41, 43, 45]
Improved stakeholder satisfaction	[1, 3, 5, 10, 21, 26, 27, 47]
Created common language/understanding	[2, 10, 19, 27, 28, 31–34, 36–38, 42, 43, 45]
Facilitated collaboration and communication	
Enhanced cross-functional team collaboration	[2, 9, 10, 18–20, 24, 27, 28, 32, 34, 37, 39, 40, 43, 48]
Increased development flexibility and adaptability	
Rapid response to changes	[6, 16, 18–20, 23, 27, 30, 34, 35, 40, 41, 43]
Support different development scales	[35]
Facilitated continuous delivery	[28, 35, 49]
Enhanced requirements traceability	[1, 3, 4, 10, 14, 19, 24, 27, 34, 36, 41, 45, 48]
Cost and resource optimization	
Reduced development costs	[9, 10, 12, 23, 35]
Optimized use of resources	[6, 10, 12, 16, 17, 39, 40, 44, 48]
Minimized development waste	[10, 29, 31, 35, 39]
Improved project management	
Enhanced project planning	[10, 35]
Better risk management	[10, 35]
Effective integration testing	[14, 19, 41, 48, 49]
Facilitated progress tracking	[5, 9, 10, 24, 27–29, 34]

Tool and technology integration is also problematic, especially when MDD tools lack compatibility with Agile toolchains or do not support required practices [4, 44]. Moreover, some tools are context-dependent and limit general applicability.

Table 7 Main challenges of MDD and Agile integration, organized by theme

Learning curve and adoption barriers	
Overcome steep learning curve	[2–4, 6, 9, 17, 20, 23, 24, 28, 39, 42, 44, 47–49]
Tool and technology integration	
Ensure compatibility between tools	[4, 5, 12, 27, 28, 33, 34, 37, 48, 49]
Overcome technology limitations	[4, 10, 12, 16, 23, 44, 49]
Manage project complexity	
Ensure requirements and model consistency	[3, 10, 18, 20, 21, 30, 31, 46–48]
Increased documentation complexity	[31, 39, 45]
Increased conflict resolution time	[10, 16]
Collaboration issues	
Ensure stakeholder commitment	[18]
Difficult communication across disciplines/teams	[19, 42]
Avoid miscommunications	[16]
Scalability and performance concerns	
Prevent performance issues	[6, 23]
Manage integration in complex projects	[12, 35–37, 49]

Table 8 Main Limitations of MDD and Agile Integration, Organized by Theme

Generalization and applicability limitations	
Difficulty generalizing methodologies	[2, 6, 13, 14, 18, 21, 23, 28, 39, 41, 42, 45]
Limited applicability in large projects	[19, 20, 23, 36]
Challenges in adapting in specific contexts	[12, 21, 49]
Tool and technology dependency	
Reliance on specific tools and technologies/possible lack of tool support	[1, 2, 4–7, 9, 12–14, 17, 21, 23, 24, 26–29, 33, 34, 37–40, 42, 44, 46, 48, 49]
Trade-off between flexibility and stability	
Balance adaptability and stability	[16, 35, 45]
Resource and cost management	
Inefficient resource allocation	[6]
Limited short-term or small-scale project ROI	[48, 49]
Reliance on expertise	[10, 17, 20, 23, 26, 41, 44, 45]

Managing model complexity poses another difficulty. Inconsistent or outdated models can disrupt development [10, 30, 31], especially in dynamic Agile settings. Ensuring model-code alignment and managing documentation burden require continuous effort [46].

Less frequently reported challenges include collaboration issues, stakeholder disengagement, and performance concerns in large-scale systems.

Limitations. Three main limitations are consistently reported. First, Agile MDD methodologies often lack generalizability across contexts; their success is highly case-dependent and not easily replicable [18, 39].

Second, the reliance on specific tools is a recurring limitation. Many implementations depend on tool support that may be immature or incompatible with Agile practices [24, 44].

Third, Agile MDD requires modeling expertise, which can lead to bottlenecks when few individuals possess the necessary skills [20]. This dependency can affect team agility and long-term sustainability.

3.5 Implications and Recommendations (SRQ5)

The integration of MDD with AM shows strong potential to improve software quality and development efficiency. Automated code generation enhances robustness and reduces defects, while Agile practices provide iterative cycles, continuous feedback, and a collaborative environment that supports modeling [6, 38, 49].

However, realizing these benefits requires addressing context-specific challenges. Implementations often depend on particular tools and require customization to suit project needs. Tool maturity and compatibility remain critical concerns.

Moreover, modeling expertise is essential, and teams must be trained to manage the associated learning curve. Success depends on investing in structured onboarding, continuous feedback, and methodological alignment.

For organizations prepared to overcome these hurdles, Agile MDD can yield substantial returns in productivity, adaptability, and code quality.

4 Discussion

This findings depicted in the previous section are further discussed in this section. Specifically, we structure around the SRQs and draw broader insights from the evidence collected in the papers.

4.1 SRQ1: Current Implementations of MDD in AM

The literature reports on diverse strategies for integrating MDD into Agile environments. The most common implementations involve combinations of MDD with Scrum and iterative AM. These integrations essentially focus on the design and development stages of the development process but requirements specification and continuous feedback are also recurrent themes. This confirms that even if MDD is traditionally perceived as a technique to be integrated in plan-driven methodologies, it can be adapted to iterative and flexible development cycles. This is especially true when complemented by appropriate tool support and stakeholder involvement.

This suggests a trend towards the "hybridization" of methods meaning that practice-oriented adaptations often blend formal modeling with agile delivery. However, some studies hint at integration challenges, particularly where modeling cycles are perceived as too rigid for rapid iteration. A trade-off space is present that yet needs to be fully understood.

4.2 SRQ2: Relative Importance of Implementations

The frequency analysis shows that *MDD with Scrum* is the most referenced implementation, followed by *MDD with Agile Iterations* and *MDD with XP*. This reflects the popularity of Scrum in contemporary Agile practices and suggests that modeling activities are compatible with its structure (the by default defined roles and ceremonies). We can also not the implementations using *Agile Mockups* and *Agile User Stories*; the are much less frequent but highlight the importance of front-end and requirement-driven modeling efforts.

The reviewed studies do not uniformly point Scrum out as the most effective way to use MDD in agile environments. Critical frictions are indeed reported (e.g. the misalignment between sprint planning and model stabilization) which points to a need for more nuanced assessments of fit-for-purpose rather than fit-for-use.

4.3 SRQ3: Key Motivations and Benefits

Key motivations for combining MDD with Agile approaches include the willingness to dispose of the complementary strengths of both paradigms. Agile provides responsiveness and stakeholder collaboration. MDD offers formalization, automation, and traceability. When we have synthesized the motivations we could clearly notice that the combination is pursued to enhance software quality and productivity but also improve traceability, stakeholder involvement, and the ability to respond to change.

The benefits observed align with these motivations. They include improved code quality, accelerated development, and increased stakeholder satisfaction. We can notice that these benefits are reported across multiple implementations, which suggests that the outcomes often converge on similar positive impacts even if the mode of implementation varies.

4.4 SRQ4: Main Challenges and Limitations

Several challenges can be point out. The integration of MDD and AM leads to a steep learning curve in tooling and modeling techniques. This represents a significant barrier to adoption. Also, compatibility issues among tools and the lack of mature integrations hinder the seamless fusion of MDD into Agile workflows.

Other limitations include reduced generalizability across domains and insufficient empirical evaluations. These challenges underscore the need for context-aware tailoring of Agile MDD practices and suggest a demand for more empirical studies to validate proposed benefits and address observed drawbacks. A key pattern across studies is the context-dependence of the challenges: some teams benefit from tight modeling and code alignment but others find it burdensome. This variability reinforces the idea that MDD integration must be customized rather than standardized.

4.5 SRQ5: Implications and Recommendations

The integration of MDD into AM offers substantial promises but requires careful consideration. Organizations should invest in training to overcome learning barriers, adopt tools that support collaborative modeling and code generation, and adapt MDD practices to fit their Agile workflows. They should not enforce rigid adherence to either approach.

Furthermore, researchers are encouraged to explore domain-specific implementations, evaluate long-term outcomes, and develop frameworks that guide the integration process across varying project contexts. This will help bridge the gap between theory and practice and enhance the generalizability and applicability of Agile MDD solutions.

5 Conclusion

The SLR that we have presented in this paper has explored the integration of MDD with AM. We can highlight that this integration excels in three major ways. First, it significantly improves software quality because precise models reduce errors and simplify updates. Second, it allows to accelerate the software development by automating repetitive tasks. Third, it allows to strengthen collaboration by using models as shared tools that bridge technical developers and business stakeholders. This also ensures that projects remain closely aligned with user needs. These strengths show how MDD can help in agile development settings.

Nevertheless, we also need to highlight three persistent challenges complicate adoption. First, mastering modeling tools demands substantial training. A steep learning curve may thus delay progress for unprepared teams. Second, the reliance on specific tools frequently leads to compatibility issues. Indeed, the integration of some tools might appear to be incompatible with Agile workflows. Third, successful integration remains highly project-specific. Solutions effective in one context (e.g. web development) may not be in other domains, and this particularly in complex enterprise environments. These limitations show the necessity of customization and adaptation when effectively integrated MDD with AM in practice.

We can point out that the findings offer significant implications for both practitioners and researchers. Practitioners should invest in team training to cultivate modeling expertise, strategically select tools supportive of Agile processes, and tailor MDD approaches to align with project objectives. Lightweight models can for example be used to clarify requirements early within sprints. As another example, automated testing tools can facilitate efficient validation. Managers can also enhance team collaboration through the strategic use of models to synchronize efforts and maintain clear communication channels. Researchers are encouraged to further investigate MDD-AM performance within specific domains and develop standardized frameworks that deal with context-dependency to simplify future adoption.

We now need to point out some limitaitons. First some niche studies from specialized sources may be missing. Moreover, since many findings derive from project-specific cases, the generalizability of the results may be threatened. Future studies should concentrate on empirically validating the point out insights in diverse real-world contexts.

Declaration of AI-assisted technologies in the writing process

ChatGPT4o was used for copy-editing. After usage, the authors reviewed and edited the output and take full responsibility for the content.

References

1. Alba, B., Granda, M.F., Parra, O.: Ui-test: a model-based framework for visual UI testing-qualitative and quantitative evaluation. In: Ali, R., Kaindl, H., Maciaszek, L.A. (eds.) Evaluation of Novel Approaches to Software Engineering - 16th International Conference, ENASE 2021, Virtual Event, April 26-27, 2021, Revised Selected Papers. Communications in Computer and Information Science, vol. 1556, pp. 328–355. Springer (2021)
2. Alfraihi, H.: Towards improving agility in model-driven development. In: Dubois, C., Parisi-Presicce, F., Kolovos, D.S., Matragkas, N. (eds.) Joint Proceedings of the Doctoral Symposium and Projects Showcase Held as Part of STAF 2016 co-located with Software Technologies: Applications and Foundations (STAF 2016), Vienna, Austria, July 4–7, 2016. CEUR Workshop Proceedings, vol. 1675, pp. 2–10. CEUR-WS.org (2016)
3. Alfraihi, H., Lano, K.: The integration of agile development and model driven development - a systematic literature review. In: Pires, L.F., Hammoudi, S., Selic, B. (eds.) Proceedings of the 5th International Conference on Model-Driven Engineering and Software Development, MODELSWARD 2017, Porto, Portugal, Feb 19–21, 2017. pp. 451–458. SciTePress (2017)

4. Alfraihi, H., Lano, K.: Practical aspects of the integration of agile development and model-driven development: An exploratory study. In: Burgueño, L., Corley, J., Bencomo, N., Clarke, P.J., Collet, P., Famelis, M., Ghosh, S., Gogolla, M., Greenyer, J., Guerra, E., Kokaly, S., Pierantonio, A., Rubin, J., Ruscio, D.D. (eds.) Proceedings of MODELS 2017 Satellite Event: Workshops (ModComp, ME, EXE, COMMitMDE, MRT, MULTI, GEMOC, MoDeVVa, MDETools, FlexMDE, MDEbug), Posters, Doctoral Symposium, Educator Symposium, ACM Student Research Competition, and Tools and Demonstrations co-located with ACM/IEEE 20th International Conference on Model Driven Engineering Languages and Systems (MODELS 2017), Austin, TX, USA, Sept 17, 2017. CEUR Workshop Proceedings, vol. 2019, pp. 399–404. CEUR-WS.org (2017)
5. Alfraihi, H., Lano, K.: A process for integrating agile software development and model-driven development. In: Burgueño, L., Corley, J., Bencomo, N., Clarke, P.J., Collet, P., Famelis, M., Ghosh, S., Gogolla, M., Greenyer, J., Guerra, E., Kokaly, S., Pierantonio, A., Rubin, J., Ruscio, D.D. (eds.) Proceedings of MODELS 2017 Satellite Event: Workshops (ModComp, ME, EXE, COMMitMDE, MRT, MULTI, GEMOC, MoDeVVa, MDETools, FlexMDE, MDEbug), Posters, Doctoral Symposium, Educator Symposium, ACM Student Research Competition, and Tools and Demonstrations co-located with ACM/IEEE 20th International Conference on Model Driven Engineering Languages and Systems (MODELS 2017), Austin, TX, USA, Sept 17, 2017. CEUR Workshop Proceedings, vol. 2019, pp. 412–417. CEUR-WS.org (2017)
6. Alfraihi, H., Lano, K., Rahimi, S.K., Sharbaf, M., Haughton, H.P.: The impact of integrating agile software development and model-driven development: A comparative case study. In: Khendek, F., Gotzhein, R. (eds.) System Analysis and Modeling. Languages, Methods, and Tools for Systems Engineering - 10th International Conference, SAM 2018, Copenhagen, Denmark, October 15-16, 2018, Proceedings. Lecture Notes in Computer Science, vol. 11150, pp. 229–245. Springer (2018). https://doi.org/10.1007/978-3-030-01042-3_14
7. Ambler, S.: Agile model-driven development: Just-in-time modeling. In: Proceedings of the Agile 2007 Conference, pp. 24–31 (2007)
8. Arias, J.M., Calvache, C.P.: Systematic review about the integration of model-driven software development and agile methodologies. Informador Técnico. UPTC **80**(1), 87 (2016)
9. Azmoodeh, M., Georgalas, N., Fisher, S.: Model-driven systems development and integration environment. BT Technol. J. **23**(3), 96–110 (2005)
10. Bisikirskienė, L., Čeponienė, L., Jurgelaitis, M., Ablonskis, L., Grigonytė, E.: Compiling requirements from models for early phase scope estimation in agile software development projects. Appl. Sci. **13**(22), 12353 (2023)
11. Brereton, O.P., Kitchenham, B.A., Budgen, D., Turner, M., Khalil, M.: Lessons from applying the systematic literature review process in software engineering. J. Syst. Softw. **80**(4), 571–583 (2007). https://doi.org/10.1016/j.jss.2006.07.009
12. Cortiñas, A., Bernaschina, C., Luaces, M.R., Fraternali, P.: Enabling agile web development through in-browser code generation and evaluation. In: Ouhammou, Y., Ivanovic, M., Abelló, A., Bellatreche, L. (eds.) Model and Data Engineering - 7th International Conference, MEDI 2017, Barcelona, Spain, Oct 4–6, 2017, Proceedings. Lecture Notes in Computer Science, vol. 10563, pp. 311–323. Springer (2017)
13. Essebaa, H., Chantit, K.: Combining model driven development and agile software development. In: Proceedings of the 18th International Conference on Software Engineering, pp. 92–100 (2018)
14. Essebaa, I., Chantit, S.: Scrum and V lifecycle combined with model-based testing and model driven architecture to deal with evolutionary system issues. In: Abdelwahed, E.H., Bellatreche, L., Golfarelli, M., Méry, D., Ordonez, C. (eds.) Model and Data Engineering - 8th International Conference, MEDI 2018, Marrakesh, Morocco, Oct 24–26, 2018, Proceedings. Lecture Notes in Computer Science, vol. 11163, pp. 77–91. Springer (2018)
15. Ex Libris: An overview of the ex libris central discovery index (cdi). https://knowledge.exlibrisgroup.com/Primo/Content_Corner/Central_Discovery_Index/Documentation_and_Training/Documentation_and_Training_(English)/CDI_-_The_Central_Discovery_Index/

010An_Overview_of_the_Ex_Libris_Central_Discovery_Index_(CDI), last accessed: July 2025
16. Falzone, E., Bernaschina, C.: Intelligent code generation for model driven web development. In: Pautasso, C., Sánchez-Figueroa, F., Systä, K., Rodríguez, J.M.M. (eds.) Current Trends in Web Engineering - ICWE 2018 International Workshops, MATWEP, EnWot, KD-WEB, WEOD, TourismKG, Cáceres, Spain, June 5, 2018, Revised Selected Papers. Lecture Notes in Computer Science, vol. 11153, pp. 5–13. Springer (2018)
17. Giachetti, G., de la Vara, J.L., Marín, B.: Mastering agile practice adoption through a model-driven approach for the combination of development methods. Bus. Inf. Syst. Eng. **65**(2), 103–125 (2023)
18. Grangel, R., Campos, C.: Agile model-driven methodology to implement corporate social responsibility. Comput. Ind. Eng. **127**, 116–128 (2019)
19. Guerrero-Ulloa, G., Hornos, M.J., Rodríguez-Domínguez, C.: Tddm4iots: a test-driven development methodology for internet of things (iot)-based systems. In: International Conference on Applied Technologies, pp. 41–55. Springer (2019)
20. Hansson, S., Zhao, Y., Burden, H.: How MAD are we? empirical evidence for model-driven agile development. In: Ruscio, D.D., de Lara, J., Pierantonio, A. (eds.) Proceedings of the 3rd Workshop on Extreme Modeling co-located with ACM/IEEE 17th International Conference on Model Driven Engineering Languages & Systems, XM@MoDELS 2014, Valencia, Spain, Sept 29, 2014. CEUR Workshop Proceedings, vol. 1239, pp. 2–11. CEUR-WS.org (2014), https://ceur-ws.org/Vol-1239/xm14_submission_4.pdf
21. Hermawan, Hastarista, F.: Colaborated architechture framework for composition uml 2.0 in zachman framework. In: IOP Conference Series: Materials Science and Engineering, vol. 105, p. 012037. IOP Publishing (2016)
22. Kitchenham, B., Charters, S.: Guidelines for Performing Systematic Literature Reviews in Software Engineering. EBSE Technical Report (2007)
23. Kobayashi, S., Kimura, E., Ishihara, K.: Archetype model-driven development framework for ehr web system. Healthc. Inform. Res. **19**(4), 271–277 (2013)
24. Kolahdouz Rahimi, S., Lano, K., Alfraihi, H., P Haughton, H.: extreme modeling: an approach to agile model-based development. J. Comput. Secur. **6**(2), 43–52 (2019)
25. KU Leuven Libraries: What is limo? https://bib.kuleuven.be/english/collections-access-borrowing/what-is-limo, last accessed: July 2025
26. Kulkarni, V., Barat, S., Ramteerthkar, U.: Early experience with agile methodology in a model-driven approach. In: Whittle, J., Clark, T., Kühne, T. (eds.) Model Driven Engineering Languages and Systems, 14th International Conference, MODELS 2011, Wellington, New Zealand, Oct 16–21, 2011. Proceedings. Lecture Notes in Computer Science, vol. 6981, pp. 578–590. Springer (2011). https://doi.org/10.1007/978-3-642-24485-8_42
27. de Lange, P., Nicolaescu, P., Klamma, R., Jarke, M.: Engineering web applications using real-time collaborative modeling. In: Gutwin, C., Ochoa, S.F., Vassileva, J., Inoue, T. (eds.) Collaboration and Technology - 23rd International Conference, CRIWG 2017, Saskatoon, SK, Canada, Aug 9–11, 2017, Proceedings. Lecture Notes in Computer Science, vol. 10391, pp. 213–228. Springer (2017)
28. de Lange, P., Nicolaescu, P., Rosenstengel, M., Klamma, R.: Collaborative wireframing for model-driven web engineering. In: Cheng, R., Mamoulis, N., Sun, Y., Huang, X. (eds.) Web Information Systems Engineering - WISE 2019 - 20th International Conference, Hong Kong, China, Nov 26–30, 2019, Proceedings. Lecture Notes in Computer Science, vol. 11881, pp. 373–388. Springer (2019)
29. Lano, K., Alfraihi, H., Yassipour Tehrani, S., Haughton, H.: Improving the application of agile model-based development: experiences from case studies. In: The Tenth International Conference on Software Engineering Advances, pp. 213–219. IARIA Spain (2015)
30. Lano, K., Rahimi, S.K., Troya, J., Alfraihi, H.: Introduction to the theme section on agile model-driven engineering. Softw. Syst. Model. **21**(4), 1465–1467 (2022). https://doi.org/10.1007/s10270-022-01016-x

31. Losada, B., Urretavizcaya, M., Fernández-Castro, I.: A guide to agile development of interactive software with a "user objectives"-driven methodology. Sci. Comput. Program. **78**(11), 2268–2281 (2013)
32. Mairon, K., Buchheit, M., Knahl, M., Atkinson, S., Furnell, S., Schreier, U.: Agile limitations and model-driven opportunities for software development projects. In: Proceedings of the Sixth Collaborative Research Symposium on Security, E-learning, Internet and Networking, p. 43 (2010)
33. Mognon, F., Stadzisz, P.C.: Modeling in agile software development: a systematic literature review. In: da Silva, T.S., Estácio, B., Kroll, J., Fontana, R.M. (eds.) Agile Methods - 7th Brazilian Workshop, WBMA 2016, Curitiba, Brazil, Nov 7–9, 2016, Revised Selected Papers. Communications in Computer and Information Science, vol. 680, pp. 50–59 (2016)
34. Pang, C.: An agile architecture for a legacy enterprise IT system. Int. J. Organ. Collect. Intell. **6**(4), 65–97 (2016)
35. Popa, S.C., Avornicului, M.C., Besfelean, V.P.: Using amdd method for database design in mobile cloud computing systems. Inform. Econ. **17**(1) (2013)
36. Rivero, J.M., Grigera, J., Distante, D., Simarro, F.M., Rossi, G.: Datamock: an agile approach for building data models from user interface mockups. Softw. Syst. Model. **18**(1), 663–690 (2019). https://doi.org/10.1007/s10270-017-0586-9
37. Rivero, J.M., Heil, S., Grigera, J., Luna, E.R., Gaedke, M.: An extensible, model-driven and end-user centric approach for API building. In: Casteleyn, S., Rossi, G., Winckler, M. (eds.) Web Engineering, 14th International Conference, ICWE 2014, Toulouse, France, July 1–4, 2014. Proceedings. Lecture Notes in Computer Science, vol. 8541, pp. 494–497. Springer (2014)
38. Rivero, J.M., Rossi, G.: Mockupdd: Facilitating agile support for model-driven web engineering. In: Sheng, Q.Z., Kjeldskov, J. (eds.) Current Trends in Web Engineering - ICWE 2013 International Workshops ComposableWeb, QWE, MDWE, DMSSW, EMotions, CSE, SSN, and PhD Symposium, Aalborg, Denmark, July 8–12, 2013. Revised Selected Papers. Lecture Notes in Computer Science, vol. 8295, pp. 325–329. Springer (2013)
39. Romano, B.L.: A framework for web applications using an agile and collaborative model driven development (ac-mdd). Acta Scientiarum. Technol. **41**, e38349 (2019)
40. Shafiee, S., Wautelet, Y., Poelmans, S., Heng, S.: An empirical evaluation of scrum training's suitability for the model-driven development of knowledge-intensive software systems. Data Knowl. Eng. **146**, 102195 (2023)
41. Snoeck, M., Wautelet, Y.: Agile MERODE: a model-driven software engineering method for user-centric and value-based development. Softw. Syst. Model. **21**(4), 1469–1494 (2022)
42. Sorgalla, J., Wizenty, P., Rademacher, F., Sachweh, S., Zündorf, A.: Applying model-driven engineering to stimulate the adoption of devops processes in small and medium-sized development organizations. SN Comput. Sci. **2**(6), 459 (2021)
43. Stojanovic, Z., Dahanayake, A.N.W., Sol, H.G.: Component-oriented agile software development. In: Marchesi, M., Succi, G. (eds.) Extreme Programming and Agile Processes in Software Engineering, 4th International Conference, XP 2003, Genova, Italy, May 25–29, 2003 Proceedings. Lecture Notes in Computer Science, vol. 2675, pp. 315–318. Springer (2003). https://doi.org/10.1007/3-540-44870-5_38
44. Süß, J.G., Swift, S., Escott, E.: Using devops toolchains in agile model-driven engineering. Softw. Syst. Model. **21**(4), 1495–1510 (2022)
45. Valente, P., Silva, T.R., Winckler, M., Nunes, N.J.: The goals approach: Enterprise model-driven agile human-centered software engineering. In: Bogdan, C., Gulliksen, J., Sauer, S., Forbrig, P., Winckler, M., Johnson, C.W., Palanque, P.A., Bernhaupt, R., Kis, F. (eds.) Human-Centered and Error-Resilient Systems Development - IFIP WG 13.2/13.5 Joint Working Conference 6th International Conference on Human-Centered Software Engineering, HCSE 2016, and 8th International Conference on Human Error, Safety, and System Development, HESSD 2016 Stockholm, Sweden, Aug 29–31, 2016, Proceedings. Lecture Notes in Computer Science, vol. 9856, pp. 261–280. Springer (2016)

46. Wegener, H.: Agility in model-driven software development? implications for organization, process, and architecture. In: OOPSLA 2002 Workshop on Generative Techniques in the Context of Model Driven Architecture, vol. 23 (2002)
47. Wijekoon, H., Merunka, V.: Combining model driven development and agile software development. In: The Sixteenth International Conference on Digital Society. IARIA, pp. 20–2 (2022)
48. Zhang, X.: Test-driven modeling for model-driven development. IEEE Softw. **21**(3), 28–35 (2004)
49. Zhang, Y., Patel, S.: Agile model-driven development in practice. IEEE Softw. **28**(2), 84–91 (2011)

Implementing DEA with Gaussian AHP and Pearson Correction for Efficiency Benchmarking in a Food Retail Chain

Levy Galas Jacob, Plácido Rogério Pinheiro, and João Batista Furlan Duarte

Abstract The manuscript investigates the application of a hybrid approach combining Data Envelopment Analysis (DEA), Gaussian Analytic Hierarchy Process (AHP), and Pearson correlation adjustments to evaluate and optimise the operational efficiency of a food retail chain of restaurants. This methodology tries to identify and rank efficiency across the organisation's stores. Inputs evaluated include product costs and sales volume, while the outputs are revenue and profit. The Gaussian AHP was used to calculate inputs and output weights, reflecting Gaussian statistical analysis, which were then adjusted based on Pearson correlation to ensure greater alignment with the data correlations. DEA was implemented using Python, and Power BI was utilised for interactive result visualisation. This combination of tools provided a comprehensive approach. The findings highlight the effectiveness of this hybrid method for evaluating efficiency discrepancies and promoting optimisation across the organisation. Identifying units with the most significant potential for improvement and providing actionable insights for decision-making was possible. This work underscores the value of hybrid methodologies for internal benchmarking and qualitative techniques in operational efficiency studies.

Keywords Data envelopment analysis · Gaussian AHP · Pearson correlation · Benchmarking · Operational efficiency

1 Introduction

The food retail sector increasingly demands data-driven solutions to optimise processes and sustain operational efficiency. The European Full-Service Restaurants Market is projected to grow at a compound annual growth rate (CAGR) of

L. G. Jacob · P. R. Pinheiro (✉) · J. B. F. Duarte
University of Fortaleza (UNIFOR), Graduate Program in Applied Informatics, Fortaleza, CE, Brazil
e-mail: placido@unifor.br

J. B. F. Duarte
e-mail: furlan@unifor.br

A. Visvizi et al. (eds.), *Research and Innovation Forum 2025 Volume 1*,
Springer Proceedings in Complexity, https://doi.org/10.1007/978-3-032-09066-9_6

1.9% from 2024 to 2031 [1]. The global socioeconomic landscape presents significant challenges to businesses, making it increasingly difficult to control costs and maintain competitive margins in a dynamic and ever-evolving market.

Many restaurants have implemented process improvement techniques and technological innovations to optimise their operations in response to these pressures. Technological tools, such as advanced management systems and data analytics, have played a crucial role in enhancing operational efficiency and reducing costs, as noted by [2]. However, adopting these technologies remains uneven across the sector: large chains and well-structured businesses successfully incorporate these innovations, while smaller establishments face financial, technical, and cultural barriers [3], making it challenging to implement significant changes. Recent studies highlight that this disparity persists, particularly in emerging markets [2, 3], underscoring the need for scalable solutions that integrate statistical rigour with practical applicability.

In this context, multicriteria decision-making methods serve as invaluable tools for decision-making. Data Envelopment Analysis (DEA) stands out as a robust methodology for decision-making and benchmarking, based on operational efficiency, which compares the relative performance of decision-making units (DMUs) [4]. While DEA is well-established, its integration with the Gaussian Analytic Hierarchy Process (AHP), which replaces pairwise comparisons with statistical weight estimation [5] and Pearson correlation (to mitigate outlier sensitivity [6]), remains underexplored in the food retail sector. This gap motivates our research question:

> How can a hybrid DEA-Gaussian AHP framework, adjusted via Pearson correlation, improve efficiency benchmarking in restaurant chains compared to conventional methods?

Addressing the above context, we propose leveraging the statistical robustness of Gaussian AHP to handle large, noisy datasets standard in retail [5]. Applying Pearson's correction aligns weights with empirical data patterns, reducing spurious correlations [7]. This approach provides a scalable solution adaptable to both large chains and smaller establishments.

This paper is structured as follows: Sect. 2 provides an overview of the fundamentals of DEA. Furthermore, Sect. 3 introduces Gaussian AHP, and Sect. 4 details Pearson's correction. On the other hand, Sect. 5 applies the hybrid method to restaurant data. Section 6 discusses the conclusions and limitations. By integrating these methods, we aim to advance both academic discourse (e.g., bridging DEA and statistical decision-making [Gomes et al., 106]) and practical efficiency benchmarking in the food retail sector.

2 Data Envelopment Analysis (DEA)

Data Envelopment Analysis (DEA) is a non-parametric method that evaluates the relative efficiency of decision-making units (DMUs) through linear programming. Developed by Charnes, Cooper, and Rhodes [8], DEA has become a cornerstone of operational benchmarking, particularly in service industries like food retail [9].

Unlike parametric approaches, DEA constructs an empirical efficiency frontier from observed data, avoiding restrictive assumptions about functional forms [4]. This flexibility makes it ideal for comparing heterogeneous DMUs—critical need in restaurant chains where units often operate under varying conditions.

The DEA calculates efficiency as the ratio of weighted outputs to weighted inputs [10]. Two primary models dominate applications:

- CRS (Constant Returns to Scale): Assumes proportional input–output relationships, suitable for ideal-scale operations [9].
- VRS (Variable Returns to Scale): Accounts for scale inefficiencies, making it more applicable to real-world scenarios, such as food retail, where DMUs often operate at suboptimal scales due to market constraints [9, 11].

Mathematically, DEA solves for each DMU:

$$max\theta_k = \frac{\sum_{r=1}^{s} u_r y_{rk}}{\sum_{i=1}^{m} v_i x_{ik}}$$

Subject to:

$$\frac{\sum_{r=1}^{s} u_r y_{rj}}{\sum_{i=1}^{m} v_i x_{ij}} \leq 1 \forall j, u_i, v_i \geq 0$$

where x_{ij} and y_{rj} are inputs and outputs, and u_r, v_i are weights. Efficient DMUs ($\theta = 1$) form the frontier, while $\theta < 1$ indicates inefficiency [11].

Orientation Strategies [9]:

- Input-oriented: Minimises resource use while maintaining output levels. critical for cost-sensitive sectors like food retail, where waste reduction directly impacts margins.
- Output-oriented: Maximises output without increasing inputs. Useful for scaling production in established chains.

Why hybridise DEA and AHP-GS for this study? While DEA is well-documented, its integration with Gaussian AHP and Pearson correction addresses two key limitations in food retail applications:

1. Subjectivity in weight allocation: Traditional DEA allows unrealistic weight distributions [12]; our hybrid method constrains weights via statistical rigour (Sect. 3).
2. Outlier sensitivity: Pearson correction mitigates distortions from anomalous data (Sect. 4), a known issue in restaurant efficiency studies [6].

3 Gaussian AHP

The Analytic Hierarchy Process (AHP) is a well-established method for multicriteria decision-making that structures problems hierarchically through a series of pairwise comparisons [13]. However, its reliance on subjective judgments makes it prone to cognitive biases—a challenge documented in both software project management [14] and user experience design [15]—and inconsistencies, particularly with large datasets, a critical challenge in food retail efficiency analysis [5].

The Gaussian AHP addresses these limitations by:

- Replacing pairwise comparisons with statistical metrics (standard deviation, mean)
- Incorporating normal distribution to handle data uncertainty [5, 16]
- Adopting a co-evolutionary approach inspired by HCI research, where iterative statistical refinement replaces subjective user testing [15].

Methodological Framework:

In Gaussian AHP, weights are derived objectively through:

$$\text{Gaussian Factor} = \frac{\sigma_i}{\mu_i}$$

where:

- σ_i = standard deviation of criterion i
- μ_i = mean of criterion i.

Key steps:

1. Calculate Gaussian factors for all criteria
2. Normalise factors to obtain weights:

$$w_i = \frac{\left({}^{\sigma_i}/_{\mu_i}\right)}{\sum_{j=1}^{n}\left({}^{\sigma_j}/_{\mu_j}\right)}$$

Advantages of Food Retail Benchmarking:

- Reduces subjectivity by eliminating manual pairwise comparisons [5],
- Automatically prioritises volatile metrics (e.g., sales fluctuations) through Gaussian-based weighting,
- Aligns with DEA integration by providing statistically rigorous weighting for efficiency analysis (Sect. 5).

Application:

In our restaurant chain study (Sect. 6), Gaussian AHP assigned higher weights to:

- Gross profit (σ = R$908,706.00) over sales volume (σ = R$52,740)
- Validated through Pearson's correction to mitigate outlier effects [7].

4 Pearson Correction

The Pearson product-moment correlation coefficient, formalised by Karl Pearson (1896) [6], quantifies linear relationships between variables ($-1 \leq r \leq +1$), with $|r| < 0.2$ indicating negligible correlation. While foundational, this method faces two critical challenges in efficiency benchmarking:

1. Outlier Sensitivity:
   - Example: A study of income versus blood pressure (n = 120) showed the correlation coefficient (r) dropping from 0.88 to 0.7 after adding one outlier [6].
   - Implications for DEA: Outliers in restaurant data (e.g., holiday sales spikes) may distort efficiency rankings if unaddressed.
2. Assumption of Linearity:
   - Fails to capture the non-linear relationships common in operational metrics, such as cost-volume-profit dynamics [17].

Our approach leverages Pearson correlation as a weight-adjustment tool to detect and downweight spurious relationships between inputs (e.g., sales volume) and outputs (e.g., profit) [7] and enhance Gaussian AHP weights by multiplying them by the global mean of correlation coefficients (Sect. 5):

$$w_i^{corrected} = w_i^{Gaussian} \cdot \overline{r_i}$$

where $\overline{r_i}$ is the mean correlation of criterion i with other variables.

Given this context, Pearson's coefficient can be an optimisation criterion for developing noise reduction mechanisms [7] in various applications. Documented uses of this technique include image stabilisation systems [18] and transformer algorithms [17].

5 Methods

The development of the proposed hybrid multicriteria method, as illustrated in Fig. 1, which integrates DEA with Gaussian AHP and Pearson correction, follows steps similar to classical DEA, with the only modification occurring in the weight

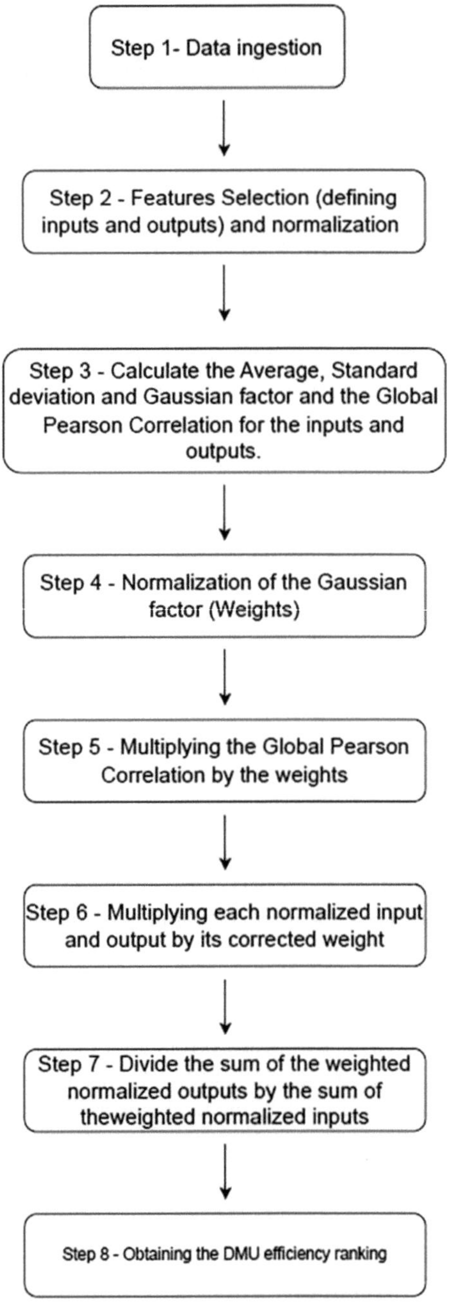

Fig. 1 Steps for DEA-AHP/GS. *Source* The Author

Table 1 Food retail chain data

DMU	Inputs		Output
Restaurant	Sales amount	Total cost (R$)	Gross profit (R$)
Restaurant A	233,655	1,673,115.11	3,241,039.50
Restaurant B	165,480	1,112,546.22	1,538,799.72
Restaurant C	137,177	953,469.73	1,564,098.94
Restaurant D	111,187	918,028.02	1,239,982.62

Source The Author

Table 2 Weights calculation detailed

Feature	Average	Standard deviation	Gaussian factor	Gaussian factor normalized	Global Pearson correlation coefficient	Weights
Sales amount	161874.75	52740.33552	0.32580952	0.679789	0.979546927	0.665885
Total cost (R$)	1164289.8	349604.8056	0.30027302	0.626508	0.979546927	0.613694
Gross profit (R$)	1895980.2	908706.1152	0.47928038	1.000000	0.979546927	0.979547

Source The Author

calculation stage. To determine the weights, statistical analysis involving the mean, standard deviation, and Gaussian factor involves normalising to generate weights, which are then incorporated into the Gaussian AHP. Once the weights are obtained, Pearson's correlation is applied by multiplying each weight by the global mean of the correlation, which involves determining the coefficients for the input and output criteria.

The hybrid method was applied to a dataset from a restaurant chain, covering the period from August 2023 to April 2024, where each business unit represents a Decision-Making Unit (DMU). The selected inputs were the total sales volume and total cost, while the output variable was the Gross Profit of each DMU (Table 1).

Next, the mean, standard deviation, Gaussian factor, normalised Gaussian factor, and the global Pearson correlation coefficient were computed. The normalised Gaussian factor was then multiplied by the global Pearson correlation coefficient to obtain the final weights (Table 2).

The standard DEA procedure was executed using the calculated weights following these steps. The inputs and outputs were normalised, and the efficiency score for each DMU was obtained by dividing the weighted sum of normalised outputs by the weighted sum of normalised inputs. The chosen normalisation method was (Value/

maxValue) to maintain data proportionality and preserve statistical properties such as mean, median, and standard deviation (Tables 3, 4 and 5).

After completing the procedures described, an efficiency ranking of all restaurants within the business group was obtained.

Access to the Python notebook developed in this work is via the QR code in Fig. 2.

Table 3 Inputs and outputs normalised

DMU	Inputs normalised		Output normalised
Restaurant	Sales amount	Total cost (R$)	Gross profit (R$)
Restaurant A	1.000000	1.000000	1.000000
Restaurant B	0.708224	0.664955	0.474786
Restaurant C	0.587092	0.569877	0.482592
Restaurant D	0.475860	0.548694	0.382588

Source The Author

Table 4 Inputs and outputs normalised and weighted

DMU	Inputs normalized and weighted		Output normalized and weighted
Restaurant	Sales amount	Total cost (R$)	Gross profit (R$)
Restaurant A	0.665885	0.613694	0.979547
Restaurant B	0.472898	0.409206	0.466360
Restaurant C	0.392016	0.350696	0.474027
Restaurant D	0.317743	0.337660	0.375798

Source The Author

Table 5 DEA—AHP/GS results detailed

DMU	Input normalized and weighted	Output normalized and weighted	Efficiency	Ranking
Restaurant A	1.283113	0.982252	0.765523	1
Restaurant C	0.742712	0.474027	0.638238	2
Restaurant D	0.655403	0.375798	0.573384	3
Restaurant B	0.882104	0.466360	0.528690	4

Source The Author

Fig. 2 Link for the DEA-AHP/GS GitHub repository. *Source* The Author

6 Conclusion

This study proposes and implements a hybrid multicriteria decision support method that combines Data Envelopment Analysis (DEA), Gaussian Analytic Hierarchy Process (Gaussian AHP), and Pearson correlation. The methodology followed the classical DEA structure, differing in the weight calculation step, where weight determination methods from Gaussian AHP were incorporated. Pearson's correction was applied to adjust the weights according to the global correlation of input and output criteria, ensuring greater robustness and alignment of the results with the relationships observed in the data.

By addressing the research question—"How can a hybrid DEA–Gaussian AHP framework, adjusted via Pearson correlation, improve efficiency benchmarking in restaurant chains compared to conventional methods?"—the study demonstrates that the proposed method enhances traditional benchmarking by reducing subjectivity and incorporating statistical rigour. This strengthens both the analytical robustness and the practical relevance of efficiency evaluations.

Applying the hybrid DEA–AHP/GS method to a restaurant chain demonstrated its technical and practical feasibility, highlighting its potential to generate strategic insights. The methodology enabled the identification of the most efficient units and those with the most significant potential for improvement, revealing opportunities for targeted operational optimization. Moreover, this approach can be easily integrated into management cycles, such as the PDCA (Plan, Do, Check, Act) cycle, and widely adopted corporate management tools, thereby enhancing objective and systematic decision-making, as well as efficiency monitoring. Another distinguishing feature of the method is its independence from subjective judgments, common in traditional AHP and DEA, as it eliminates cognitive recurrence bias by relying solely on statistical analysis and linear programming.

Theoretically, integrating Gaussian AHP and Pearson correlation into the DEA framework contributes to the academic literature by bridging the gaps between deterministic models and statistical validation, offering a replicable and adaptable approach for efficiency analysis across diverse sectors. This hybridization addresses long-standing criticisms about subjectivity in multicriteria decision-making and provides a data-driven alternative that may inspire new methodological pathways in performance evaluation research [19, 20].

For future work, it is intended to expand the approach to sectors beyond food retail, exploring its applicability in areas with distinct operational structures, such as manufacturing industries, public management, and marketing. Additionally, incorporating machine learning and artificial intelligence techniques could improve the method, making its application more versatile and valuable to organisations. Investigating the integration of this model with advanced visualisation tools, such as interactive dashboards, would also be beneficial in facilitating the interpretation and practical use of the results.

Despite the promising results, this research and its development present some limitations that should be taken into consideration. The application of the method was limited to a specific restaurant chain within a specified timeframe. Although the input and output criteria were suitable for the given context, they may not encompass all variables influencing operational efficiency in different organisational and production environments. Lastly, while Pearson's correction was applied to mitigate distortions, spurious correlations may still affect the analysis, indicating the need for further investigations employing methods that account for such aspects.

Acknowledgements The first author thanks the Coordination of Superior Level Staff Improvement (CAPES) through grant # 88887.798679/2022-00. The second and third authors thank the National Council for Scientific and Technological Development (CNPq) through grant #306435/2023-3.

References

1. Dharmadhikari, S.: Europe full-service restaurants market report 2025. Cognitive market research (2025). www.cognitivemarketresearch.com/regional-analysis/europe-full-service-restaurants-market-report. Accessed 13 Jan 2025
2. Nadai, S.: Ganhos com Implantação de ERP no Setor de Compras de uma Empresa do Setor Alimentício. X Congresso de Logística das Faculdades de Tecnologia do Centro Paula Souza - FatecLog Guarulhos (2019)
3. Lima, M.S., Carrieri, A., Pimentel, T.D.: Resistência à Mudança Gerada pela Implementação de Sistemas de Gestão Integrada (ERP): Um Estudo de Caso. Rev. Gestão Planejamento **8**(1), 89–105 (2007). revistas.unifacs.br/index.php/rgb/article/view/258.
4. Cooper, W.W., Seiford, L.M., Tone, K.: Data Envelopment Analysis: A Comprehensive Text with Models, Applications, References and DEA-Solver Software, 2nd edn. Springer (2007)
5. Dos Santos, M., De Araújo Costa, I.P., Gomes, C.F.S.: Multicriteria decision-making in the selection of warships: a new approach to the AHP method. Int. J. Anal. Hierarchy Process. **13**(1) (2021)
6. Okwonu, F.Z., Asaju, B.L., Arunaye, F.I.: Breakdown analysis of Pearson correlation coefficient and robust correlation methods. IOP Conf. Ser. Mater. Sci. Eng. 012065 (2020)
7. Cohen, I., et al.: Pearson correlation coefficient. In: Noise Reduction in Speech Processing, pp. 1–4 (2009)
8. Charnes, A., Cooper, W.W., Rhodes, E.: Measuring the efficiency of decision-making units. Eur. J. Oper. Res. **2**(6), 429–444 (1978)
9. Huguenin, J.-M.: Data Envelopment Analysis (DEA): a pedagogical guide for decision makers in the public sector. Swiss Graduate School of Public Administration (2012)
10. Gomes, C.F.S., et al.: Ordering of warships for the brazilian navy using the new method: AHP-Gaussian with Pearson's correlation. In: Modern Management Based on Big Data II and Machine Learning and Intelligent Systems III, pp. 104–111. IOS Press (2021)

11. Meza, L.A., Gomes, E.G., Neto, L.B.: Curso de Análise de Envoltória de Dados. In: XXXVII Simpósio Brasileiro de Pesquisa Operacional, pp. 2520–2547 (2005)
12. Dyson, R.G., Thanassoulis, E.: Reducing weight flexibility in data envelopment analysis. J. Oper. Res. Soc. **39**(6), 563–576 (1988)
13. Saaty, T.L.: Relative measurement and its generalisation in decision making are why pairwise comparisons are central in mathematics for measuring intangible factors in the analytic hierarchy/network process. RACSAM-Rev. Real Acad. Ciencias Exactas Físicas Nat. Serie A Matemáticas **102**(2), 251–318 (2008)
14. Almeida, L.H., et al.: A multicriteria model for planning and fine-tuning distributed scrum projects. In: 2011 IEEE Sixth International Conference on Global Software Engineering, pp. 75–83. IEEE (2011). https://doi.org/10.1109/ICGSE.2011.36
15. Mendes, M., et al.: A co-evolutionary interaction design of digital TV applications based on verbal decision analysis of user experiences. In: Lytras, M.D., et al. (eds.) The Open Knowledge Society: A Computer Science and Information Systems Manifesto. Communications in Computer and Information Science, vol. 19, pp. 732–741. Springer (2008). https://doi.org/10.1007/978-3-540-87783-7_89.
16. Rodrigues, T.S., Pinheiro, P.R.: Hyperparameter optimization in Generative Adversarial Networks (GANs) using Gaussian AHP. IEEE Access **13**, 770–788 (2025)
17. Wu, L.: Automatic scoring model for English compositions using bidirectional encoder representations from transformer with Pearson correlation coefficient. In: 2024 International Conference on Integrated Intelligence and Communication Systems (ICIICS), pp. 1–5. IEEE (2024)
18. Liu, J., Zhang, Y., Zhao, Q.: The video stabilisation algorithm is based on the Pearson correlation coefficient. In: 2019 International Conference on Advanced Mechatronic Systems (ICAMechS), pp. 289–293. IEEE (2019)
19. Pinheiro, P.R., Rodrigues, T.S.: Optimisation of regression models for rental price recommendations: an integrated approach using PyCaret and Gaussian analytic hierarchy process. In: Silhavy, R., Silhavy, P. (eds.) Lecture Notes in Networks and Systems, vol. 1119, pp. 504–512. Springer (2024)
20. Saaty, T.L.: Highlights and critical points in the theory and application of the analytic hierarchy process. Eur. J. Oper. Res. **74**(3), 426–447 (1994)

Systematic Literature Reviews (SLRs) in Business, Management and Economics: Trends, Gaps, and Methodological Rigor

Radosław Malik, Anna Visvizi, and Roman Wosiek

Abstract This paper examines how and to what extent Systematic Literature Reviews (SLRs) and the Preferred Reporting Items for Systematic Reviews and Meta-Analyses (PRISMA) protocol have been endorsed by researchers conducting research in economics, business and management, and what implications it bears for academic debates unfolding in respective areas of research. The findings reveal an increase (14.415%) SLR adoption over the period 2005–2024 in the body of Web of Science (WoS) indexed publications. Management and business emerge as leading areas of SLR adoption, with economics lagging behind. A comparison with other areas of research reveals that economics, business and management fall behind as regards PRISMA/SLR adoption overall. The variability of SLR and PRISMA adoption suggests that different approaches to specific aspects of methodology across disciplines exist, which in turn raises questions about the prospects of inter- and multi-disciplinary research as well as about comparability of research output, also in context of quality assurance. From a different angle, while management and business have embraced SLRs as a tool for enhancing research quality, economics presents opportunities for improvement. The study concludes that a more structured approach to SLRs, such as the application of PRISMA in business and management, could further enhance the reliability and impact of research in these fields.

Keywords Systematic literature review · SLR · PRISMA · Research methodology · Economics · Business · Management · Quality assurance · Evaluation

R. Malik (✉) · A. Visvizi · R. Wosiek
SGH Warsaw School of Economics, Warsaw, Poland
e-mail: rmalik@sgh.waw.pl

A. Visvizi
Effat University, Jeddah, Saudi Arabia

A. Visvizi et al. (eds.), *Research and Innovation Forum 2025 Volume 1*,
Springer Proceedings in Complexity, https://doi.org/10.1007/978-3-032-09066-9_7

1 Introduction

The number of academic publications has seen a consistent rise over the past years [6]. Several factors have contributed to that, including the evolution of the publishing market—a trend best described as moving away from oligopoly to (fierce, very scarcely regulated) competition; technology-driven relative decline in the cost of the publishing process; and pressures on new, more flexible, more equitable modes of access to published research. The trend towards quantified scientific outputs [1], growing international competition among universities [7] and the "publish or perish" attitude prevalent in the research community [8] are equally important in this context. That being said, the critical question that remains is that whether quantity is commensurate with quality [19], whether increased research output does indeed translate in new discoveries, in line with the imperative that science should be emancipatory [21–23], and research-driven innovations based on practical applications of research outcomes [11].

Systematic Literature Reviews (SLR) have grown in importance in recent years, being recognized as a valuable tool to stimulate the progress and quality of scientific research [9]. A great number of studies delving on questions of methods and methodology highlighted the latter's role and salience in view of advancing research and scientific discovery, in terms of quality enhancement, new avenues of research [18], building bridges within and across disciplines, and avoiding cul-de-sacs of excessive specialization. Accordingly, also in relation to SLR, it has been stressed that SLRs serve as a tool for synthesizing existing knowledge, providing insights in areas of consensus or disagreement within specific research domains [24]. Moreover, it was shown that SLR facilitate the identification of research gaps, thereby informing decision-making about the direction of future research [20]. Furthermore, in an era characterized by high specialization in research and the often relatively limited scale and scope of several publications, systematic literature reviews enable the integration of disparate studies into a more coherent and comprehensive picture of the current state of the art in a given research field [5]. Additionally, SLR are viewed as a means to enhance methodological rigor, as they often follow specific procedures and are replicable [16]. In this view, SLRs, fulfilling the requirements of good science, i.e. empirical verifiability; ethical neutrality; malleability; objectivity; observability; replicability; systematic reliability, may serve as the condition sine qua non for establishing a sound point of departure for any research endeavor. For the sake of objectivity and neutrality, SLRs and the PRISMA protocol do display shortcomings [23].

The clear benefits of SLRs have led to their enhancement and are reflected in the development and spread of specific protocols in the domain of SLRs [4], such as for instance the Preferred Reporting Items for Systematic Reviews and Meta-Analyses (PRISMA) protocol, the less known Methodi Ordinatio [12], and other. These protocols aim to further structure the application of SLRs [10], boost their benefits, and increase scientific rigor [15]. The most popular of these protocols, i.e. PRISMA provides a standardized framework for reporting systematic reviews and meta-analyses, thus ensuring transparency and completeness of the literature review process [13].

The application of SLRs and their further enhancement through the use of protocols, as above, and guidelines, add rigor and transparency of literature reviews in general. Therefore, if properly utilized they also enhance the quality and reliability of academic research across various disciplines [14]. These structured approaches to literature review ensure that the examination of existing research is comprehensive, methodologically sound, replicable and fit-for-purpose, which is crucial for advancing scientific knowledge [2]. However, despite these significant benefits, the application of SLRs and the tools that further strengthen their for progress in studies appears to be unevenly distributed across different research areas, and so also among different academic disciplines [3]. This variability suggests that while some research communities have fully embraced and integrated SLRs into their research practices, others may lag behind, potentially missing out on the advantages that systematic reviews offer in terms of evidence synthesis, gap identification, and informed decision-making. Against this backdrop, by zooming in the art and practice of economics, business and management, the objective of this paper is to examine the extent to which the benefits of SLRs have been recognized and thus utilized by respective research communities. In other words, the two research questions that this paper addresses are:

- RQ1: What is the extent of the application of SLRs in economics, business and management as compared to other research areas?
- RQ2: What is the extent of the adoption and implementation of the PRISMA protocol in SLR conducted in economics, business and management as compared to other research areas?

To address these questions, the remainder of this paper is structured as follows: the Methods and Materials section outlines the approach taken to investigate the research questions, detailing the data collection and analysis procedures. Findings are presented in the following section. Discussion and Conclusions follow.

2 Methods and Materials

In view of addressing the research questions outlined above, the objective of this section is (i) to shed light on the number and share of academic publications the authors of which have employed SLRs percentage of scientific texts that applied systematic literature reviews across various disciplines, and (ii) and to examine the prevalence of this method in publications classified as belonging to debates in economics, business and management. To add to the conversation, (iii) the use of the PRISMA protocol, will also be examined. To this end, a comprehensive and systematic approach was employed, involving five key steps: (i) data collection, (ii) selection criteria, (iii) data extraction, (iv) data analysis, (v) comparative analysis.

In the Data Collection stage, the Web of Science (WoS) database, a widely recognized and reliable source for academic publications across multiple disciplines, was employed. The WoS database was chosen for its extensive coverage of peer-reviewed literature, which thus ensured a robust and representative sample [17] for the analysis in this paper. In the second step, i.e. Selection Criteria, specific criteria were applied to ensure the relevance and quality of the publications included in the sample. Specifically, to maintain consistency across disciplines, only articles and review articles were considered in the analysis. The time frame of the research was limited to 2005–2024 to balance the scope of the analysis with not so recent publications with an up-to-date dataset. Additionally, only texts written in English were included in the scope of the analysis to ensure comparability and accessibility.

To identify publications that applied SLRs as the primary research method, the search criterion the keyword "systematic literature review" in the Title section of WoS search engine was used. It was assumed that texts containing this keyword in the title utilized SLR as the primary research method. Furthermore, to identify articles that applied the PRISMA protocol within SLRs, the search criteria included "systematic literature review" in the Title and "PRISMA" in All Fields. This approach was based on the assumption that if PRISMA was used, this information would be indicated in the title or keywords, or abstract of the text, and all them are included as All Field section of WoS search engine.

In the third step, Data Extraction (from WoS), was performed on April 16, 2025. The selection criteria indicated above were applied for three separate searches to extract the number of texts across all disciplines. The first search covered texts that utilized SLRs. The second search covered texts that employed both SLRs and the PRISMA protocol. The third search covered the total number of texts per discipline to provide the background for frequency analysis per discipline.

The fourth step, Data Analysis, implies the analysis of the collected data to determine the number and percentage of publications that applied systematic literature reviews (SLRs) across various disciplines. Descriptive statistics, including frequencies and percentages, were calculated to quantify the prevalence of SLRs in different fields. The frequency of SLR use in economics, business and management was then compared to that in other disciplines as measured by the WoS' categories to identify trends and disparities. Similarly, the adoption and implementation of the PRISMA protocol, as a part of SLRs, were assessed by analyzing the subset of publications that explicitly mentioned PRISMA. This analysis included evaluating the proportion of SLRs that adhered to the PRISMA guidelines and comparing this adoption rate across disciplines.

In the fifth, and the final step, Comparative Analysis is applied to gain deeper insights, a comparative analysis was performed to identify trends and gaps in the application of SLRs and the PRISMA protocol across different disciplines. This analysis involved comparing the frequency of SLR usage and PRISMA adoption in economics, business, and management with other fields, such as life sciences, earth sciences, and other social science.

3 Findings

3.1 An Overview

The application of the research methodology outlined in the previous sections yielded the following results. A total of 15,260 papers written in English and published between 2005 and 2024 were identified as having applied SLR. This subset represents a specific focus within the broader dataset of 39,953,013 texts published and indexed in the WoS database that fulfilled the research criteria. The results indicated a growing number of SLRs that have been published over this period (see Fig. 1 for details).

The analysis of publications' trend based on the above figure reveals a remarkable evolution in the adoption of SLRs in academic research over the past two decades. The data demonstrates an extraordinary growth trajectory, with the number of SLRs increasing from just 20 in 2005 to 2,883 in 2024, representing a 14.415% increase over the period 2005–2024. This exponential growth significantly outpaces the 281% increase observed in the total number of publications indexed in the WoS database during the same timeframe. Despite this substantial expansion, SLRs still constitute only 0.041% of all published texts in the analyzed dataset, possibly highlighting their specialized nature within the broader academic landscape, or other issues, as discussed in conclusions.

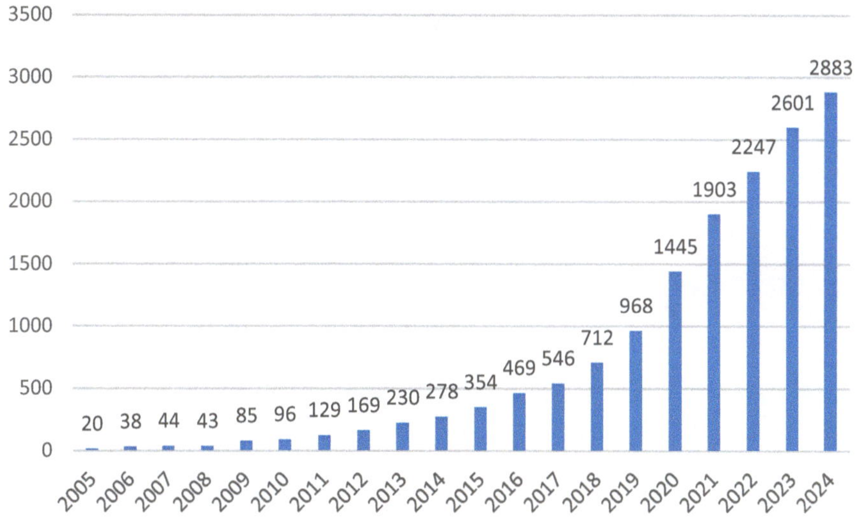

Fig. 1 Total number of SLRs published in 2005–2024 in WoS. *Source* Authors

3.2 SLR Adoption Across Research Areas

The published SLRs were attributed to 223 WoS' categories based on the predefined labels provided by the WoS database. The results reveal a considerable distribution of SLRs across various disciplines over 2005–2024. The top five categories with the highest number of SLRs were Management (1,268), Environmental Sciences (1,012), Computer Science Information Systems (992), Business (888), and Green Sustainable Science Technology (804). This distribution highlights the widespread adoption of SLRs in diverse fields, reflecting their relevance and perceived value in synthesizing research across different areas of research. Figure 2 depicts the 20 top WoS categories in terms of the number of SLRs published. It thus provides a visual representation of the disciplines where SLRs have been most frequently applied.

The results highlight the important role of SLRs in Management and Business, which ranked among the top categories with 1,268 and 888 publications, respectively. In contrast, SLRs appear to play a less important role in economics, with only 280 publications, placing it in the 26th position compared to other research categories. This disparity suggests that while Management and Business have embraced SLRs as methodological tool to a greater extent than economics, which seems to lag behind in adopting this approach to literature review.

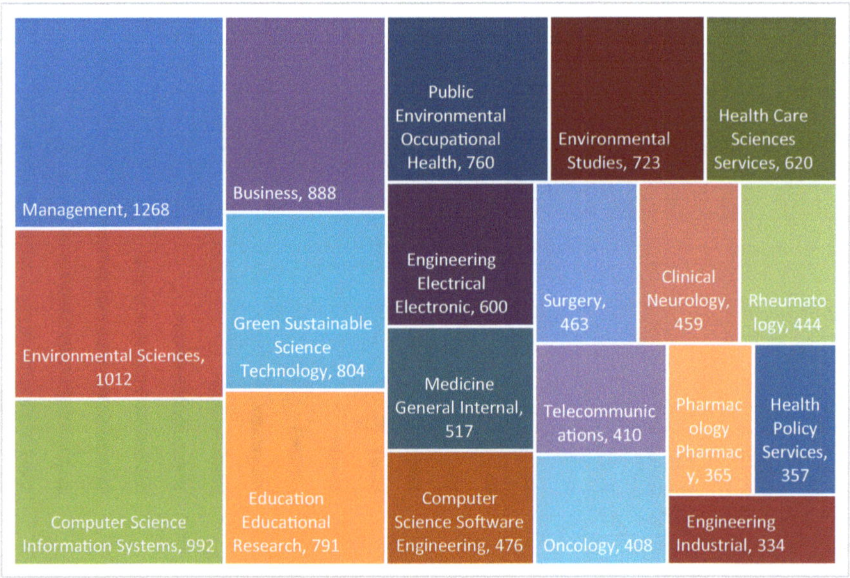

Fig. 2 Total number of SLRs published in 2005–2024 in per WoS category. *Source* Authors

3.3 *SLR Adoption in Economics, Business and Management*

The analysis of SLR adoption across different disciplines reveals significant variability, which in itself asks for a closer examination. Given that the WoS categories differ considerably in terms of their total research output, which naturally influences the absolute number of SLRs published, it becomes essential to analyze the relative frequency of SLR adoption per category. This normalized approach provides a more accurate comparison of how different fields utilize systematic review methodologies. Figure 3 represents the results for the top 20 categories out of 223 where at least one SLR was published, in terms of the percentage of texts employing SLRs.

The analysis presented in Fig. 3 reveals significant variability in the adoption of SLRs across different WoS' categories. Among the 223 categories examined, Management and Business emerged as leading disciplines in SLR adoption, with frequencies of 0.375 and 0.341%, securing the 2nd and 3rd positions, respectively. In contrast, Economics ranked 66th, with a considerably lower adoption rate of 0.053%.

Notably, all categories included in the analysis demonstrated a higher frequency of SLR usage than the overall average of 0.041% for all publications in the research scope. Furthermore, specific subcategories within Business and Management, such as Business Finance and Operations Research Management Science, stood out with particularly high adoption rates of 0.136 and 0.130%, respectively, placing them

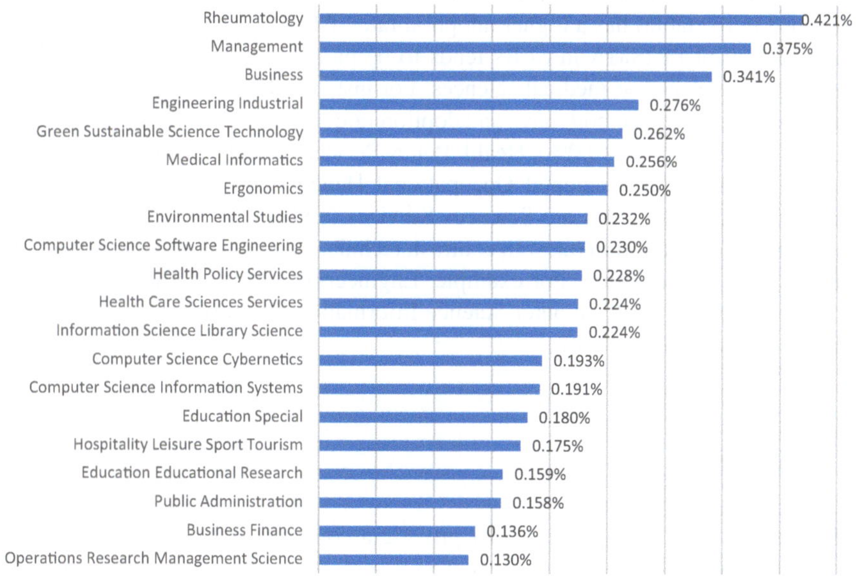

Fig. 3 Top 20 WoS categories by percentage of texts using SLRs (2005–2024). *Source* Authors

among the top 20 categories. Interestingly, politics, international studies, international relations, have not been included in the results. Please, see the concluding section for a few points on that.

3.4 The PRISMA Protocol Adoption: An Overview

To address the second research question, i.e. the extent of PRISMA protocol adoption in SLRs in economics, business and management, compared to other disciplines, a targeted analysis was conducted. The findings indicate that, within the examined dataset of 15,260 SLR articles, the PRISMA protocol was applied in 2,170 texts, therefore in 14.2% of the SLRs. To ensure a robust and unbiased analysis, this presented data focuses on the top 20 WoS categories ranked by the total number of SLRs published previously shown in the Fig. 2. This approach mitigates potential bias that could arise from including categories with a very low number of SLRs but disproportionately high PRISMA adoption rates (e.g., cases where only a few SLRs exist, all of which use PRISMA). By prioritizing categories with a substantial volume of SLRs, this analysis provides a more accurate and meaningful assessment of PRISMA's significance and adoption trends across disciplines as detailed in Table 1.

The data (Table 1) shows that the adoption of PRISMA in systematic literature reviews (SLRs) varies significantly across disciplines. In Management, the adoption rate stands at 11.0%, while in Business, it is 8.2%. Notably, Economics (not included in the table) has a higher adoption rate of 13.6%. The results indicate that the adoption of PRISMA in SLRs tends to be particularly high in studies related to the environment and medical sciences. For instance, Environmental Studies and Environmental Science both have an adoption rate of 20.2%, while various medical fields, such as Surgery (19.0%), Health Policy Services (17.6%), Clinical Neurology (17.4%), Public Environmental Occupational Health (17.4%), Medicine General Internal (17.2%), Health Care Sciences Services (16.0%), and Oncology (13.2%), also exhibit high adoption rates. In contrast, technical studies show a more moderate rate of PRISMA adoption. For example, Engineering Electrical Electronic has an adoption rate of 9.8%, Computer Science Information Systems stands at 9.2%, and Telecommunications at 8.8%.

4 Discussion

The findings presented in this paper provide a comprehensive answer to the first research question regarding the extent of SLR application in economics, business and management research compared to other research areas. The results reveal a growing trend in SLRs as a distinct genre of academic publications, with their increase in the 2005–2024 period significantly outpacing the overall growth rate of all articles

Table 1 Percentage adoption of PRISMA in SLRs across top 20 WoS categories (2005–2024)

Environmental sciences	20.7%
Environmental studies	20.2%
Surgery	19.0%
Green sustainable science technology	18.3%
Health policy services	17.6%
Education educational research	17.4%
Clinical neurology	17.4%
Public environmental occupational health	17.4%
Medicine general internal	17.2%
Health care sciences services	16.0%
Oncology	13.2%
Pharmacology pharmacy	11.8%
Management	11.0%
Engineering electrical electronic	9.8%
Computer science information systems	9.2%
Telecommunications	8.8%
Business	8.2%
Rheumatology	6.5%
Engineering industrial	6.0%
Computer science software engineering	2.3%

Source Authors

published. This suggests that SLRs are becoming an increasingly valued methodological approach in academic research across disciplines, albeit with varying degrees of utilization.

The use of SLRs is particularly prominent in management and business, both in terms of absolute publication numbers and relative frequency of SLRs' use. Management leads with 1,268 SLR publications, followed by business with 888, representing the adoption rates of 0.375 and 0.341%, respectively. These figures are substantially higher than the overall average of 0.041% across all fields of research. This may indicate that researchers in management and business recognize the value of SLRs in structuring knowledge, synthesizing research findings, and identifying gaps in the literature to a greater extent than most other fields. Certainly, other factors may be at play too.

In contrast, economics exhibits a more moderate adoption of SLRs, with 280 publications and a frequency of 0.053%, i.e. only slightly above the overall average. While this suggests some recognition of SLRs as a valuable research tool, the relatively lower adoption rate implies that economics lags behind management and business in leveraging systematic reviews for knowledge synthesis. This presents an opportunity for progress, particularly in enhancing the systematic structuring of research outcomes, identifying consensus and disagreements in the literature, and mapping future research directions.

The second research question was designed to address the extent of PRISMA protocol adoption in SLRs in economics, business and management, as compared to other areas of research. The findings reveal notable variability in adoption rates across these fields. Specifically, in management the adoption rate of PRISMA stands at 11.0%, while in business, it is 8.2%. Interestingly, economics displays a higher adoption rate of 13.6%. However, all three research areas fall below the overall average adoption rate of 14.2% observed in the analyzed sample. This suggests that while economics demonstrates a relatively stronger uptake of PRISMA compared to management and business, all three fields still lag behind the broader academic average.

When compared to other disciplines, the adoption of PRISMA in economics, management, and business is lower than in environmental and medical sciences, where structured review protocols have been more widely embraced by respective research communities. However, these fields exhibit slightly higher adoption rates than technical studies identified in our sample of top 20 WoS categories, where methodological standardization is often less emphasized. This variation underscores differing levels of emphasis on methodological rigor across academic domains.

These findings suggest that while economics, management, and business communities recognize the value of PRISMA, significant potential exists for a more frequent adoption of the latter as a research tool. Encouraging a more frequent use of PRISMA in these fields of research could enhance transparency, reproducibility, and rigor of systematic reviews, ultimately strengthening the quality of research in these fields. Future efforts should focus on promoting awareness of PRISMA guidelines and their benefits, particularly in disciplines where adoption remains below the broader academic standard.

5 Conclusions and Recommendations

The consistent year-over-year increase in SLR publications underscores their growing recognition as a valuable research methodology. This trend reflects the academic community's increasing appreciation for the rigorous, evidence-based approach that SLRs provide in synthesizing existing knowledge, identifying research gaps, and informing future research directions. The disproportionate growth rate between SLRs and general publications suggests that while SLRs remain a niche methodology, their importance and application are expanding at an unprecedented pace across various disciplines. This development points to a shifting paradigm in research practices, where systematic approaches to literature review are becoming increasingly integral to advancing scientific knowledge.

The differences in the frequency of SLR adoption across these disciplines underscore varying levels of methodological rigor, evidence-based research practices, as well as prevailing tradition of narration in a given area of research and, even specific research community. The analysis presented in this paper reveals that researchers in management and business have effectively embraced SLRs as a tool for enhancing research quality, while the research community dwelling on diverse aspects of economics has been slower to adopt this approach. These findings suggest that increased use of SLRs in economics might improve the field's capacity for knowledge synthesis, gap identification, and methodological strengthening of future studies. While management and Business currently lead in SLR adoption, economics presents substantial opportunities for growth. By promoting greater adoption of SLRs in economics, the discipline could enhance its ability to conduct systematic reviews and synthesize research findings, thereby fostering more robust and evidence-based academic discourse.

Recognizing that PRISMA is not the only tool used to enhance the quality of SLRs, in this paper, its application is viewed as an indicator of methodological rigor that influences the overall academic quality of SLRs, and by default adds to quality of research published. This is the function of the features of good science. This paper concludes with a point that the adoption of this PRISMA protocol in management, business, and economics remains below the average, as compared to other fields of research, with business and management falling particularly short of expectations. While scholars in these fields have embraced the use of SLRs, the lower-than-average adoption of PRISMA suggests that the rigor and quality of their application may still require improvement. This finding highlights an important opportunity for enhancement, emphasizing the need for greater emphasis on structured review methodologies to strengthen the reliability and impact of SLRs in these disciplines. Future efforts should focus on promoting best practices in systematic reviewing to ensure that research in Management, Business, and Economics meets the highest standards of academic rigor.

While this study provides valuable insights into the adoption of SLRs and the PRISMA protocol across disciplines, it is not without limitations; these will be addressed in our future research. Specifically, first, this study relies exclusively on data from the WoS database, which may be the source of selection bias. Future studies could incorporate additional datasets, such as Scopus, to validate and compare findings, ensuring a more comprehensive analysis of global research trends. Second, the selection inclusion criteria have been designed in a manner that exclude all those papers that do use SLR as one of the methods employed, yet, as one of many, not as the key one [23]. Therefore, future research should focus on bibliometric analysis engaging with content extraction as well.

Third, the typology of research disciplines/areas in this study follows the WoS classification system, which may be overly granular in some cases. Also, the WoS typology it does not reflect the disciplinary boundaries typical to specific markets, or countries, say regions, where disciplines, such as economics, politics, and management make serious claims to their territoriality. The WoS typology is good in identifying leading and emerging areas of research, yet it also falls short of clustering inter-disciplinary research that defies the dominant categories. This may suggest that research otherwise locating itself in economics and politics, has not been included in either of categories depicted above. Accordingly, future research could explore alternative approaches, such as broader disciplinary groupings, to mitigate the influence of narrowly defined categories with limited publication volumes. This adjustment could provide a more generalized overview of SLR adoption trends.

Finally, while PRISMA is a widely recognized protocol for enhancing the quality of SLRs, it is not the only methodological framework available. Other tools, such as A MeaSurement Tool to Assess Systematic Reviews (AMSTAR), Risk Of Bias In Systematic Reviews (ROBIS), Synthesis Without Meta-analysis (SWiM), or Methodi Ordinatio, also play crucial roles in ensuring the rigor of systematic reviews. Future studies should evaluate the adoption and impact of these alternative protocols to provide a more holistic understanding of methodological standards in SLRs.

This study makes a threefold contribution to the understanding of SLRs and the PRISMA protocol across research areas. First, we conducted an analysis of SLR and PRISMA adoption across various fields, offering a novel methodological approach that can be applied to diverse sets of disciplines. This cross-disciplinary perspective provides a foundation for future comparative studies in research methodologies. Second, we provided new insights into the differential adoption of SLRs and PRISMA between Economics, Management, and Business studies. Our findings reveal distinct patterns of methodological rigor, with Management and Business demonstrating higher adoption rates than Economics, highlighting varying levels of engagement with systematic review practices. Third, our study underscores the need for increased SLR adoption in Economics and a more structured approach to SLRs in Business and Management. By identifying these gaps, we emphasize opportunities for improvement, encouraging researchers in these fields to enhance the scientific rigor and quality of their systematic reviews.

References

1. Abramo, G., D'Angelo, C.A.: Hyperprolific authorship: unveiling the extent of extreme publishing in the 'publish or perish' era. J. Inf. **19**(2), 101658 (2025). Elsevier, https://doi.org/10.1016/J.JOI.2025.101658.
2. Akl, E.A., et al.: Extension of the PRISMA 2020 statement for living systematic reviews (PRISMA-LSR): checklist and explanation. BMJ **387**, e079183 (2024). British Medical Journal Publishing Group, https://doi.org/10.1136/BMJ-2024-079183.
3. Azarian, M., et al.: Do we perform systematic literature review right? A scientific mapping and methodological assessment. Logistics **7**(4), 89 (2023). Multidisciplinary Digital Publishing Institute, https://doi.org/10.3390/LOGISTICS7040089
4. Bandara, W., Syed., R.: The role of a protocol in a systematic literature review. J. Decis. Syst. (2024). Taylor & Francis, https://doi.org/10.1080/12460125.2023.2217567
5. Burgers, C., et al.: Systematic literature reviews: four applications for interdisciplinary research. J. Pragmat. **145**, 102–109 (2019). North-Holland, https://doi.org/10.1016/J.PRAGMA.2019.04.004
6. Hanson, M.A., et al.: The Strain on Scientific Publishing (2024). https://doi.org/10.1162/qss_a
7. Hart, P.F., Rodgers, W.: Competition, competitiveness, and competitive advantage in higher education institutions: a systematic literature review. Stud. High. Educ. (2024). Routledge, https://doi.org/10.1080/03075079.2023.2293926

8. Ioannidis, J.P.A., et al.: Evolving patterns of extreme publishing behavior across science. Scientometrics **129**(9), 5783–5796 (2024). Akademiai Kiado ZRt., https://doi.org/10.1007/S11192-024-05117-W/METRICS
9. Kraus, S., et al.: The burgeoning role of literature review articles in management research: an introduction and outlook. Rev. Manag. Sci. **18**(2), 299–314 (2024). Springer Science and Business Media Deutschland GmbH, https://doi.org/10.1007/S11846-024-00729-1/METRICS
10. Malik, R., et al.: Smart services in smart cities: insights from science mapping analysis. Sustainability (Switzerland) **14**(11) (2022). Springer Science and Business Media Deutschland GmbH, https://doi.org/10.3390/SU14116506
11. Munari, F., et al.: Public research funding and science-based innovation: an analysis of ERC research grants, publications and patents. In: Research Evaluation. Oxford University Press (OUP) (2024). https://doi.org/10.1093/RESEVAL/RVAE012
12. Pagani, R.N., Kovaleski, J.L., Resende, L.M.: Methodi Ordinatio: a proposed methodology to select and rank relevant scientific papers encompassing the impact factor, number of citation, and year of publication. Scientometrics **105**, 2109–2135 (2015). https://doi.org/10.1007/s11192-015-1744-x
13. Page, M.J., et al.: The PRISMA 2020 statement: an updated guideline for reporting systematic reviews. BMJ **372** (2021). British Medical Journal Publishing Group, https://doi.org/10.1136/BMJ.N71
14. Sarkis-Onofre, R., et al.: How to properly use the PRISMA statement. Syst. Rev. **10**(1), 1–3 (2021). BioMed Central Ltd., https://doi.org/10.1186/S13643-021-01671-Z/METRICS.
15. Sauer, P.C., Seuring, S.: How to conduct systematic literature reviews in management research: a guide in 6 steps and 14 decisions. Rev. Manag. Sci. **17**(5), 1899–1933 (2023). Springer Science and Business Media Deutschland GmbH, https://doi.org/10.1007/S11846-023-00668-3/FIGURES/2
16. Simsek, Z., et al.: Methodological rigor in management research reviews. J. Manag. (2024). SAGE Publications Inc., https://doi.org/10.1177/01492063241237222/SUPPL_FILE/SJ-DOCX-1-JOM-10.1177_01492063241237222.DOCX.
17. Skrzek-Lubasińska, M., Malik, R.: Is critical thinking a future skill for business success: science mapping and literature review. Cent. Eur. Manag. J. **31**(1), 48–63 (2023). Emerald Publishing, https://doi.org/10.1108/CEMJ-09-2021-0110/FULL/PDF
18. Snyder, H.: Literature review as a research methodology: an overview and guidelines. J. Bus. Res. **104**, 333–339 (2019). Elsevier, https://doi.org/10.1016/J.JBUSRES.2019.07.039.
19. Trueblood, J.S., et al.: The misalignment of incentives in academic publishing and implications for journal reform. Proc. Natl. Acad. Sci. **122**(5), e2401231121 (2025). National Academy of Sciences, https://doi.org/10.1073/PNAS.2401231121
20. Varsha, P.S., et al.: How to undertake an impactful literature review: understanding review approaches and guidelines for high-impact systematic literature reviews. South Asian J. Bus. Manag. Cases **13**(1), 18–35 (2024). SAGE Publications Ltd., https://doi.org/10.1177/22779779241227654/ASSET/D7CA6F24-4EF9-497C-8A87-B87D4ADE69DC/ASSETS/IMAGES/LARGE/10.1177_22779779241227654-FIG3.JPG
21. Visvizi, A.: Computers and human behavior in the smart city: issues, topics, and new research directions. Comput. Hum. Behav. **140**, 107596 (2023). Pergamon, https://doi.org/10.1016/J.CHB.2022.107596
22. Visvizi, A., Troisi, O., Grimaldi, M., Kozłowski, K.: Getting things right: ontology and epistemology in smart cities research. In: Visvizi, A., Troisi, O., Grimaldi, M. (eds.) Research and Innovation Forum 2022. RIIFORUM 2022. Springer Proceedings in Complexity. Springer, Cham (2023). https://doi.org/10.1007/978-3-031-19560-0_14
23. Visvizi, A., Malik, R., Guazzo, G.M., Çekani, V.: The Industry 5.0 (I50) paradigm, blockchain-based applications and the smart city. Eur. J. Innov. Manag. **28**(1), 5–26 (2025). https://doi.org/10.1108/EJIM-09-2023-0826
24. Williams, R.I., et al.: Re-examining systematic literature review in management research: additional benefits and execution protocols. Eur. Manag. J. **39**(4), 521–533 (2021). Pergamon, https://doi.org/10.1016/J.EMJ.2020.09.007

Artificial Intelligence (AI) Across Issues and Domains

Why Artificial Intelligence Needs Sociology?

Pablo de-Gracia-Soriano, Francisco Javier Jiménez-Loaisa, and Kateryna Mamchur

Abstract Artificial intelligence (AI) is rapidly reshaping key dimensions of contemporary society, including education, healthcare, labor, and social relations. However, the development and deployment of AI systems has been largely driven by technical priorities, frequently neglecting their broader social, political, and ethical implications. This chapter contends that sociology, as a discipline rooted in the critical analysis of institutions, power dynamics, and social structures, is indispensable to understand and guide the evolution of AI technologies. Methodologically, the chapter presents a case study conducted at the University of Alicante, involving a socio-educational innovation project based on a mixed-methods design and participatory workshops with students and faculty. The findings underscore the importance of integrating sociological perspectives into the design, implementation, and regulation of AI systems, particularly in educational and professional settings. Sociologists are especially well placed to contribute to the responsible development of AI by identifying risks and opportunities associated with these technologies, analyzing social biases and structural inequalities, and anticipating their unintended consequences. In doing so, they play a vital role in promoting sustainable, equitable, and socially informed technological innovation.

Keywords STS studies · Sociology · AI · Data analysis · Trandisciplinarity · SDG

P. de-Gracia-Soriano (✉) · F. J. Jiménez-Loaisa
Department of Sociology I, University of Alicante, San Vicente del Raspeig, Spain
e-mail: pablo.dgsoriano@ua.es

F. J. Jiménez-Loaisa
e-mail: jimenez.loaisa@ua.es

K. Mamchur
Department of Computer Languages and Systems, University of Alicante, San Vicente del Raspeig, Spain
e-mail: kateryna.mamchur@ua.es

A. Visvizi et al. (eds.), *Research and Innovation Forum 2025 Volume 1*,
Springer Proceedings in Complexity, https://doi.org/10.1007/978-3-032-09066-9_8

1 Introduction: The Sociotechnical Challenge of AI

In 1956, John McCarthy convened the Dartmouth Conference, where the term *Artificial Intelligence* (AI) was coined. This milestone marked the beginning of a technological field whose evolution has since shaped and been shaped by the social, political, and economic contexts in which it is embedded. Far from being merely a technical phenomenon, AI has become a core element of what some call the *fourth industrial revolution*, restructuring the boundaries between humans and machines and transforming labor, communication, and knowledge production.

Although AI is often framed as a neutral or autonomous process, its development has always responded to specific historical and institutional logics—from Cold War military priorities to the commercial imperatives of today's platform capitalism. These historical trajectories, as well as current applications, show that AI reflects and also reproduces and amplifies existing social structures and inequalities.

This raises a central research question. What role can sociology play in the development and governance of AI? This question is increasingly relevant, given the growing social impact of algorithmic systems and the need for interdisciplinary approaches that go beyond technical performance to consider issues such as bias, power concentration, and ethical accountability [1].

This chapter presents a case study conducted at the University of Alicante (Spain) as part of a socio-educational innovation project that explores the relationship between sociology and AI. The study pursues two main objectives:

First, to analyze the ways in which generative AI alters how individuals—particularly students—organize their academic and professional activities, with an emphasis on time management.

Second, to examine the potential of sociology as a discipline capable of contributing theoretical and methodological tools for the responsible use of AI in educational and professional contexts.

The chapter is structured as follows. It begins with a conceptual and historical overview of AI, situating its development within broader sociotechnical and political contexts, and outlining the relevance of sociological inquiry into its implications. This initial section also explores the key literature at the intersection of AI and the social sciences. Next, we present the methodological framework of the study, including the design and implementation of a series of participatory workshops conducted with students and faculty. The fourth section provides an analysis of the results derived from these workshops. We then move to a discussion of the broader implications of these findings, particularly regarding the integration of sociological perspectives into the development, evaluation, and governance of AI systems. Finally, the chapter concludes with reflections on future research directions and educational strategies.

2 The Development of AI and the Emergence of a Sociology of AI

The emergence of sociological interest in AI connects with broader traditions within Science and Technology Studies (STS), actor-network theory, and the sociology of innovation [2]. Recent contributions in this field have examined the role of AI in reproducing social inequalities, strengthening power asymmetries, and creating new epistemological challenges around explainability and accountability. This chapter draws on these perspectives to propose a case-based, pedagogically grounded approach to the sociological analysis of AI.

AI remains a fluid and evolving concept, shaped by continuous technological advancements that challenge any stable or unified definition. As Russell and Norvig outline [3], AI has historically been categorized through four distinct approaches: systems that think like humans, act like humans, think rationally and act rationally. In parallel, a functional classification based on capability distinguishes among Artificial Narrow Intelligence (ANI), which performs specific tasks (e.g., chatbots); Artificial General Intelligence (AGI), which hypothetically matches human cognitive abilities; and Artificial Super Intelligence (ASI), which would theoretically surpass human intelligence in all respects.

Thus, AI is framed as part of the "Fourth Industrial Revolution" [4], a period marked by the convergence of nanotechnology, biotechnology, information technology and cognitive science (NBIC). This convergence has accelerated the integration of human and technological spheres, transforming political, economic, and social systems.

Historically, early explorations of AI date back to the 1950s, both in academic and cultural domains. Alan Turing's seminal work, *Computing Machinery and Intelligence* [5], posed the foundational question: *Can machines think?*—a query that led to the formulation of the Turing test. Cultural works such as *I, Robot*, 1984, and *Metropolis* had already imagined machine-human interactions.

Subsequent developments included MIT's ELIZA in the 1960s, one of the first natural language processing programs, and several limited AI experiments in the 1970s–1990s, such as Japan's Fifth Generation Project. A renewed surge began in the 1990s, fueled by personal computing and internet expansion. The launch of OpenAI in 2015 and ChatGPT in 2022 further accelerated interest in AI. Since then, major corporations such as Microsoft, Google, and Amazon have invested heavily in machine learning, neural networks, and data mining to develop applications for productivity, automation, and recommendation systems. However, this growth has outpaced the establishment of adequate regulatory frameworks, highlighting the urgent need for social, ethical, and legal oversight.

The history of AI cannot be understood without considering the actors who drove it, who, in turn, were shaped by the intellectual debates and institutional frameworks of their time [6, 7]. That is, it would be imprudent to view AI solely through the lens of technological or economic milestones. From its inception during the Cold War to its consolidation in contemporary digital capitalism, AI has been a battleground where scientific, political, and corporate interests converge.

The early steps of AI were marked by the logic of geopolitical conflict. In the 1940s, World War II had highlighted the need to develop automated information processing systems for cryptography, ballistic prediction, and real-time decision making. In this context, Norbert Wiener articulated the concept of cybernetics, a theory aimed at explaining communication and control in machines and living organisms through feedback mechanisms [8]. His work not only influenced the early computational models of the human mind but also attracted the interest of the US government, which saw in these ideas a potential strategic advantage over the Soviet Union.

It was in this climate of strong state investment in technology that in 1956, John McCarthy organized the Dartmouth Conference, where the term "AI" was coined. McCarthy, along with Marvin Minsky, Allen Newell, and Herbert Simon laid the foundations of the field with the belief that computers could simulate human cognitive processes [9]. However, its development was not without disputes: while Simon and Newell promoted symbolic models of thought, Minsky advocated for a vision based more on networks of neural interconnection. These conceptual differences were framed in a context where funding came from DARPA, the U.S. Department of Defense's research agency, which prioritized military and strategic applications.

The initial enthusiasm for AI suffered significant setbacks in the 1970s and 1980s when the promises of the pioneers did not materialize at the expected pace. The lack of progress in natural language processing and the limited capacity of computers to generalize knowledge led to the cancellation of numerous projects funded by the U.S. and U.K. governments. This period was referred to as the "AI winter," a time when funding was redirected toward other fields like conventional computing and expert systems.

However, in the 1990s, with the expansion of personal computing and widespread access to the Internet, AI found a renewed space for development within the emerging framework of digital capitalism. Companies such as IBM and Microsoft began developing algorithms designed to improve the management of large volumes of data, giving rise to the era of machine learning. During this period, Geoffrey Hinton revived the concept of artificial neural networks, which had been dismissed by Marvin Minsky decades earlier [10]. His work laid the foundations for deep learning, an approach that, thanks to the exponential growth in computing power, would enable the development of voice and image recognition systems with direct commercial applications.

The new century marked the consolidation of AI as the cornerstone of platform capitalism. Tech giants like Google, Facebook, and Amazon have invested heavily in big data-based predictive models, reshaping the digital economy. Figures such as Demis Hassabis, the founder of DeepMind, rose to prominence through developments in AI for gaming and biomedical challenges.

Today, AI functions not only as a tool for business efficiency, but also as a mechanism that reproduces social inequalities. Research by Joy Buolamwini and Timnit Gebru has exposed racial and gender biases in facial recognition and algorithmic decision making, undermining claims of technological neutrality [11]. Trained in historical data, AI tends to replicate and amplify existing societal hierarchies.

These concerns have brought new actors into the AI landscape. Governments and institutions such as the European Union have begun developing regulations to curb algorithmic threats to privacy and civil rights [12]. Parallel to this, AI ethics movements advocate for more transparent and democratic models of development.

This section proposes a sociological analysis of AI that situates it within its broader social, political, and economic context, encouraging a deeper understanding beyond technical narratives. From its roots in the Cold War to its inflection into digital capitalism, AI has evolved through the influence of key actors and institutional incentives.

The trajectories of figures like Wiener, McCarthy, Hinton, and Buolamwini must be viewed within the power structures that shaped their work. AI is neither an inevitable outcome nor a neutral technology, but a contested domain where economic interests, ideologies, and visions of the future converge [13].

3 Methods and Materials

Unlike most studies that approach AI from a purely technical or abstract ethical perspective, this project offers an empirical, classroom-based exploration of how sociological reflection on AI can be fostered through participatory pedagogy. In a context marked by accelerating technological change, this chapter presents a case study that explores the intersections between sociology and AI through a socio-educational innovation project conducted at the University of Alicante (Spain). The study examines how AI is used and perceived in educational and professional contexts, while advancing the idea that sociology can play a proactive role in shaping AI systems through critical, methodological and ethical frameworks.

The project is based on the assumption that, despite the growing presence of AI in academic and everyday life, sociologists—and sociology as a field—have not yet fully recognized their potential to become key actors in this domain [14]. Rather than limiting itself to critical discourse, sociology has the ability to contribute to the responsible development of AI by offering methodological, analytical, and ethical frameworks that consider its social implications.

To address these questions, a case study was conducted using a qualitative approach grounded in participatory pedagogy. The fieldwork consisted of four workshops held within the university environment: three with students enrolled in sociology programs and one with faculty members from the same department. The student group comprised a total of 22 participants—12 women and 10 men—who voluntarily participated in the activity. Their ages ranged from 20 to 21. The faculty workshop involved eight professors, each bringing different perspectives based on their experience and disciplinary subfields. The sessions were structured to encourage reflection on the uses, perceptions, and ethical challenges associated with AI, with a particular

Table 1 Demographic profile of workshop participants

Group	Gender (F/M)	Age range
Students	12/10	20–21
Faculty	4/4	29–65

Source Authors

focus on how these technologies influence time management, academic performance, and professional development (Table 1).

The methodological design followed a mixed strategy of dialogue and observation. Each workshop had a duration of approximately 90 min and was organized around collective discussions, guided questions, and reflective activities that facilitated the emergence of shared narratives. The workshops created a space not only to identify the ways in which AI is being integrated into university life, but also to question the assumptions behind its use and the opportunities it opens or closes within academic and labor contexts.

Data collection was carried out through direct observation and systematic note taking by the research team. These notes allowed the team to capture verbal interactions, expressions of agreement or resistance, and significant moments of learning or tension. No audio recordings or formal questionnaires were used, as the objective was to maintain fluid and open dialogue within a context of pedagogical experimentation rather than empirical generalization. All participants were informed about the purpose of the project and participated voluntarily. The workshops were conceived as spaces of mutual learning, where both students and faculty could explore and redefine their relationship with AI from a sociological perspective.

Case study research is particularly suited for examining complex social processes in context, allowing the researcher to explore meaning-making practices, lived experiences, and the interplay between individuals and institutions [15, 16]. The strength of case study methodology lies in its ability to capture the richness and specificity of a case as it unfolds in its natural setting. In this study, the case represents not only a pedagogical intervention but also an instance of disciplinary positioning—how sociology responds to the rise of AI in higher education.

Given the exploratory nature of the project and its emphasis on participant reflexivity, the use of participatory and dialogical methods is especially appropriate [17]. Participatory approaches help bridge the gap between academic research and social practice by fostering collective inquiry and encouraging horizontal knowledge production. In the context of this research, they allow both students and faculty to reflect critically on their own engagements with AI in academic and professional settings.

4 Results: Insights from Participatory Workshops

The first workshop aimed to establish an initial contact and understand the uses that students make of AI in their academic life. Through an open dialogue, the main purposes for which they employ these tools were identified, highlighting the improvement of text writing, clarifying doubts, and the preparation of academic papers.

Students indicated that they use AI to correct grammar and spelling, rephrase sentences, and better structure ideas, resulting in clearer, more cohesive texts. Some also mentioned using these tools to translate or adapt their writing to different linguistic registers depending on the context.

Concerning the resolution of doubts, AI has become a common resource to obtain quick answers to specific concepts, define terms, or dive into certain topics without the need to consult multiple sources. However, during the debate, concerns about the reliability of the information generated by these systems emerged, as well as the importance of cross-referencing it with academic sources.

It was also observed that there seems to be a tendency to use AI as a substitute for traditional search engines (for them, it is not a replacement, but rather their "natural" way of proceeding).

In preparing academic documents, students reported using AI to generate outline, suggest references, and organize ideas prior to writing. This helped them structure texts, connect concepts, and improve argument coherence. Some admitted using AI to summarize articles or extract key information from long texts, *optimizing* study time, and aiding comprehension of complex material.

However, during the discussion, a phenomenon emerged that could be considered a form of "academic trickery": the use of AI to prepare documents almost entirely, with minimal student intervention. Some acknowledged having resorted to these tools to generate complete texts, limiting themselves to making small modifications to prevent the result from appearing entirely automated. In some cases, they pointed out that AI provides written content and can also generate citations and references, although not always verifiable.

This led to a debate on the ethical and academic implications of outsourcing work to AI. Although students valued efficiency and reduced workload, concerns were raised about how such dependence could hinder the development of critical thinking, analytical, and creative skills. The risk of uncritically trusting AI output was also observed, potentially resulting in inaccurate or poorly substantiated documents.

The discussion also addressed the role of educators and institutions in regulating AI use. There was consensus on the need for clear criteria defining ethical academic practices, as well as on the importance of training in the critical and responsible use of these tools.

These discussions reflect the growing integration of AI into academic activities, both as a support and as a potential substitute for student work. The workshop provided space to reflect on related challenges, overreliance, diminished critical skills, and the need for ethical strategies.

The second workshop introduced a sociologist involved in AI research, allowing students to see the relevance of sociology in the shaping and evaluating of AI models. Key analytical distinctions, such as machine learning versus deep learning, and methodologies such as supervised, unsupervised, and reinforcement learning, along with techniques including regression, clustering, factor analysis, and neural networks, were discussed.

The third workshop was structured as a focus group, with the aim of gathering feedback from students after the previous sessions. One of the most significant aspects that emerged from the discussion was the participants' surprise at realizing that they had been able to understand the content covered in the second workshop. This recognition allowed them to perceive that they possess the necessary skills to actively participate in the creation and evaluation of AI-related processes.

In the session with the teaching staff, the focus was on how AI is transforming education. Opinions varied: some saw AI as a threat to student engagement, while others saw potential benefits if used appropriately. Assignment production was a key concern; some feared that students would completely delegate their work to AI, while others believed that it could assist in organizing ideas and improving writing. This sparked a discussion on rethinking assessment strategies to promote deeper involvement and critical thinking.

Challenges around AI integration in the classroom were also raised: how much teaching should adapt and what role educators should play in guiding students. The need to train educators in digital competencies was emphasized, as students often appear to be more familiar with AI tools.

At the end of the workshop, there was a general consensus on the importance of critically addressing these changes without resorting to outright rejection. AI should not be seen as a threat, but as a tool that, when used correctly, can enrich higher education. The key, as discussed, is to learn to integrate it in a way that enhances learning without replacing the students' effort and creativity.

Studies suggest that in e-learning, combining pedagogical technologies and AI improves student outcomes [18, 19]. Used properly, these tools can personalize learning, simplify content and improve accessibility, for example, for students with intellectual disabilities.

5 Discussion: The Meeting Point Between Sociology as a Discipline and AI

The case study reveals that both sociology students and faculty tend to use AI instrumentally, particularly generative tools, without initially perceiving themselves as professionals capable of contributing to its design or governance. However, as widely recognized, AI is not merely a technical development; it is a social phenomenon that changes the way knowledge is produced, decisions are made, and societal life is organized [20]. From this point of view, sociology is well equipped to analyze the impacts of AI and should also play an active role in its development and regulation.

Participants in the reflective workshops identified key intersections between AI and sociology in areas such as data production and interpretation, ethical dilemmas, labor transformations, and social inequality. These themes align closely with sociological concerns and demonstrate the potential of the discipline to contribute meaningfully to AI governance.

Sociology offers tools to interrogate how data, far from being neutral, are socially constructed and selectively prioritized, potentially reinforcing structural inequalities [21]. For example, algorithms used in hiring processes may replicate historical patterns of exclusion. By questioning which data are included, who defines them, and with what consequences, sociology can intervene critically in the logic and design of AI systems. Machine learning labels are not just technical categories, but embed social judgments.

The ethical challenges posed by AI require regulatory frameworks that go beyond technical parameters. Participants recognized the role sociology can play in analyzing social impacts and informing responsible regulation, particularly in relation to transparency and explainability, which are especially urgent in contexts like justice, healthcare, or employment. Although commercial interests often hinder such initiatives, proposals such as external audits or certification systems could improve public oversight and accountability.

Finally, the workshops addressed how AI-driven automation is transforming labor: not only by displacing jobs, but by changing the structure of employment, redefining required skills, and exacerbating existing inequalities [22]. Automation does not affect everyone equally. It reinforces gender, economic, and educational disparities. At the same time, the rise of new roles related to AI development and oversight points to the need for inclusive training policies that support a fair transition in the evolving labor market.

6 Conclusion: Why Does AI Need Sociology?

As shown throughout this chapter, AI is not an isolated or purely technical phenomenon, but a technology whose design, implementation, and effects are deeply embedded in society. While this may appear self-evident, current practices suggest that it is far from fully acknowledged or addressed. In this context, sociology and sociologists emerge as crucial actors in ensuring that AI evolves in equitable, transparent, and democratic directions.

AI relies on massive data processing, but its effectiveness is not based on quantity alone. The quality of data, the design of systems, the methods and the visualization of information are critical. Sociology, with its tradition of integrated and mixed methodologies, is particularly well suited to interpret data in relation to the social processes that generate them. Moreover, it is well equipped to detect biases in both data and analysis, which influence the design of algorithms and their social consequences.

This links sociological work with the evaluation of how technological artifacts, as Langdon Winner argued, shape individuals, structures, and social dynamics [23].

Given the significant social impacts of AI, it is problematic to leave its development solely in the hands of technical experts, especially when engineering fields often operate in isolation from the social sciences. The danger lies in turning AI-based decision making into a technocratic process that neglects how societies function.

AI also reflects existing power structures and dominant values [24]. From training data selection to automated decision making, every phase of development is shaped by social, political, and economic forces. Sociology offers analytical and critical tools to understand and influence these processes, helping to ensure that AI increases efficiency and also respects equity, transparency, and responsibility.

The future of AI depends on technological advancement and the integration of diverse disciplinary perspectives. Interestingly, we see increasing collaboration through multidisciplinary teams and growing institutional support, including legislative efforts.

Ultimately, the central question is not merely how to make AI more efficient but how to ensure its development aligns with the values of a democratic and just society [25]. Sociology and AI should not remain isolated; they should be integrated to guide innovation in a way that is both effective and socially grounded [26]. The future depends on this convergence.

The case study at the University of Alicante illustrates how participatory pedagogical approaches can foster critical reflection on AI among students and faculty, especially around issues such as academic integrity, inequality, and ethics. Although participants were aware of the potential and risks of AI, they often lacked conceptual tools for critical engagement beyond practical use.

Although the study provides valuable information on the pedagogical integration of AI in higher education, its scope is limited to a specific institutional and cultural context. Future research could explore similar initiatives in other disciplines or regions or adopt longitudinal approaches to track changes over time.

Further studies should also examine the role of sociology in interdisciplinary teams working on AI development, regulation, and evaluation, particularly in areas such as algorithmic governance, educational technology, and digital labor. The design of training programs and curricular frameworks to prepare sociologists for these roles remains a crucial challenge. This chapter has argued that integrating sociological expertise into AI is desirable and essential to ensure that technological progress supports democratic and socially just principles.

References

1. Joyce, K., Smith-Doerr, L., Alegria, S., Bell, S., Cruz, T., Hoffman, S.G., Noble, S.U., Shestakofsky, B.: Toward a sociology of artificial intelligence: a call for research on inequalities and structural change. Socius **7** (2021). https://doi.org/10.1177/2378023121999581
2. Liu, Z.: Sociological perspectives on artificial intelligence: a typological reading. Sociol. Compass **15**, e12851 (2021). https://doi.org/10.1111/soc4.12851
3. Russell, S.J., Norvig, P.: AI: A Modern Approach, 4th edn. Pearson (2020)
4. Ausín, T.: ¿Por qué ética para la Inteligencia Artificial? Lo viejo, lo nuevo y lo espurio. Sociol. tecnociencia **11**(2), 1–16 (2021)
5. Turing, A.M.: Computing machinery and intelligence. Mind **49**, 433–460 (1950)
6. Latour, B.: Science in Action: How to Follow Scientists and Engineers Through Society. Harvard University Press (1987)
7. Callon, M.: Techno-economic networks and irreversibility. Sociol. Rev. **38**(1_suppl), 132–161 (1990). https://doi.org/10.1111/j.1467-954X.1990.tb03351.x
8. Wiener, N.: Cybernetics: Or Control and Communication in the Animal and the Machine. MIT Press (1948). https://doi.org/10.7551/mitpress/11810.001.0001
9. McCarthy, J., Minsky, M.L., Rochester, N., Shannon, C.E.: A proposal for the Dartmouth summer research project on AI. AI Mag. **27**(4), 12 (2006). https://doi.org/10.1609/aimag.v27i4.1904
10. Hinton, G.E., Osindero, S., Teh, Y.W.: A fast learning algorithm for deep belief nets. Neural Comput. **18**(7), 1527–1554 (2006). https://doi.org/10.1162/neco.2006.18.7.1527
11. Buolamwini, J., Gebru, T.: Gender shades: intersectional accuracy disparities in commercial gender classification. Proc. Mach. Learn. Res. **81**, 77–91 (2018). Proceedings of the 1st Conference on Fairness, Accountability and Transparency, https://proceedings.mlr.press/v81/buolamwini18a.html
12. Zuboff, S.: The Age of Surveillance Capitalism: The Fight for a Human Future at the New Frontier of Power. PublicAffairs, New York (2019)
13. Sarirete, A., Balfagih, Z., Brahimi, T., Lytras, M.D., Visvizi, A.: AI and machine learning research: towards digital transformation at a global scale. J. Ambient. Intell. Humaniz. Comput. **13**(7), 3319–3321 (2022). https://doi.org/10.1007/s12652-021-03168-y
14. Jiménez-Loaisa, F.J., de-Gracia-Soriano, P., Jareño-Ruiz, D.: La inteligencia artificial en el contexto de la educación superior en la Unión Europea: Un análisis bibliométrico y de redes sociales. In: Caldevilla Domínguez, D., Barrientos Báez, A. (eds.) Internet al servicio de la modernidad en red, pp. 183–195. Peter Lang (2024)
15. Yin, R.K.: Case Study Research: Design and Methods, 4th edn. SAGE, Thousand Oaks, CA (2009)
16. Stake, R.E.: The Art of Case Study Research. SAGE, Thousand Oaks, CA (1995)
17. Rodríguez-Villasante Prieto, T., Montañés, M., Martí Olivé, J.: Participatory social research: Building citizenship [La investigación social participativa: construyendo ciudadanía]. El Viejo Topo, Barcelona (2010)
18. Castillo-Merino, D., Serradell-López, E.: An analysis of the determinants of students' performance in e-learning. Comput. Hum. Behav. **30**, 476–484 (2014)
19. Espinosa-Izquierdo, J.G., Espinosa-Figueroa, J.A., Espinosa-Arreaga, G.B.: E-learning: A necessary tool for learning [E-learning una herramienta necesaria para el aprendizaje]. Polo del Conocimiento **6**(3), 659–669 (2021). https://doi.org/10.23857/pc.v6i3.2394
20. Schwartz, R.D.: Artificial intelligence as a sociological phenomenon. Can. J. Sociol./Cah. Can. Sociol. **14**(2), 179–202 (1989). https://doi.org/10.2307/3341290
21. Joyce, K., Cruz, T.M.: A sociology of artificial intelligence: inequalities, power, and data justice. Socius **10** (2024). https://doi.org/10.1177/23780231241275393
22. Zajko, M.: Artificial intelligence, algorithms, and social inequality: sociological contributions to contemporary debates. Sociol. Compass **16**(3), e12962 (2022). https://doi.org/10.1111/soc4.12962

23. Winner, L.: Do artifacts have politics? Daedalus **109** (1), 121–136 (1980). https://www.jstor.org/stable/20024652
24. Svetlova, E.: AI meets narratives: the state and future of research on expectation formation in economics and sociology. Soc. Econ. Rev. **20**(2), 841–861 (2022). https://doi.org/10.1093/ser/mwab033
25. Bosco, G., Riccardi, V., Sciarrone, A., D'Amore, R., Visvizi, A.: AI-driven innovation in smart city governance: Achieving human-centric and sustainable outcomes. In: Transforming Government: People, Process and Policy (2024). Ahead-of-print, https://doi.org/10.1108/TG-04-2024-0096
26. Visvizi, A., Calzada, I., Kozłowski, K., Troisi, O.: Road-mapping generative AI (Gen-AI) and its impact: AI for whom and for what? In: Visvizi, A., Kozłowski, K., Calzada, I., Troisi, O. (eds.) Multidisciplinary Movements in AI and Generative AI, Society, Business, Education. Elgar (2025). https://doi.org/10.4337/9781035358663.00007

The (Dis)Illusion of the Web3 Decentralization for Global Governance in the Age of GenAI

Igor Calzada

Abstract This paper critically examines both the promises and pitfalls associated with Web3 decentralization within the evolving landscape of global governance, increasingly influenced by Generative AI (GenAI). It questions whether Web3 technologies meaningfully redistribute power or primarily strengthen a new cyber-libertarian elite composed of technologically skilled individuals with significant digital capital. Drawing from extensive fieldwork action research conducted in Silicon Valley since August 2022, as well as direct interactions with scholars and practitioners—including detailed dialogues held in Washington, DC, in April 2025—the analysis explores the underlying techno-political beliefs driving the global Web3 movement. Advocates commonly depict Web3 as an effective means to break centralized data monopolies and challenge traditional state institutions. However, this paper argues that decentralization does not inherently lead to democratization. Instead, Web3 potentially reproduces or even deepens existing hierarchies rooted in algorithmic control and digital privilege, especially when integrated with GenAI systems. To illustrate this point, the paper delineates three emerging paradigms shaped by Web3 logic: (i) Network States, promoting crypto-libertarian models of governance; (ii) Network Sovereignties, enabling collective self-determination for marginalized groups; and (iii) Algorithmic Nations, constructing decentralized, data-driven infrastructures aimed at global collaboration. While innovative and provocative, these paradigms simultaneously raise significant concerns about inclusivity, equity, and the concentration of technocratic influence. The paper concludes by proposing hybrid policy frameworks that acknowledge global power imbalances and

I. Calzada (✉)
University of the Basque Country, Public Policy & Economic History Department, Donostia-San Sebastian, Spain
e-mail: igor.calzada@ehu.eus

Basque Foundation for Science, Ikerbasque, Bilbao, Spain

Cardiff University, WISERD, Cardiff, UK

Decentralization Research Centre, Toronto, Canada

US-UK Fulbright Commission, London, UK

A. Visvizi et al. (eds.), *Research and Innovation Forum 2025 Volume 1*,
Springer Proceedings in Complexity, https://doi.org/10.1007/978-3-032-09066-9_9

local contextual realities, highlighting the urgency of pursuing digital justice and fostering international cooperation within a GenAI-powered Web3 environment.

Keywords Global governance · Web3 · Decentralization · GenAI · SDG-16: peace · Justice and strong institutions

1 Introduction: Cyberlibertarianism and Global Governance in the Age of GenAI

The Westphalian system, foundational to the modern international order since 1648, defines sovereignty through the territorial authority of the state [25, 42]. Yet in the digital age, this model faces unprecedented disruption. The confluence of decentralized Web3 technologies and the exponential development of Generative Artificial Intelligence (GenAI) is not merely challenging traditional governance—it is reconfiguring it [29, 46]. Emerging techno-political imaginaries blur the boundaries between states and platforms, citizenship and protocol, public legitimacy and algorithmic authority. In this paper, we critically examine how *cyberlibertarianism*—a dominant ideological thread within the Web3 movement—is shaping these transformations and reinforcing new forms of technocratic control under the guise of decentralization. This paper examines three post-Westphalian global paradigms shaped by decentralized Web3 policy trends: (i) Network States [52], (ii) Network Sovereignties [26], and (iii) Algorithmic Nations [10].

While these paradigms propose alternatives to centralized governance, they also reveal the illusion of decentralization [17], where the promise of distributing power may instead elevate a new, libertarian, tech-savvy elite [48]. By critically analyzing these trends, this paper seeks to contribute to the debate on the evolving notions of sovereignty, governance, and citizenship in the digital age, emphasizing the need for inclusive and equitable policy frameworks in the Web3 era [51]. While decentralized technologies like blockchain and Decentralized Autonomous Organizations (DAOs) are heralded as the harbingers of digital democratization, there is a paradoxical risk that they may reinforce existing power structures or create new forms of centralization [16]. The governance models of many decentralized systems often concentrate power among a small group of technically proficient actors, undermining the democratic ideals they purport to uphold [38]. Despite the promise of distributed, participatory governance, blockchain-based systems and DAOs frequently require technical expertise that is not widely accessible, thus creating new gatekeepers in the form of developers, validators, and major token holders [1, 31, 40, 43, 44]. This dynamic raises the research question of this paper: *Are we truly decentralizing power, or are we merely shifting it into the hands of a new elite under the guise of technological innovation?* [7, 28, 38].

Cyberlibertarianism, as advocated by figures such as Hope and Ludlow [32], promotes a worldview in which governance can be autonomously organized through cryptographic systems, market incentives, and voluntary association [12, 13, 24]. It valorizes individual sovereignty and distrusts collective institutions. Golumbia [30] argues, however, that such positions often conceal their political underpinnings, advancing a deeply ideological project of deregulated, technocratic governance. Web3 technologies, including blockchain, peer-to-peer infrastructures, and Decentralized Autonomous Organizations (DAOs), are frequently framed as emancipatory [14]. Yet they risk reproducing centralized authority—now exercised through code and capital rather than law and representation.

The integration of GenAI into these decentralized systems amplifies the paradox. GenAI tools—ranging from large language models to synthetic decision agents—promise democratization through access to knowledge, creativity, and automation. But they also introduce opaque computational infrastructures, governed by those who possess the resources and expertise to develop, train, and control them. These actors, operating under the banner of decentralization, may in fact constitute a new elite: a technocratic class with disproportionate influence over the future of governance, identity, and sovereignty.

To interrogate this dynamic, the paper analyzes three post-Westphalian paradigms of governance that have emerged from the Web3 and GenAI nexus:

1. *Network States* [52], which embody a market-driven, crypto-libertarian vision of political formation detached from territorial legitimacy;
2. *Network Sovereignties* [26], which pursue digital self-determination for marginalized groups, but must navigate the risk of technocratic capture;
3. *Algorithmic Nations* [10], which seek to orchestrate distributed global collaboration through algorithmic coordination mechanisms [35, 57].

These paradigms reflect competing visions of how decentralization might reimagine sovereignty. Yet they also raise critical concerns about the concentration of power, equity of participation, and legitimacy of algorithmic rule. As ethnographic and empirical studies [40, 43] increasingly show, access to these systems is often mediated by technical knowledge, economic capital, and privileged social networks. What is claimed as horizontal governance frequently becomes a vertical hierarchy of influence.

The implications are profound. If GenAI-enhanced Web3 systems continue to evolve under cyberlibertarian ideals, we risk constructing a post-Westphalian order governed not by states, but by platforms, protocols, and proprietary models. In this scenario, algorithmic power substitutes political accountability, and digital infrastructures become the primary means of inclusion—or exclusion—from global citizenship [3, 17, 18, 38]. Golumbia's [30] critique is thus more urgent than ever: the rhetoric of decentralization may be a discursive smokescreen for a deeply hierarchical transformation of global governance.

Web3 has been celebrated as a technological revolution capable of redistributing power from centralized institutions to individuals and communities. Anchored in a (crypto-)libertarian ethos prevalent in Silicon Valley [22], the movement embraces individual sovereignty, transparency, and minimal state intervention [6–8, 15, 27, 36]. Figures such as Barlow [6] and May [39] envisioned digital technologies as a means of resisting coercion and securing freedom through cryptographic autonomy and algorithmic coordination. This libertarian vision positions decentralization as both a technical innovation and a political project aimed at undermining centralized governance structures [7, 27].

The research question of this paper becomes even more urgent with the integration of GenAI, which adds layers of algorithmic complexity and opacity. The governance of these systems increasingly depends on closed-source models, proprietary data pipelines, and elite expertise. Instead of dismantling centralized hierarchies, GenAI-enhanced Web3 architectures may solidify them in new, less visible forms.

In sum, while Web3 and GenAI technologies hold potential for reimagining sovereignty and global governance, they also risk reproducing exclusionary systems under the banner of innovation. Recognizing and confronting this paradox is essential if we are to move toward more inclusive, democratic, and equitable digital futures. The next section will map the state of the art in decentralized governance, comparing the models of Network States, Network Sovereignties, and Algorithmic Nations to further assess the implications of these paradigms.

The structure of the paper is as follows: Sect. 2 outlines the methodological and empirical approach. Section 3 examines the current landscape of Web3 governance paradigms. Section 4 evaluates their implications for digital sovereignty and global justice. The conclusion offers reflections on future policy trajectories, calling for governance models that resist cyberlibertarian capture and instead prioritize equity, accountability, and democratic participation in an increasingly post-Westphalian world.

2 Method and Findings: Fieldwork Action Research in April 2025, Washington, D.C.

This paper employs fieldwork action research conducted in Washington, D.C., in April 2025, as a critical methodological component of an ongoing iterative research process. Action research, in this context, extends beyond merely adding isolated events or conversational snapshots; rather, it actively reshapes, refines, and dynamically updates previously published insights and theoretical frameworks within the ongoing research program. This iterative process involves systematic cycles of planning, acting, observing, and reflecting [49], continuously recontextualizing existing data and findings with new empirical observations.

The fieldwork conducted in Washington, D.C., sought to explore and deepen understanding around emerging discussions on Web3 decentralization and GenAI, specifically scrutinizing their implications for governance and digital citizenship. This stage of action research not only provided fresh empirical data through participatory workshops and collaborative events but also critically re-examined and reshuffled earlier theoretical and empirical contributions.

The methodological rigor of this action research lies in its reflexivity and adaptability, as the author demonstrated over the years through his scholarship production. It explicitly recognizes previous publications and findings as provisional, subject to continuous interrogation and reshaping by newly emerging data and insights. Thus, the fieldwork in Washington, D.C., did not simply generate new data points in isolation; it actively influenced the analytical and theoretical orientation of the overall AI4SI action research program [12–14].

This section clearly emphasizes the methodological relevance of action research by demonstrating how each iterative cycle enhances the robustness and depth of the ongoing scholarly inquiry. Such methodological stance underscores action research as particularly suitable for exploring dynamic socio-technical phenomena such as Web3 and GenAI, where both technological developments and social implications evolve rapidly and unpredictably, particularly as directly noted in timely discussions that this paper focuses on [34]. In sum, the fieldwork action research conducted in Washington, D.C., in April 2025, contributes significantly to this paper by providing empirical richness and by iteratively refining existing scholarly narratives, ultimately enhancing the coherence, relevance, and critical rigor of the broader research agenda.

This action research investigates how decentralized technologies—such as blockchain, DAOs, and data cooperatives—are mobilized by diverse actors seeking alternatives to dominant governance models. These actors collectively challenge the enduring Westphalian paradigm of sovereignty characterized by territorial borders, centralized authority, and surveillance-capitalist dynamics. In contrast, the emergent ethos underpinning Web3 technologies embraces borderless coordination, algorithmic trust, and participatory infrastructures.

This paper identifies and analyzes three paradigmatic responses to this shift—*Network States*, *Network Sovereignties*, and *Algorithmic Nations*. These paradigms are treated not as ideal types but as heuristic lenses through which to assess how governance is being reimagined in a post-Westphalian context. Each reflects distinct motivations, organizational logics, and techno-political imaginaries that shape the evolving global debate on decentralization and digital sovereignty [12, 16, 55].

2.1 Network State

As conceptualized by Srinivasan [52], the Network State paradigm exemplifies a post-Westphalian experiment in sovereignty, wherein decentralized digital communities assert governance beyond territorial borders. Grounded in blockchain and Web3 infrastructure, this model constructs legitimacy through consent-based participation and cryptographic systems, bypassing conventional state institutions. Its underlying ethos reflects a cyberlibertarian ideology that valorizes individual autonomy, minimal intervention, and market-centric coordination [50].

These "startup societies" operate under lean governance structures, where influence often correlates with financial input and tokenized engagement. Internally, governance mixes centralization around founders with decentralized consensus mechanisms [37]. Externally, Network States pursue international recognition while navigating legal ambiguities—challenging the normative coherence of state sovereignty.

Technologically deterministic and economically transactional, this paradigm prioritizes efficiency, scalability, and financial viability. However, its reliance on token economies and digital infrastructures risks reinforcing socio-economic exclusion and deepening the digital divide [41]. By merging technological solutionism with libertarian logics, the Network State risks reproducing elite enclaves under the guise of decentralized governance.

2.2 Network Sovereignties

As conceptualized by De Filippi et al. [26], the Network Sovereignties paradigm represents a commons-centric model of post-Westphalian governance, in which decentralized communities assert sovereignty through blockchain and related technologies. These fluid, transnational jurisdictions reject both statist centralization and the market-driven individualism of the Network State [26].

Rooted in collective stewardship, this paradigm prioritizes ethical and inclusive governance over financialization. It draws on the legacy of commons theory and collaborative governance [2, 45], emphasizing shared responsibility for knowledge, infrastructure, and public goods. Decision-making is decentralized and participatory, guided by the principle of subsidiarity and grounded in moral legitimacy rather than capital investment.

Economically, Network Sovereignties promote mutualism and sustainability, favoring cooperation over competition and exploring non-monetary systems of value exchange. Technologically, they adopt a pragmatic and ethical approach, rejecting techno-solutionism in favor of tools that serve equity and community well-being.

Rather than seeking recognition from nation-states, these communities leverage existing legal frameworks to construct autonomous zones. Their territoriality is relational and ecological, focusing on stewardship and local engagement. By balancing autonomy with participation in global governance, Network Sovereignties articulate a transformative alternative to both neoliberal and statist models.

2.3 Algorithmic Nations

As conceptualized by Calzada [10, 19], the Algorithmic Nations are emergent political entities that use algorithms, data-driven technologies, and digital platforms to reconfigure governance beyond the nation-state [4]. Rooted in emancipatory and transnational ideals, they aim to empower marginalized groups—such as indigenous peoples, e-diasporas, and stateless nations—through culturally grounded, ethical uses of technology [21, 23, 33, 58].

These nations emphasize self-determination, cultural preservation, and social justice. Governance is participatory and decentralized, guided by transparency, inclusivity, and community consent [53, 54]. Legitimacy derives from local impact rather than state recognition, with power distributed through culturally sensitive, algorithmically mediated structures.

Economically, Algorithmic Nations prioritize mutual aid and endogenous development, resisting market logic in favor of ethical, community-led value systems. Data cooperatives exemplify this, aligning exchange with social priorities rather than profit.

Their techno-emancipatory approach resists both state centralism and techno-solutionism, using digital platforms to support solidarity, heritage, and cooperation. Grounded in territorial and data sovereignty, they advocate for governance rescaling via city-regional frameworks and devolved, community-led control [9, 11].

As these decentralized governance paradigms continue to evolve, future research should explore how Network States, Network Sovereignties, and Algorithmic Nations can address issues of inclusivity, equity, and global cooperation. Policymakers will need to engage with these models to ensure that decentralized governance contributes to a more just and equitable global order.

Consequently, Table 1 depicts how each paradigm would respond to the research question fixed in this paper.

3 Final Remarks: The (Dis)Illusion of Web3 Decentralization

This paper has interrogated the (Dis)Illusion of Web3 decentralization, particularly as it converges with Generative AI (GenAI) to reshape the contours of global governance. While the rhetoric surrounding Web3 often invokes empowerment, transparency, and participation, the empirical and conceptual analysis presented here reveals a more ambivalent reality. Rather than dismantling traditional hierarchies, decentralized technologies risk reassembling them—under new guises of algorithmic authority, technical literacy, and digital capital [16, 38, 43].

The integration of GenAI further intensifies these tensions. As algorithmic systems mediate decision-making processes and knowledge production, the locus of sovereignty shifts from nation-states to opaque technical infrastructures governed by a narrow stratum of actors with privileged access to data, expertise, and computational resources. Without deliberate safeguards, this fusion of GenAI and Web3 threatens to solidify a new technocratic class under the guise of decentralization [1, 44].

The three paradigms analyzed—*Network States, Network Sovereignties, and Algorithmic Nations*—offer divergent responses to this evolving landscape. *Network States* [52], shaped by a crypto-libertarian ethos, concentrate power among founders, developers, and early investors. Despite the promise of borderless innovation,

Table 1 Responding to the research question

Research question: distributing power or creating a new tech-savvy elite?			
Criteria	Network states [52]	Network sovereignties [26]	Algorithmic nations [10]
Centralized versus decentralized	Centralized but with an illusion of decentralization. Key decision-making remains with the founder or core developers, leading to elite-driven power structures	Decentralized and participatory. Governance is rooted in collective decision-making, aligning with ethical and commons-based principles	Decentralized, algorithm-driven governance. Governance through data-driven algorithms, emphasizing cultural sensitivity and inclusivity
Membership and access	Exclusive. Membership often requires financial contribution or technological literacy, favoring those who are tech-savvy or economically capable	Voluntary and value-driven. Based on commons-driven alignment. Members share responsibilities, not absent of small groups co-opting the power	Voluntary and culturally rooted. Members are aligned with shared cultural, ethical values, emphasizing belonging and self-determination over financial input
Power distribution	Centralized among tech-savvy elites. Power distribution is skewed towards those with technical expertise or financial resources, clearly reinforcing a new elite class	Distributed among community members. Power is decentralized, fostering collective governance and minimizing the rise of a centralized elite, although not absent of manipulation	Algorithmically mediated. Power distribution is balanced through algorithms, ensuring transparency and participatory decision-making
Cultural integration	Techno-utopian. Focuses more on technological innovation and efficiency, often at the expense of cultural or community-based governance models	Commons-centric and community-based. Emphasizes libertarian digital nomadic patterns before localism and roots, which contradicts the sustainable commons view that seems to aim to advocate	Culturally rooted. Promotes the preservation of cultural identity, transnational cooperation, and social justice through technology. Not absent of local-first judgment, excluding newcomers, with high entrance barriers

(continued)

Table 1 (continued)

Research question: distributing power or creating a new tech-savvy elite?			
Criteria	Network states [52]	Network sovereignties [26]	Algorithmic nations [10]
Economic Model	Financialized. Driven by market efficiency and on-chain monetarism (cryptocurrencies), often exacerbating economic inequalities and fostering elite control	Commons-driven. Focus on reciprocity, mutual aid, and the stewardship of public goods, challenging financialization. Still too anchored in state-driven mindsets rather than nation-building processes	Cooperative and ethical. Data and digital economies are driven by cultural and ethical considerations, prioritizing social well-being over financial gain. Federated internationalization is the antidote to real distribution of power
Risk of creating a new elite	High. By requiring technological and financial access, Network States may inadvertently create a privileged class, limiting broader participation and equity	Moderate. Collective decision-making and a focus on mutual benefits reduce the likelihood of elite formation. Although the digital nomadic trend shows that power can be co-opted by a small group	Moderate. While algorithms promote transparency, their reliance on data and technology might still favor tech-savvy groups if not carefully managed
Potential for inclusivity	Low. The reliance on technological proficiency and financial contributions limits inclusivity, potentially excluding marginalized groups	Moderate. The emphasis on community-driven governance ensures that all members have a say, fostering inclusivity across diverse backgrounds. Although the digital nomadic patterns are still bound to a privileged group of *world citizens*	Moderate to high. Inclusivity is encouraged through the ethical use of technology and algorithmic governance, though data access remains a challenge. Inclusive internationalist nation-building is a precondition to encourage inclusive and highly solidaristic communities

Source Calzada

these models often reinforce exclusion through financialization and technological gatekeeping [15, 48]. Their framing of autonomy, rooted in market logic, risks undermining public accountability and democratic legitimacy.

In contrast, *Network Sovereignties* [26, 45] build upon commons-based traditions of self-governance and mutual aid. Emphasizing subsidiarity and participatory ethics, they resist both statist centralism and cyberlibertarian individualism. Yet even these models face challenges: internal power asymmetries, digital nomadism, and the difficulty of scaling without compromising inclusivity [2].

Algorithmic Nations [10, 19] offer perhaps the most compelling vision for equitable, post-Westphalian governance in the GenAI era. By centering marginalized communities—such as e-diasporas, indigenous groups, and stateless nations—and integrating culturally grounded algorithmic infrastructures, these formations challenge both centralized state power and market radicalism [21, 23, 33, 58]. However, their viability depends on addressing the risks of data sovereignty erosion, algorithmic opacity, and unequal access to digital infrastructures [4, 20, 53].

In response to the research question—whether Web3 genuinely redistributes power or merely shifts it into the hands of a new, technologically adept elite—this paper demonstrates that decentralization alone does not inherently democratize power structures. The analysis of three paradigms—*Network States*, *Network Sovereignties*, and *Algorithmic Nations*—clearly illustrates that outcomes vary significantly based on underlying governance models and their embedded techno-political imaginaries. *Network States* predominantly reinforce elite-driven centralization through financialized participation and technical gatekeeping, while *Network Sovereignties* and *Algorithmic Nations* offer more inclusive possibilities, provided they actively address internal power asymmetries and ensure equitable participation. Therefore, the paper's findings emphasize that realizing the democratic and inclusive potential of Web3 decentralization requires deliberate, context-sensitive policy interventions designed to mitigate technocratic capture, promote digital justice, and foster genuine cross-border cooperation.

To move beyond the (dis)illusion, global governance in the age of GenAI must develop *hybrid policy frameworks* that bridge technological innovation with normative commitments to justice, transparency, and democratic participation. These frameworks should confront both global asymmetries and local specificities, resisting cyberlibertarian capture while fostering solidaristic infrastructures for collective governance [5, 47, 51].

Ultimately, if Web3 and GenAI are to fulfill their emancipatory potential, they must be steered not only by technical ingenuity but by *ethical and political imagination* [56]. Only then can we imagine and realize a post-Westphalian digital order in which power is genuinely distributed, and governance becomes inclusive, reflexive, and accountable in a world increasingly shaped by code.

Acknowledgements This paper integrates insights from transatlantic action research fieldwork supported by the US-UK Fulbright Commission (https://fulbright.org.uk/people-search/igor-calzada/) and the Economic and Social Research Council (ESRC) (https://wiserd.ac.uk/blog/international-comparative-and-action-research-triangulating-wales-with-the-basque-country-and-california/) across North America and Europe. The author gratefully acknowledges the support of these institutions, as well as the contributions of participants engaged during the fieldwork conducted in Washington, D.C., in April 2025. These activities were conducted as part of an iterative, longitudinal inquiry within the AI4SI research program.

This Springer Nature paper is part of a broader AI4SI action research programme and builds upon preliminary findings published in *Future Internet* journal and action research fieldwork initiated in August 2022 [14], which provided a framework for this AI4SI research programme co-funded by the European Commission via (i) Ikerbasque, (ii) US-UK Fulbright Commission, Astera, and Decentralization Research Centre in the USA, and (iii) Cardiff University/WISERD and ESRC in the UK. Through this paper, consequently, the author updates its findings up to April 2025, including participatory observation at the Data Cooperatives Workshop (Georgetown University) and the Decentralized Tech Summit 2025 (National Press Club, Washington, D.C.). These engagements, along with earlier work at Stanford DAO Workshops and initiatives such as SOAM, BlockchainGov, and Decentralization Research Centre, constitute a longitudinal action research study into the techno-political ethos of the global Web3 movement. It should add that action research process does not simply add events or conversations, but reshuffles iteratively previous published research by providing new flavour and updates to the current research programme.

References

1. Abdi, I.: Digital capital and the territorialization of virtual communities: an analysis of Web3 governance and network sovereignty (2023). https://www.researchgate.net/publication/370057091
2. Aligica, P.D., Tarko, V.: Policentricity: from Polanyi to Ostrom, and beyond (2012). https://papers.ssrn.com/sol3/papers.cfm?abstract_id=2149165
3. Allen, D., Frankel, E., Lim, W., Siddarth, D., Simons, J., Weyl, E.G.: Ethics of decentralized social technologies: lessons from Web3, the fediverse, and beyond. In: The Edmond & Lily Safra Center for Ethics. Harvard University (2023)
4. Amoore, L.: Cloud geographies: computing, data, sovereignty. Prog. Hum. Geogr. **42**(1), 4–24 (2016). https://doi.org/10.1177/0309132516662147
5. Arendt, H.: The rights of man: what are they? Mod. Rev. **3**, 4–37 (1949)
6. Barlow, J.P.: A declaration of the independence of cyberspace (1996). https://vimeo.com/111576518?ref=tw-v-share. Accessed 22 Feb 2018
7. Bodó, B., Brekke, J.K., Hoepman, J.-H.: Decentralisation: a multidisciplinary perspective. Internet Policy Rev. **10**(2) (2021). https://doi.org/10.14763/2021.2.1563
8. Bratton, B.H.: The Stack: On Software and Sovereignty. The MIT Press, Boston (2017)
9. Calzada, I.: Benchmarking future city-regions beyond nation-states. RSRS Reg. Stud. Reg. Sci. **2**(1), 350–361 (2015). https://doi.org/10.1080/21681376.2015.1046908
10. Calzada, I.: Algorithmic nations: seeing like a city-regional and techno-political conceptual assemblage. Reg. Stud. Reg. Sci. **5**(1), 267–289 (2018). https://doi.org/10.1080/21681376.2018.1507754
11. Calzada, I.: Emerging Digital Citizenship Regimes: Postpandemic Technopolitical Democracies. Emerald Points Series. Emerald Publishing Limited, Bingley (2022). https://doi.org/10.1108/9781803823317, ISBN: 9781803823324
12. Calzada, I.: Disruptive technologies for e-Diasporas: blockchain, DAOs, data cooperatives, metaverse, and ChatGPT. Futures **154**(C), 103258 (2023). https://doi.org/10.1016/j.futures.2023.103258
13. Calzada, I.: Blockchain-driven digital nomadism in the Basque e-Diaspora. Globalizations **21**(5), 777–802 (2023). https://doi.org/10.1080/14747731.2023.2271216
14. Calzada, I.: Data (un)sustainability: navigating utopan resistance while tracing emancipatory datafication strategies. In: Certomá, C., Martelozzo, F., Iapaolo, F. (eds.) Digital Technologies for Sustainable Futures: Promises and Pitfalls. Routledge, Oxon (2024). https://doi.org/10.4324/9781003441311-11
15. Calzada, I.: Artificial intelligence for social innovation: beyond the noise of algorithms and datafication. Sustainability **16**(19), 8638 (2024). https://doi.org/10.3390/su16198638
16. Calzada, I.: Decentralized Web3 reshaping internet governance: towards the emergence of new forms of nation-statehood? Future Internet **16**(10), 361 (2024). https://doi.org/10.3390/fi16100361
17. Calzada, I.: Special issue (2025) Transforming government: people, process, and policy. generative AI and urban AI policy challenges ahead: trustworthy AI for whom? (2025). https://www.emeraldgrouppublishing.com/calls-for-papers/generative-ai-and-urban-ai-policy-challenges-ahead-trustworthy-ai-whom
18. Calzada, I.: Special issue (2024) Discover cities. In: Digital Inclusion & Urban AI (2025). https://link.springer.com/collections/bhgeijbhib

19. Calzada, I., Bustard, J.: The dilemmas around digital citizenship in a post-Brexit and post-pandemic northern Ireland: towards an algorithmic nation? Citizsh. Stud. **27**(2), 271–292 (2022). https://doi.org/10.1080/13621025.2022.2026565
20. Calzada, I., Németh, G., Al-Radhi, M.S.: Trustworthy AI for whom? GenAI detection techniques of trust through decentralized Web3 ecosystems. Big Data Cogn. Comput. **9**, 62 (2025). https://doi.org/10.3390/bdcc9030062
21. Carroll, S.R., Rodriguez-Lonebear, D., Martinez, A.: (2019). Indigenous data governance: Strategies from United States native nations. Data Sci. J. **18**(1), Chapter 31. https://doi.org/10.5334/dsj-2019-031
22. Cheney-Lippold, J.: The silicon future. New Media Soc. **27**(7), 4164–4180 (2024). https://doi.org/10.1177/14614448241234864
23. Cook, D.: Breaking the contract: digital nomads and the state. Crit. Anthropol. **42**(3), 304–323 (2022). https://doi.org/10.1177/0308275X221120172
24. D'Andrea, A.: Neo-nomadism: a theory of post-identitarian mobility in the global age. Mobilities **1**(1), 95–119 (2006). https://doi.org/10.1080/17450100500489148
25. Davidson, J.D., Rees-Mogg, W.: The Sovereign Individual: Mastering the Transition to the Information Age. Touchstone, London (1999)
26. De Filippi, P., Reijers, W., Mannan, M.: Blockchain Governance. MIT Press, Boston (2024)
27. DuPont, Q.: A progressive Web3: from social coproduction to digital polycentric governance (2023). SSRN. https://doi.org/10.2139/ssrn.4320959, https://ssrn.com/abstract=4320959. Accessed 9 Jan 2023
28. Galloway, A.R., Thacker, E.: The Exploit: A Theory of Networks. University of Minnesota Press, Minnesota (2007)
29. Gray, J.E., Hutchinson, J., Stilinovic, M., Tjahja, N.: The pursuit of 'good' Internet policy. Policy Internet **16**, 480–484 (2024). https://doi.org/10.1002/poi3.423
30. Golumbia, D.: Cyberlibertarianism: The Right-Wing Politics of Digital Technology. Minnesota University Press, Minnesota (2024)
31. G'sell, F.: Regulating under uncertainty: governance options for generative AI. Stanford GenAI report, Florence-GSELL (2024)
32. Hope, J., Ludlow, P.: Farewell to Westphalia: crypto sovereignty and post-nation-state governance (2024). https://logos.co/farewell-to-westphalia/
33. Kannisto, P.: Global Nomads and Extreme Mobilities. Routledge, Oxon (2016)
34. Kemmis, S., McTaggart, R.: Participatory action research: communicative action and the public sphere. In: Denzin, N.K., Lincoln, Y.S. (eds.) The Sage Handbook of Qualitative Research, 3rd edn., pp. 559–603. Sage Publications Ltd., London (2005)
35. Kukutai, T., Cormack, D.: "Pushing the space": data sovereignty and self-determination in Aotearoa NZ. In: Walter, M., Kukutai, T., Carroll, S.R., Rodriguez-Lonebear, D. (eds.) Indigenous Data Sovereignty and Policy, pp. 21–35. Routledge, London (2020)
36. Lehdonvirta, V.: Cloud Empires: How Digital Platforms Are Overtaking the State and How We Can Regain Control. MIT Press, Boston (2022)
37. Lessig, L.: Code: And Other Laws of Cyberspace. Basic Books, London (1999)
38. Mathew, A.J.: The myth of the decentralised internet. Internet Policy Rev. **5**(3) (2016). https://policyreview.info/chapters/analysis/myth-decentralised-internet
39. May, T.: The crypto anarchist manifesto (1992). https://groups.csail.mit.edu/mac/classes/6.805/chapters/crypto/cypherpunks/may-crypto-manifesto.html
40. Merk, T.: The unusual DAO: An ethnography of building trust in "trustless" spaces. Internet Policy Rev. **13**(3) (2024). https://doi.org/10.14763/2024.3.1795
41. Morozov, E.: Web3: a map in search of territory (2022). https://the-crypto-syllabus.com/web3-a-map-in-search-of-territory/
42. Nanni, R., Bizzaro, P.G., Napolitano, M.: The false promise of individual digital sovereignty in Europe: comparing artificial intelligence and data regulations in China and the European Union. Policy Internet **16**, 1–16 (2024). https://doi.org/10.1002/poi3.424
43. Nabben, K., De Filippi, P.: Accountability protocols? On-chain dynamics in blockchain governance. Internet Policy Rev. **13**(4) (2024). https://doi.org/10.14763/2024.4.1807

44. Ohlhaver, P., Weyl, E.G., Buterin, V.: Decentralized Society: Finding Web3's Soul. SSRN (2022). https://ssrn.com/abstract=4105763
45. Ostrom, E.: Governing the Commons. Cambridge University Press, Cambridge, UK (1990)
46. Pohle, J., Santaniello, M.: From multistakeholderism to digital sovereignty: toward a new discursive order in internet governance? Policy Internet **16**, 1–20 (2024). https://doi.org/10.1002/poi3.426
47. Polanyi, K.: The Great Transformation: The Political and Economic Origins of Our Time. Beacon Press, Boston (2001)
48. Rauchfleisch, A., Jungherr, A.: Blame and obligation: the importance of libertarianism and political orientation in the public assessment of disinformation in the United States. Policy Internet **16**, 1–17 (2024). https://doi.org/10.1002/poi3.407
49. Reason, P., Bradbury, H. (eds.): The SAGE Handbook of Action Research: Participative Inquiry and Practice, 2nd edn. Sage, London (2013)
50. Singh, R.: Give me a database and I will raise the nation-state. South Asia J. South Asian Stud. **42**(3), 501–518 (2019). https://doi.org/10.1080/00856401.2019.1602810
51. Spelliscy, C., Hubbard, S., Schneider, N., Vance-Law, S.: Toward Equitable Ownership and Governance in the Digital Public Sphere. Belfer Center, Harvard (2023)
52. Srinivasan, B.: The Network State: How to Start a New Country (2022). www.thenetworkstate.com
53. Stucke, M.: Breaking Away: How to Regain Control Over Our Data, Privacy, and Autonomy. Oxford University Press, Oxford (2022)
54. Tucker, A., Piero de Bellis, G.: Panarchy: Political Theories of Non-Territorial States. Routledge, Oxon (2016)
55. Utrata, A.: Engineering territory: space and colonies in silicon valley. Am. Polit. Sci. Rev. **118**(3), 1097–1109 (2024). https://doi.org/10.1017/S0003055423001156
56. Visvizi, A., Kozlowski, K., Calzada, I., Troisi, O.: Multidisciplinary Movements in AI and Generative AI: Society, Business, Education. Edward Elgar, Chentelham (2025). https://doi.org/10.4337/9781035358663
57. Walter, M., Lovett, R., Maher, B., Williamson, B., Prehn, J., Bodkin-Andrews, G., Lee, V.: Indigenous data sovereignty in the era of big data and open data. Aust. J. Soc. Issues **56**(2), 143–156 (2021). https://doi.org/10.1002/ajs4.141
58. Walter, M., Kukutai, T., Carroll, S.R., Rodriguez-Lonebear, D.: Indigeneous Data Sovereignty and Policy. Routledge, Oxon (2021)

Digital Human: The New Frontier of Artificial Intelligence for Customer Experience

Claudia Spilotro, Giustina Secundo, and Felice Vitulano

Abstract Artificial Intelligence (AI) is increasingly shaping customer service in the banking sector, with a growing focus on Digital Humans—AI-powered virtual assistants that simulate human appearance, speech, and behavior. Unlike traditional chatbots, Digital Humans aim to provide more natural, empathetic, and personalized interactions by combining advanced machine learning, natural language processing, and 3D rendering technologies. This study explores how Digital Humans influence customer perceptions of digital banking services through the development and evaluation of a prototype based on Algho, a platform by QuestIT. Following a design science methodology, the research involved creating a Digital Human tailored to loan request management and assessing it through a pilot test and two qualitative surveys. Findings reveal that Digital Humans enhance engagement and innovation perception, yet barriers such as limited AI familiarity and the "Uncanny Valley" effect reduce full user acceptance. Most users expressed preference for a hybrid model that combines human empathy with AI efficiency. This study contributes to the understanding of how AI-driven interfaces reshape digital banking experiences, offering practical insights for the design of customer-centric AI solutions.

Keywords Artificial intelligence · Digital human · Customer experience · SDG-9 Industry innovation and infrastructure

C. Spilotro (✉)
Department of Management, Finance and Technology, LUM Giuseppe Degennaro University, Casamassima, Italy
e-mail: spilotro.phdstudent@lum.it

G. Secundo
Department of Engineering, LUM Giuseppe Degennaro University, Casamassima, Italy
e-mail: secundo@lum.it

F. Vitulano
QuestIT, Siena, Italy
e-mail: felice.vitulano@quest-it.com

A. Visvizi et al. (eds.), *Research and Innovation Forum 2025 Volume 1*,
Springer Proceedings in Complexity, https://doi.org/10.1007/978-3-032-09066-9_10

1 Introduction

The evolution of conversational agents has advanced significantly, leading to the emergence of "Digital Humans", a substantial shift away from traditional chatbots. Digital Humans are avatars capable of displaying comprehensive human body language, utilizing artificial intelligence (AI) to interpret user inputs and deliver both verbal and non-verbal responses [1]. By adopting humanoid forms, these Digital Humans exceed traditional limitations, fostering more empathetic and meaningful customer interactions [2] and reshaping social dynamics due to ongoing digital innovation [3]. Such anthropomorphic designs have significant psychological effects, potentially increasing user engagement [4, 5]. However, they may also have negative impacts, especially regarding transparency about their artificial nature, as human users may have biases against machines regardless of the Digital Humans' demonstrated competence [6]. Balancing the benefits of anthropomorphism against transparency challenges is essential for effective human–machine interactions in the rapidly growing Digital Human economy, projected to reach $125 billion by 2035 [7].

Despite increased interest in AI-driven customer experience enhancement, digital technologies are still in early developmental stages, providing substantial areas for exploration. Innovation management research has focused on only a few AI types [8]. AI, defined as models trained to acquire knowledge and generate predictions from data, is a critical digitalization trend with transformative potential across various industries, notably finance. Currently, AI primarily supports decision-making processes [9], with banking and financial sectors significantly investing in and benefiting from AI technologies [10]. The banking sector, in particular, is undergoing substantial digital transformation driven by emerging technologies, including Digital Humans [11]. Nevertheless, most research has overlooked consumer perspectives, necessitating deeper insights into customer motives, expectations, and technology adoption [12].

Addressing this gap, this study aims to explore the strategic potential of AI-driven Digital Humans in enhancing customer experience [13]. Digital Humans facilitate natural, empathetic two-way communication [14], significantly enhancing customer interactions. Specifically, the research investigates how Digital Humans reshape customer experience within banking by implementing a virtual assistant prototype designed to automate financial processes. Thus, the key research question driving this analysis is: *How do Digital Humans influence customers' perceptions and acceptance of digital banking services, considering both emotional engagement and perceived usability?*

The study employs a design science process [15–17] to acquire knowledge and comprehension of an issue by creating and implementing a designed artifact. Results indicate that corporate processes can benefit from the integration of AI technologies. Due to a lack of understanding about these technologies and the idea that they are hard to use, there is also resistance to the full use of these tools.

This paper is structured as follows. Section 2 reviews the literature on the rise of AI-based digital humans, with a particular focus on their role in reshaping customer experience in the banking sector. Section 3 presents the research design, detailing the methodological approach adopted in this study. Section 4 outlines the main findings, followed by a discussion in Sect. 5 that interprets the results according to existing literature. Finally, Sect. 6 concludes the paper and reflects on the broader implications for both research and practice.

2 Literature Background

2.1 The Rise of AI-Based Digital Humans

Digital transformation involves a deliberate and substantial shift in an organization's characteristics, driven by integrating information, computing, communication, and connectivity technologies [18]. This shift enables the rise of AI technologies, significantly enhancing business processes and customer interactions. Recent advancements in conversational AI, particularly chatbots, have attracted significant interest for their potential to revolutionize various industries [19]. Chatbots utilize natural language processing (NLP) to closely mimic human speech and interact with users via digital platforms [20, 21]. The evolution towards Digital Humans, characterized by anthropomorphic features, marks a substantial advancement over traditional chatbots, offering more empathetic interactions and better user experiences [2, 7]. Digital Humans leverage advanced algorithms and machine learning to handle complex queries and interpret nonverbal communication, fundamentally altering the landscape of human–machine interactions [22]. Anthropomorphism—the attribution of human characteristics to AI—significantly influences user perceptions and interactions. While moderate anthropomorphism may increase usage intention and reduce perceived risks [23], excessive human-like appearance can have the opposite effect, diminishing acceptance and negatively impacting interactions [6, 24–26]. Importantly, trust in anthropomorphic AI hinges not just on human-like traits, but also on perceived interpersonal attraction, both social and functional [27].

2.2 AI-Based Digital Humans: Reshaping Customer Experience in Banking

Customer experience remains central to marketing management, offering substantial competitive advantage [28]. Effective customer experiences involve interactions across multiple touchpoints, combining human interaction, functional utility, and mechanical elements [29]. AI-driven chatbots enhance these experiences by delivering efficient, predictive, and consistent interactions across various platforms, thus boosting satisfaction and engagement [30, 31]. However, despite these advantages, some consumers still prefer interactions with humans [32]. Emotional and functional enhancements in service robots significantly improve customer attitudes, engagement, and overall outcomes [33]. Improving machine-mediated interactions requires optimal design strategies and continuous customer feedback to minimize gaps between automated and human interactions [34]. Technological advancements significantly transform service delivery, especially in finance, as evidenced by digital payment solutions and automated financial advisory services [35]. Banks are increasingly adopting artificial intelligence-based solutions—such as robo-advisors—to lower service costs, expand access to 24/7 financial services, make financial advice more accessible and personalized through automated investment decisions [36, 37]. To fully harness the benefits of digital transformation, it's essential that leaders understand this shift from the customer's perspective [38, 39]. AI tools like chatbots have become widespread in banking to enhance customer service [40, 41], with a growing number of financial institutions deploying both chatbots [42, 43]. Chatbots are integral to banks' digital transformation initiatives, offering sustainable solutions and reshaping marketing, sales, and customer relationship strategies [44]. Ensuring chatbot reliability, responsiveness, and utility is vital to sustaining customer engagement [45]. Employing Digital Humans further enhances banking customer service by improving navigation, providing real-time insights, forecasting outcomes, and optimizing workflows, ultimately reducing operational costs [2].

3 Research Process

This study employs a design science approach [15–17], aiming to generate practical knowledge that professionals can apply to solve technical problems within their fields. Central to design science research is creating and evaluating artifacts to gain deeper understanding and address specific issues [16, 46]. This research involved developing and analyzing a Digital Human using Algho, an advanced conversational AI platform developed by QuestIT, trademarked by The Digital Box. Algho leverages proprietary technologies to create virtual assistants and Digital Humans capable of emulating human appearance, voice, and movements, fostering empathetic user interactions. Figure 1 summarizes the research process, outlining steps, key inputs, methods/tools, and outputs:

- Phase 1: Identified the research problem—growing complexity and resource-intensive nature of customer care—highlighting Digital Humans as potential solutions for process automation and improved customer experiences, leading to the research question: *"How do Digital Humans influence customers' perceptions and acceptance of digital banking services, considering both emotional engagement and perceived usability?"*

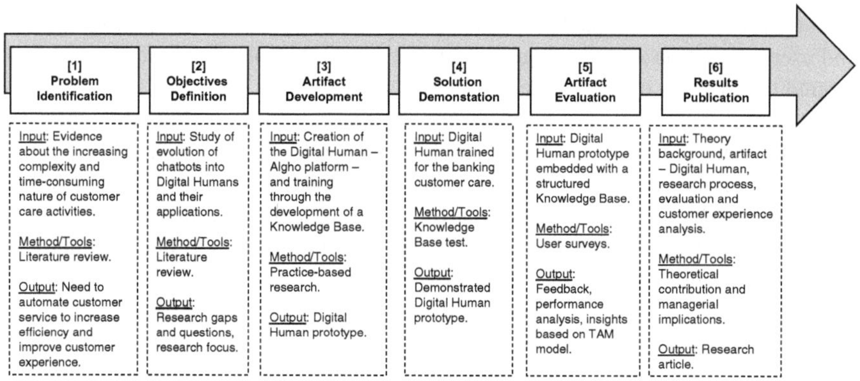

Fig. 1 Research process. *Source* Authors' own creation

- Phase 2: Defined research objectives by analyzing Digital Humans' capabilities, addressing gaps in fragmented literature, and exploring customer-oriented impacts on customer care processes.
- Phase 3: Developed the Digital Human artifact on the Algho platform, including a comprehensive knowledge base crucial for resolving user-specific issues and enhancing user trust. Practice-based research methods [47] were integral to gaining deeper insights through practical implementation.
- Phase 4: Conducted iterative testing of the Digital Human's knowledge base, optimizing its effectiveness within the Algho platform, according to prototyping design principles [48].
- Phase 5: Implemented qualitative surveys—a Performance Survey and Profiling, and a Technology Acceptance Model-based Likert-scale survey [49, 50]—to capture nuanced user feedback on experiences with the Digital Human. The study involved 25 diverse participants.
- Phase 6: Concluded with synthesizing theoretical and practical insights into Digital Humans' potential for enhancing banking customer experiences, aiming to disseminate these findings academically and professionally through publication.

Phases 1 and 2 are conceptually addressed within the *Literature Background* (Sect. 2) of the study. In particular, the identification of the research problem and the definition of the research objectives are supported by a critical examination of existing literature on Digital Humans, AI in customer service, and user perception in digital banking. This theoretical foundation allowed us to highlight existing gaps—particularly the lack of empirical studies focused on customer perceptions and psychological responses to AI interfaces—and thus justified the need for a user-centered investigation.

Phases 3, 4, and 5 are presented in the *Findings* (Sect. 4), specifically within the subsections dedicated to artifact development, solution demonstration, and artifact evaluation. These phases describe the practical implementation of the Digital Human prototype on the Algho platform, the testing of its Knowledge Base, and the collection and analysis of user feedback through qualitative surveys. The methodological rigor and user interaction at these stages ensure the empirical robustness of the study and ground the insights in real user experiences.

Phase 6 is summarized in the *Conclusions and implications* (Sect. 6), where the theoretical and practical contributions of the research are discussed. This includes reflections on the potential of Digital Humans in banking, the importance of hybrid AI-human models, and future directions for technological integration and user education. The conclusions also acknowledge current limitations and open avenues for further exploration in the AI–customer experience domain.

3.1 Research Context: Algho Platform

Algho, developed by QuestIT and trademarked by The Digital Box, is an advanced virtual assistant platform using 3D technology to enhance customer interactions. With human-like traits, it fosters empathy and emotional intelligence, making interactions more intuitive. Its Knowledge Base extends its capabilities beyond typical assistants, while its Digital Human Interface simulates gestures, expressions, and contextual responses for deeper user engagement. Algho's multimodal abilities enable seamless interaction through visual, audio, and semantic channels.

In banking, it integrates with Home Banking platforms, offering account insights, Data Narration for financial analysis, and Smart Document Access for streamlined document retrieval.

Powered by Speech Recognition, NLP, Machine Learning, and Reinforcement Learning, Algho continuously evolves, ensuring accurate, emotionally intelligent responses. Its human-like interface enhances engagement, aligning with the demand for more natural, interactive AI experiences.

3.2 Data Collection

Data collection of our study regarded the collection of documents for developing the knowledge base of the Digital Human as well as the collection of data to analyze the behavior and perception of customers with reference to the customer care service.

Knowledge Base

The creation of the knowledge base for the Digital Human on the Algho platform for banking customer care has been the foundation for ensuring effective and satisfactory interaction with users. This process began with the collection and organization of detailed information about banking products and services, as well as frequently asked questions and common problems encountered by customers. Iterative testing and user feedback are used to refine and optimize the knowledge base, ensuring that the prototype is always aligned with real customer needs.

Customer Experience

The data collection process in this study adopted a qualitative approach [51, 52] to delve into the complex meanings and experiences of users interacting with Algho. This multi-faceted strategy ensures a comprehensive understanding of user interactions by monitoring important aspects of the customer experience such as perceived ease of use and perceived usefulness, thereby enhancing the accuracy and relevance of the knowledge base. The initial phase involved a performance and profile survey, offering an in-depth examination of user engagement, by examining participants' prior experience with technology and virtual assistants, their familiarity and comfort with these tools, and their expectations and attitudes towards Digital Humans in customer care. Subsequently, a 5-point Likert scale questionnaire survey, developed according to the technology acceptance model, was administered. These surveys were designed to capture diverse facets, ranging from user sentiment to performance perceptions, and to gauge the overall readiness of participants to embrace the Algho technology. The selected sample of 25 individuals, which actively and personally tested the Digital Human created by engaging directly with Algho. This pilot test ensured that the collected data authentically reflected the experiences of individuals who had firsthand interactions with the technology.

3.3 *Data Analysis*

The data analysis phase was structured according to the results obtained from two main survey instruments:

- Performance Survey and Profiling, aimed at understanding user engagement with the Algho platform. The responses collected from this survey provided a detailed picture of users' interactions and user engagement with Digital Human, allowing them to identify patterns and trends in their experiences. Through the qualitative analysis of these responses, valuable insights into users' perceptions and interaction with technology emerged.
- Survey Based on the Technology Acceptance Model (TAM) —used as a theoretical framework—which identifies two main factors that influence an individual's intention to use a new technology: perceived ease of use and perceived usefulness. Perception of usefulness was defined by Fred Davis as "the degree to which a person believes that using a particular system would improve his or her job performance", while perceived ease of use is defined as "the degree to which a person believes that using a particular system would be effortless" [49]. Although the Likert-scale format is typically used for quantitative analysis, in this study, responses were interpreted through qualitative content analysis [53], treating each statement as a reflection of the respondent's perception. This allowed for the identification of recurrent patterns and thematic interpretation aligned with the TAM dimensions, rather than statistical aggregation.

Through a content analysis of the responses obtained, we were able to thoroughly assess users' perceptions of the ease of use and usefulness of the Digital Human Algho. This approach contributed to a detailed understanding of the factors influencing technology adoption, enabling the identification of any barriers or facilitators in its acceptance by users. So, the implementation of this methodology allowed for a comprehensive overview of user experiences and perceptions, thus contributing to an in-depth assessment of the performance and acceptance of the Algho platform in the actual context of use.

4 Findings

After identifying the problem and defining the objectives, the next step was to understand how Digital Humans can reshape customer experience in banking customer care. The problem is characterized by the increasing complexity and resource-intensive nature of customer care activities, which are often time-consuming. Through a literature analysis, it was found that integrating Digital Humans could potentially automate processes, enhance customer experience, and improve machine-human interaction. The objectives were defined by studying Digital Humans and their potential. Given the fragmented literature and limited understanding of the phenomenon from the customer's perspective, it was essential to explore how this AI tool could transform customer care processes. Findings led to the development of an artifact, represented by the prototyping of a Digital Human on the Algho platform and its training through the construction of a knowledge base for an Italian banking website (Sect. 4.1). The solution was demonstrated by testing the knowledge base of the Digital Human (Sect. 4.2). The artifact was then evaluated using a qualitative approach (Sect. 4.3), which comprised a Performance Survey and Profiling to understand the sample's background and perspectives on virtual assistants, and a survey based on the Technology Acceptance Model to gain deeper insights into the customer experience from the sample who directly interacted with the Digital Human prototype.

4.1 Artifact Development: Digital Human Prototype

The study began with the development of the Digital Human, tailored to business requirements. Key attributes such as name, voice, language, and gender were carefully planned. Since the target bank is Italian, the Digital Human was fully designed in Italian. A small Italian bank was chosen for the prototype due to its low level of digitization, offering an opportunity to integrate AI-driven solutions. The bank's website was analyzed to build the virtual assistant's knowledge base, ensuring it accurately reflected the bank's services and customer habits. This integration enhanced the Digital Human's relevance and effectiveness in user interactions. To facilitate natural

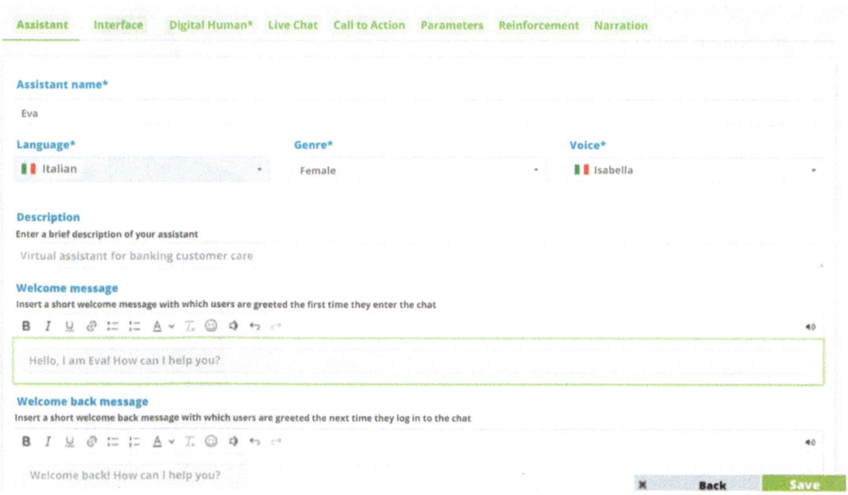

Fig. 2 Creation of the Digital Human. *Source* Algho platform (https://www.alghoncloud.com/)

communication, the assistant includes a welcome greeting, with Italian language and cultural elements reinforcing a personalized and engaging user experience (Fig. 2).

We configured the chat's colors, company logo, and the assistant's response speed. The "Live Chat" feature enables human intervention upon user request or after multiple unanswered queries. The "Parameters" section allows adjustments to flexibility, confidence, and response accuracy, optimizing syntactic and semantic emphasis. The "Reinforcement" section facilitates machine learning by enabling users to evaluate responses and improve future interactions. Once these settings are configured, the knowledge base is built. For example, a structured response to "What services do you offer?" includes details on account management, fund transfers, bill payments, loans, and investment support, ensuring 24/7 assistance. For casual conversations, pre-made basic questions are available. Custom questions require users to define both the query and a comprehensive response that covers multiple variations.

4.2 *Solution Demonstration*

The developed prototype of the Digital Human (Fig. 3) was designed to enhance banking customer care, specifically targeting the automation of procedures like loan requests. To assess and refine the customer experience derived from interactions with the trained Digital Human, we conducted three distinct but complementary performance tests. These evaluations aimed to optimize the Digital Human for practical implementation. The Knowledge Base test, in particular, was crucial in enhancing the virtual assistant's accuracy in responding to customer inquiries. Some of the questions included in the knowledge base of the Digital Human are: "What services do

Fig. 3 Digital Human prototype. *Source* Authors' own creation using Algho platform (https://www.alghoncloud.com/)

you offer?", "Who can apply for a personal loan?", "How do I check the accounting status of my card?", "How do I find the branch nearest to me?", "What credit cards do you have?", "How does early repayment of a loan work?", "What do I need to apply for a loan?", "What documents do I need to apply for a credit card?", "Do I need a guarantor to apply for a personal loan?".

The Algho platform presented the outcome as a percentage of this performance after the test, which was carried out by inserting all the questions in the virtual assistant's Knowledge Base with any alternatives that a user might ask in an Excel file submitted directly to the Algho platform. The Knowledge Base was consequently improved to achieve a 100% performance result.

4.3 Artifact Evaluation

The Digital Human—our artifact—was then evaluated through a pilot test using a qualitative approach, which involved the implementation of two complementary surveys: *the performance survey and profiling* and a questionnaire based on the *technology acceptance model*. These surveys aimed to explore the experience of customers belonging to the sample who interacted directly with the trained Digital Human prototype using an Italian bank website. A diverse sample of 25 individuals, varying in age, gender, education, and occupation, was selected. These participants actively tested the Digital Human before completing the surveys. This approach enriched the evaluation by capturing a broad range of perspectives on the Algho technology. The combination of qualitative exploration, user surveys, and a focused design science approach improved the robustness of the analysis, providing a nuanced understanding of user perceptions and Digital Human performance in real-world scenarios. The results of these surveys provided key data to refine the Digital Human prototype and optimize its integration into the banking customer service framework.

Performance Survey and Profiling Results

The first section of the artifact evaluation aimed to profile the sample and understand user engagement with virtual assistants. It assessed participants' prior experience, familiarity, and comfort with technology, as well as their expectations and attitudes toward Digital Humans in customer care. The survey explored perceptions of the benefits and drawbacks of virtual assistants in banking and their openness to adopting such technology. It also collected demographic data—age, gender, education, and occupation—to ensure a diverse sample. User experience with the Digital Human prototype was evaluated through satisfaction levels, issue identification, and performance feedback. This section provided critical insights for refining the technology and optimizing its integration into banking services. To enhance user clarity, the term "virtual assistant" was used in the questionnaire, as Digital Humans represent an advanced form of these systems.

The key findings from this section (Figs. 4 and 5) reveal a strong preference for digital banking services. While 16% prefer fully digital services and 84% favor a hybrid approach, no participants chose traditional banking. The innovative operator was preferred for three main reasons: breadth of services and products (68%); simplicity of use (44%); ease of access (40%). Additionally, 72% of participants were familiar with virtual assistants, yet only one had ever used one. 64% viewed them as useful, while 24% preferred human interaction. Regarding Digital Humans, 84% believed they would soon replace traditional customer service, and 76% considered them beneficial in banking.

To deepen their understanding, all participants engaged directly with the Digital Human before completing the second part of the survey, which assessed their experience and evaluated the assistant's performance.

The second part of the survey (Figs. 6 and 7) revealed mixed but generally positive feedback on the Digital Human experience: 20% rated it the highest score, while 44% liked it but gave a lower score; 28% were neutral, and 8% did not enjoy the experience; 72% noticed some anomalies in responses, while 28% found no issues. While no one reported completely inaccurate answers, the results indicate room for optimization based on external user inquiries. Regarding user preference: 48% were indifferent to interacting with a human or a virtual assistant; 28% preferred human interaction, while 16% favored the virtual assistant; 8% said their preference depends on the situation.

All participants agreed that the Digital Human Interface improves user interaction and comfort. Most also recognized that awareness of interacting with a non-human

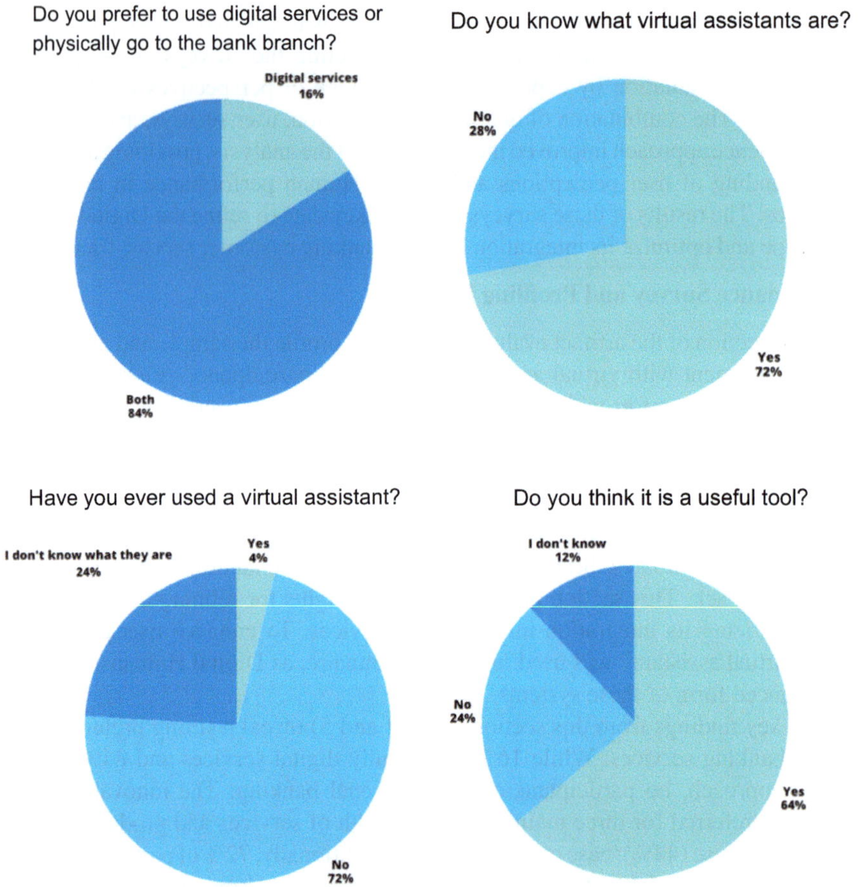

Fig. 4 Profiling survey results (1/2). *Source* Authors' own creation

entity could influence the experience. Finally, nearly all users viewed the integration of virtual assistants into customer service as beneficial, providing advantages for both businesses and customers.

Survey Based on the Technology Acceptance Model Results

The second part related to artifact evaluation focused on the analysis of the Digital Human prototype through a survey based on the Technology Acceptance Model. This section aimed to delve into the customer experience from the perspective of the sample who directly interacted with the Digital Human prototype. The TAM-based survey explored two key dimensions: perceived usefulness and perceived ease of use of the Digital Human technology by users. Participants were asked to rate their agreement with statements on these dimensions using a 5-point Likert scale. The results of this evaluation offered critical insights into the potential of Digital

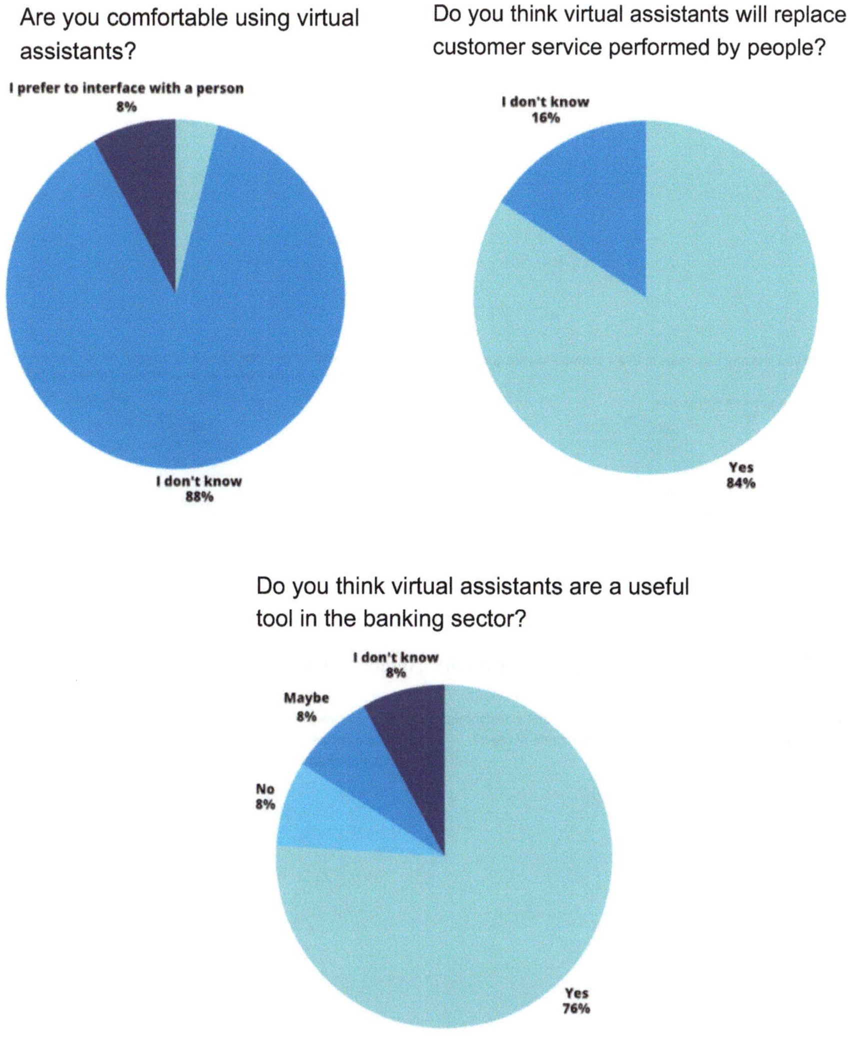

Fig. 5 Profiling survey results (2/2). *Source* Authors' own creation

Human to reshape customer service processes in the banking industry, guiding further development and optimization of the technology.

The last step saw the administration of a further survey based on the TAM, where the sample responses ranged from 1 (very little) to 5 (very much), whose findings (Figs. 8 and 9) showed a high level of perceived usefulness by the sample, indicating a digital maturity inherent in individuals and their ability to recognize the benefits derived from the implementation of such technology. However, there was a low perceived ease of use, indicating a low propensity to adopt digital tools, probably due to individuals' low knowledge of these new technologies.

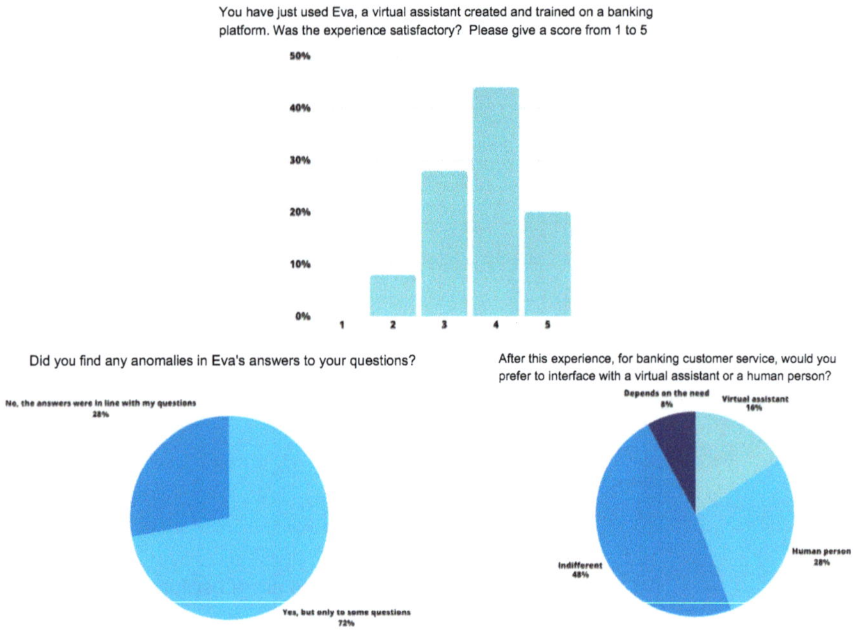

Fig. 6 Performance survey results (1/2). *Source* Authors' own creation

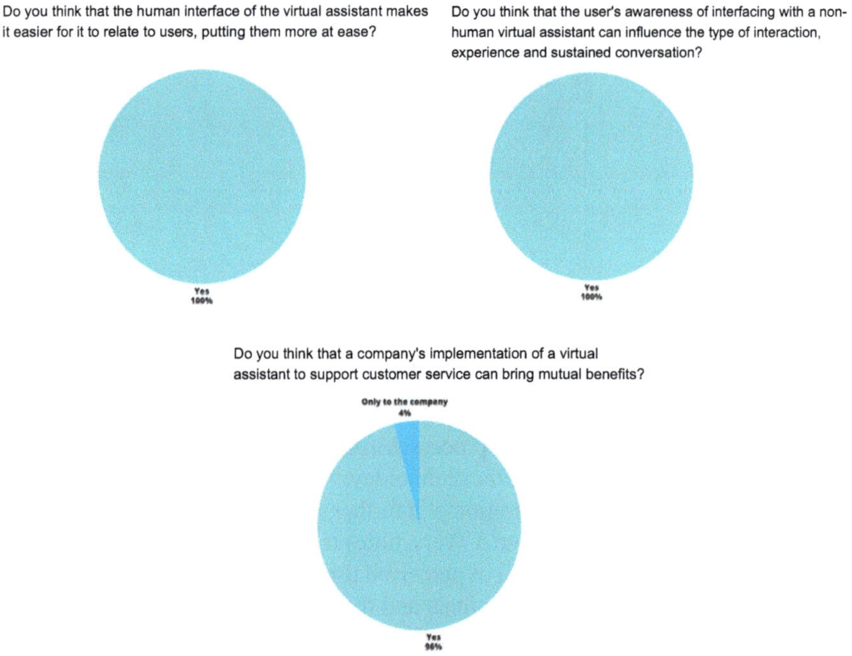

Fig. 7 Performance survey results (2/2). *Source* Authors' own creation

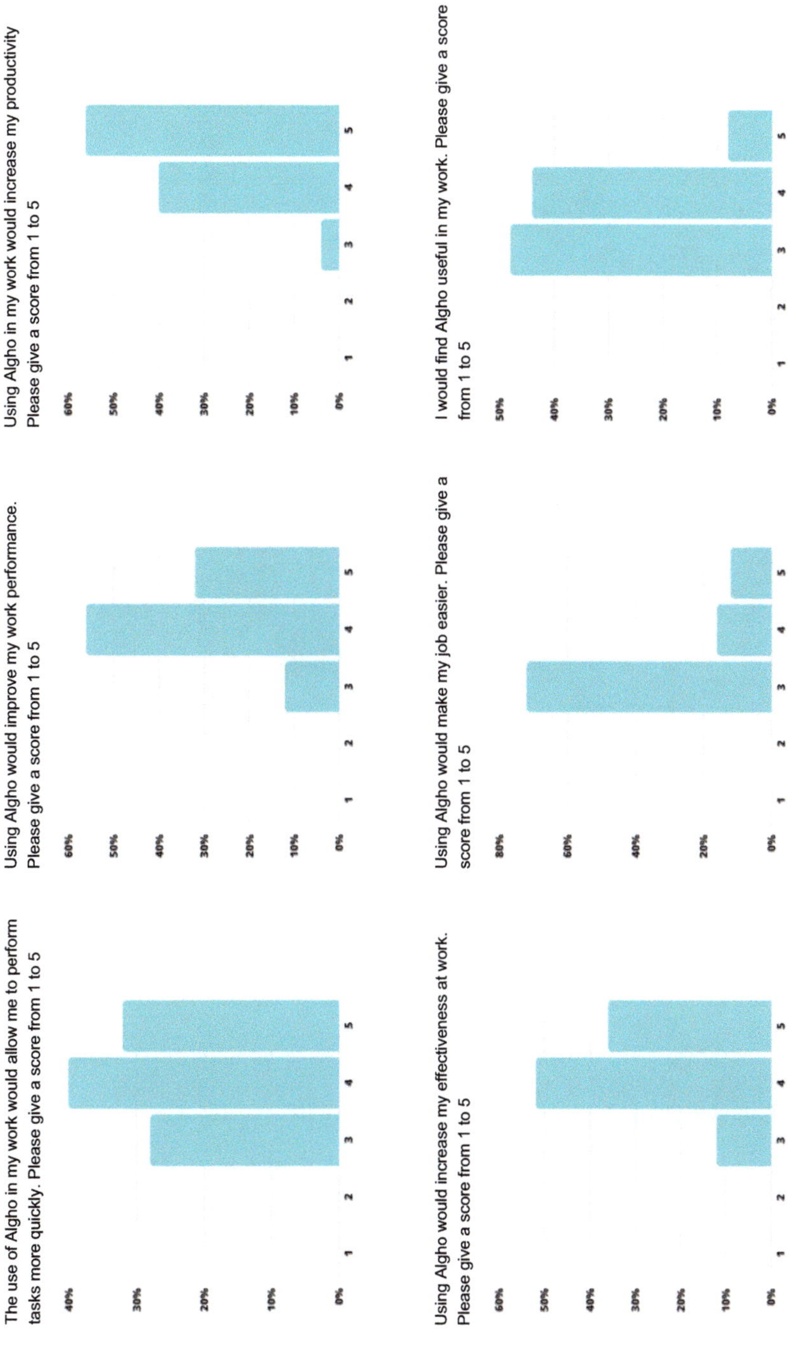

Fig. 8 TAM survey results: perceived usefulness. *Source* Authors' own creation

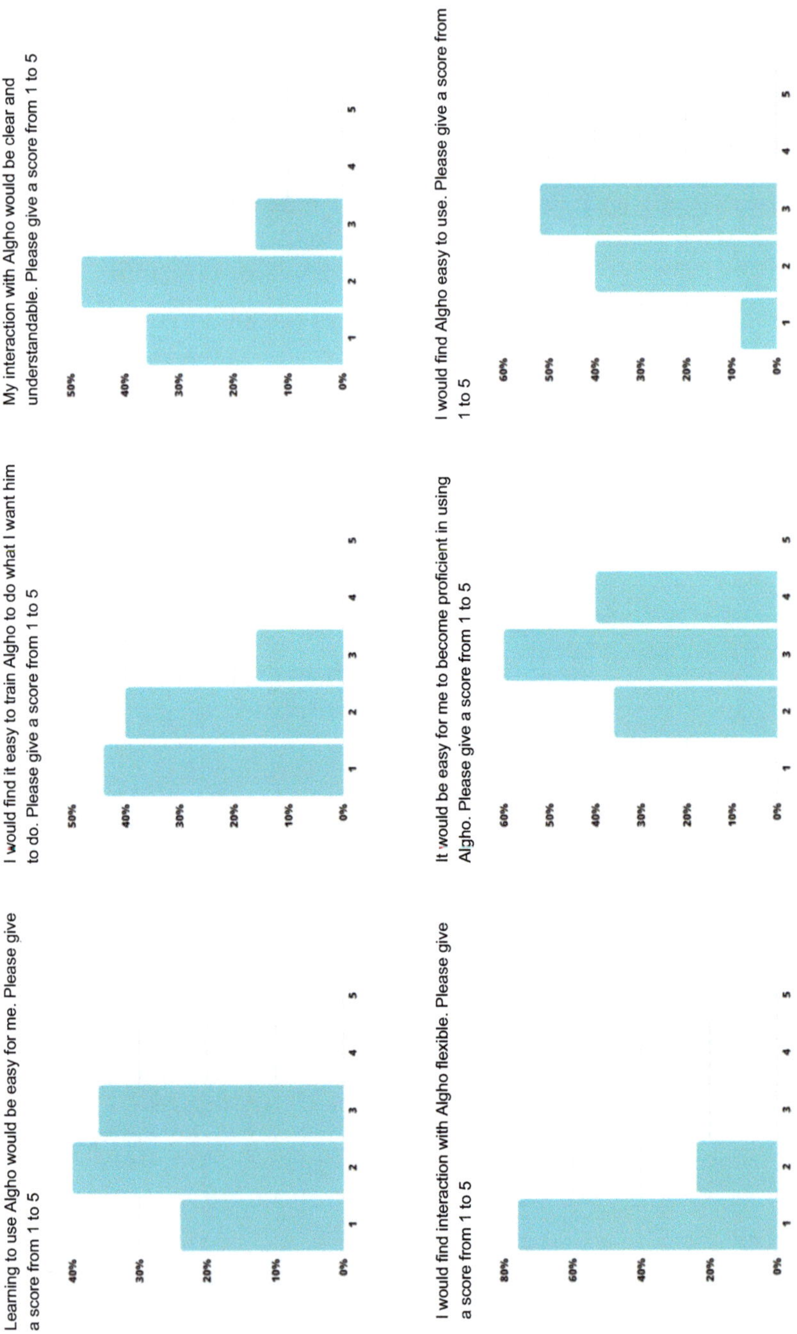

Fig. 9 TAM survey results: perceived ease of use. *Source* Authors' own creation

In conclusion, it can be stated that in a potential path of digital readiness, individuals are not yet fully aware and prepared for a radical and innovative change, probably due to an unconscious aversion to the potential risks of digital evolution. This result lies between two opposing currents of thought, the first seeing positive feedback from users [54] influencing their behavior [5, 55–57], and the second highlighting the negative impact [24].

5 Discussion

Our study focuses on Digital Humans—AI-driven conversational agents that represent an empathetic evolution of traditional virtual assistants. Specifically, we examined the Algho platform, which enables the creation and training of Digital Humans to support customer interactions in service environments. In this context, we developed and tested a Digital Human prototype designed to provide customer service in the banking sector, with particular attention to the management of loan request procedures. The goal was to enhance the overall digital banking experience by improving engagement, efficiency, and personalization. These findings directly address our research question, by showing that Digital Humans can enhance perceived usefulness, particularly through availability, speed, and automation of basic tasks. However, concerns persist around trust, emotional connection, and technological unfamiliarity, which continue to affect user acceptance. This duality highlights the need to design AI solutions that are not only operationally efficient but also emotionally intelligent—capable of generating trust and comfort in users, especially in sensitive service sectors like banking. By integrating these findings, our study contributes to understanding how Digital Humans may reshape customer expectations and service paradigms, potentially leading to increased satisfaction, loyalty, and innovation in customer care technologies.

This innovation offers a substantial market advantage by optimizing costs and reallocating human resources to more complex and unpredictable tasks. Research suggests that artificial intelligence technologies significantly improve customer interactions, as individuals tend to form stronger connections with robots compared to disembodied AI systems [58]. The Algho platform exemplifies the transition from conventional chatbots to Digital Humans, highlighting a new era of human–machine interaction. The successful deployment of a Digital Human in the banking sector demonstrates the adaptability of this technology in streamlining complex procedures such as loan applications. This advancement delivers both economic benefits and a more personalized user experience, opening new possibilities for how businesses approach customer service and banking operations.

Our analysis indicates a growing awareness of market evolution and the transformative impact of AI-driven technologies on banking operations. While most respondents favor digital tools, 28% still prefer human interaction, underscoring that AI solutions are in an early stage and that certain human experiences remain irreplaceable. This preference aligns with the "Uncanny Valley" hypothesis [24], which suggests that as robots become increasingly lifelike, initial familiarity may give way to discomfort and aversion [59]. This phenomenon presents a potential barrier to AI adoption, as reflected in our user testing results. Furthermore, the perceived complexity and limited understanding of these technologies hinder full adoption. A promising approach is a hybrid model that combines machine efficiency with human empathy, ensuring a balanced and user-friendly customer experience. Digital Humans, with their ability to evolve and emulate human interactions, hold significant potential for accurately replicating real-world customer interactions in the future [22].

From a broader perspective, the deployment of Digital Humans in banking raises important social and political considerations. On one hand, AI-driven customer service could improve access to banking services, particularly for users who face physical, geographic, or time-based barriers. On the other hand, digital exclusion remains a risk: individuals with limited digital literacy or access may be disadvantaged in fully automated environments. Moreover, regulatory frameworks for AI-driven services—particularly regarding transparency, data privacy, and accountability—are still evolving and need to adapt to the increasingly lifelike nature of Digital Humans.

For practitioners, the findings suggest that a hybrid service model—where AI supports human agents rather than replaces them—Is likely the most acceptable and effective path forward. Organizations should invest not only in technology development, but also in user education, onboarding strategies, and continuous feedback collection to ensure positive user experiences. For policymakers, the challenge lies in fostering innovation while safeguarding inclusivity, ensuring that the benefits of AI do not exacerbate existing inequalities or compromise user rights.

6 Conclusions and Implications

This study explored the development of a Digital Human prototype for banking customer care, revealing a strong user inclination toward digital banking services and virtual assistants, despite limited familiarity with such technologies. While there is growing digital readiness, the preference for a hybrid model—blending human and AI-driven interactions—highlights the enduring importance of human elements in customer experience. The transition to fully digitalized banking services requires continuous efforts in user education and technological refinement to ensure accessibility, effectiveness, and adoption.

Beyond usability, the increasing deployment of Digital Humans raises ethical concerns, particularly regarding privacy, security, and employment. The ability of AI to collect and process vast amounts of data intensifies privacy risks, while automation of routine tasks could lead to job displacement, exacerbating economic inequalities [60]. A balanced approach is necessary to support workforce transitions while leveraging AI's efficiency. Moreover, advancements in artificial intelligence, such as cognitive enhancements and emotional responsiveness in robots [61], continue to reshape the banking industry. The introduction of sophisticated AI models like ChatGPT [62] has significantly improved chatbot interactions, increasing their relevance in financial services [37] and enabling professionals to focus on strategic, high-value tasks [63].

From a managerial perspective, the integration of Digital Humans presents opportunities to automate key operations, optimize processes, and enhance customer interactions. However, technological complexity and limited user proficiency remain barriers, emphasizing the need for better user training and seamless AI-human collaboration. While users are not yet ready for full automation, a hybrid approach that augments human expertise with AI-driven tools appears to be the most viable path forward. Future research should explore the psychological factors influencing human–machine interactions to optimize user experience and reduce resistance to adoption.

A key limitation of this study is its small sample size (25 participants), which, while allowing for in-depth user engagement and detailed insights, limits the generalizability of findings. Expanding the sample in future research would provide a broader understanding of technology adoption dynamics in banking. Additionally, the convergence of generative AI and Digital Humans offers promising opportunities for disruptive innovations that optimize business activities while augmenting human intelligence and creativity [64, 65]. The continued exploration of AI-human synergies will be crucial in developing more intuitive, efficient, and personalized customer service models that enhance both user satisfaction and operational effectiveness.

References

1. Deloitte: Digital human elevating the digital human experience. https://www.deloittedigital.com/nl/en/insights/perspective/digital-human.html. Accessed 02 Aug 2024
2. QuestIT (2023). https://www.alghoncloud.com/artificial-human/
3. Kohli, R., Melville, N.P.: Digital innovation: a review and synthesis. Inf. Syst. J. **29**(1), 200–223 (2019)
4. Alabed, A., Javornik, A., Gregory-Smith, D.: AI anthropomorphism and its effect on users' self-congruence and self–AI integration: a theoretical framework and research agenda. Technol. Forecast. Soc. Change **182**, 121786 (2022)
5. Go, E., Sundar, S.S.: Humanizing chatbots: The effects of visual, identity and conversational cues on humanness perceptions. Comput. Hum. Behav. **97**, 304–316 (2019)
6. Luo, X., Tong, S., Fang, Z., Qu, Z.: Frontiers: machines vs. humans: the impact of artificial intelligence chatbot disclosure on customer purchases. Mark. Sci. **38**(6), 937–947 (2019)
7. Gartner (2022). https://www.gartner.com/en/documents/4000073
8. Mariani, M.M., Machado, I., Magrelli, V., Dwivedi, Y.K.: Artificial intelligence in innovation research: a systematic review, conceptual framework, and future research directions. Technovation **122**, 102623 (2023)
9. Cavazza, A., Dal Mas, F., Campra, M., Brescia, V.: Artificial intelligence and new business models in agriculture: the 'ZERO' case study. Manag. Decis. (2023)
10. Machkour, B., Abriane, A.: Industry 4.0 and its implications for the financial sector. Procedia Comput. Sci. **177**, 496–502 (2020)
11. Nicoletti, B.: Banking 5.0: how fintech will change traditional banks in the 'new normal' post pandemic. Springer Nature (2021)

12. Lu, V.N., et al.: Service robots, customers and service employees: what can we learn from the academic literature and where are the gaps? J. Serv. Theory Pract. **30**(3), 361–391 (2020)
13. Secundo, G., Spilotro, C., Ardito, C.A., Vitulano, F.: Artificial intelligence for the customer experience: developing a digital human for banking using algho platform. In: Presented at the IFKAD 2023 - Managing Knowledge for Sustainability Conference, pp. 1574–1586 (2023)
14. Nah, K., Oh, S., Han, B., Kim, H., Lee, A.: A study on the user experience to improve immersion as a digital human in lifestyle content. Appl. Sci. **12**(23), 12467 (2022)
15. Markus, M.L., Majchrzak, A., Gasser, L.: A design theory for systems that support emergent knowledge processes. MIS Q. 179–212 (2002)
16. Hevner, A.R., March, S.T., Park, J., Ram, S.: Design science in information systems research. MIS Q. 75–105 (2004)
17. Van Aken, J.E.: Management research as a design science: articulating the research products of mode 2 knowledge production in management. Br. J. Manag. **16**(1), 19–36 (2005)
18. Vial, G.: Understanding digital transformation: a review and a research agenda. Manag. Digit. Transform. 13–66 (2021)
19. Davenport, T.H., Ronanki, R.: Artificial intelligence for the real world. Harv. Bus. Rev. **96**(1), 108–116 (2018)
20. Ling, E.C., Tussyadiah, I., Tuomi, A., Stienmetz, J., Ioannou, A.: Factors influencing users' adoption and use of conversational agents: a systematic review. Psychol. Mark. **38**(7), 1031–1051 (2021)
21. Thomaz, F., Salge, C., Karahanna, E., Hulland, J.: Learning from the Dark Web: leveraging conversational agents in the era of hyper-privacy to enhance marketing. J. Acad. Mark. Sci. **48**, 43–63 (2020)
22. Kaczorowska-Spychalska, D.: Chatbots in marketing. Management **23**(1) (2019)
23. Yang, M., Li, J., Feng, L., Chen, S.-C., Tseng, M.-L.: Service robot anthropomorphism on consumer usage intention: curvilinear and linear effect. Ind. Manag. Data Syst. **124**(2), 612–639 (2024)
24. Mori, M.: The uncanny valley. Energy **7**(4), 33–35 (1970)
25. Ciechanowski, L., Przegalinska, A., Magnuski, M., Gloor, P.: In the shades of the uncanny valley: an experimental study of human–chatbot interaction. Future Gener. Comput. Syst. **92**, 539–548 (2019)
26. Shin, M., Kim, S.J., Biocca, F.: The uncanny valley: no need for any further judgments when an avatar looks eerie. Comput. Hum. Behav. **94**, 100–109 (2019)
27. Chen, Q.Q., Park, H.J.: How anthropomorphism affects trust in intelligent personal assistants. Ind. Manag. Data Syst. **121**(12), 2722–2737 (2021)
28. Kushwaha, A.K., Kumar, P., Kar, A.K.: What impacts customer experience for B2B enterprises on using AI-enabled chatbots? Insights from Big data analytics. Ind. Mark. Manag. **98**, 207–221 (2021)
29. Prentice, C., Nguyen, M.: Engaging and retaining customers with AI and employee service. J. Retail. Consum. Serv. **56**, 102186 (2020)
30. Cheetham, W.E., Goebel, K.: Appliance call center: a successful mixed-initiative case study. AI Mag. **28**(2), 89–89 (2007)
31. Nazir, S., Khadim, S., Asadullah, M.A., Syed, N.: Exploring the influence of artificial intelligence technology on consumer repurchase intention: the mediation and moderation approach. Technol. Soc. **72**, 102190 (2023)
32. Nanji, A.: How consumers feel about brands' use of AI (2019)
33. Hlee, S., Park, J., Park, H., Koo, C., Chang, Y.: Understanding customer's meaningful engagement with AI-powered service robots. Inf. Technol. People **36**(3), 1020–1047 (2023)
34. Novak, T.P., Hoffman, D.L., Yung, Y.-F.: Measuring the customer experience in online environments: a structural modeling approach. Mark. Sci. **19**(1), 22–42 (2000)
35. Basole, R.C., Patel, S.S.: Transformation through unbundling: visualizing the global FinTech ecosystem. Serv. Sci. **10**(4), 379–396 (2018)
36. Belanche, D., Casaló, L.V., Flavián, C.: Artificial Intelligence in FinTech: understanding robo-advisors adoption among customers. Ind. Manag. Data Syst. **119**(7), 1411–1430 (2019)

37. Barile, D., Secundo, G., Bussoli, C.: Exploring artificial intelligence robo-advisor in banking industry: a platform model. Manag. Decis. (2024)
38. Kandampully, J., Bilgihan, A., Bujisic, M., Kaplan, A., Jarvis, C.B., Shukla, Y.: Service transformation: how can it be achieved? J. Bus. Res. **136**, 219–228 (2021)
39. Lähteenmäki, I., Nätti, S., Saraniemi, S.: Digitalization-enabled evolution of customer value creation: an executive view in financial services. J. Bus. Res. **146**, 504–517 (2022)
40. Hwang, S., Kim, J.: Toward a chatbot for financial sustainability. Sustainability **13**(6), 3173 (2021)
41. Song, M., Xing, X., Duan, Y., Cohen, J., Mou, J.: Will artificial intelligence replace human customer service? The impact of communication quality and privacy risks on adoption intention. J. Retail. Consum. Serv. **66**, 102900 (2022)
42. Sarbabidya, S., Saha, T.: Role of chatbot in customer service: a study from the perspectives of the banking industry of Bangladesh. Int. Rev. Bus. Res. Pap. **16**(1), 231–248 (2020)
43. Suhel, S.F., Shukla, V.K., Vyas, S., Mishra, V.P.: Conversation to automation in banking through chatbot using artificial machine intelligence language. In: 2020 8th International Conference on Reliability, Infocom Technologies and Optimization (Trends and Future Directions) (ICRITO), pp. 611–618. IEEE (2020)
44. Eren, B.A.: Determinants of customer satisfaction in chatbot use: evidence from a banking application in Turkey. Int. J. Bank Mark. **39**(2), 294–311 (2021)
45. Nguyen, D.M., Chiu, Y.-T.H., Le, H.D.: Determinants of continuance intention towards banks' chatbot services in Vietnam: a necessity for sustainable development. Sustainability **13**(14), 7625 (2021)
46. Elia, G., Margherita, A., Secundo, G.: Project management canvas: a systems thinking framework to address project complexity. Int. J. Manag. Proj. Bus. **14**(4), 809–835 (2021)
47. Candy, L.: Practice based research: a guide. CCS Rep. **1**(2), 1–19 (2006)
48. Sadek, M., Calvo, R.A., Mougenot, C.: Co-designing conversational agents: a comprehensive review and recommendations for best practices. Des. Stud. **89**, 101230 (2023)
49. Davis, F.D.: Perceived usefulness, perceived ease of use, and user acceptance of information technology. MIS Q. 319–340 (1989)
50. Al-Rahimi, W.M., Othman, M.S., Musa, M.A.: Using TAM model to measure the use of social media for collaborative learning. Int. J. Eng. Trends Technol. IJETT **5**(2), 90–95 (2013)
51. Fink, A.: The Survey Handbook. Sage (2003)
52. Jansen, H.: The logic of qualitative survey research and its position in the field of social research methods. In: Forum Qualitative Sozialforschung/Forum: Qualitative Social Research (2010)
53. Ahmed, M., Almotairi, M.A., Ullah, S., Alam, A.: Mobile banking adoption: a qualitative approach towards the assessment of TAM model in an emerging economy. Acad. Res. Int. **5**(6), 248 (2014)
54. Tran, A.D., Pallant, J.I., Johnson, L.W.: Exploring the impact of chatbots on consumer sentiment and expectations in retail. J. Retail. Consum. Serv. **63**, 102718 (2021)
55. Zogaj, A., Mähner, P.M., Yang, L., Tscheulin, D.K.: It's a Match! The effects of chatbot anthropomorphization and chatbot gender on consumer behavior. J. Bus. Res. **155**, 113412 (2023)
56. De Cicco, R., Silva, S.C., Alparone, F.R.: Millennials' attitude toward chatbots: an experimental study in a social relationship perspective. Int. J. Retail Distrib. Manag. **48**(11), 1213–1233 (2020)
57. Sheehan, B., Jin, H.S., Gottlieb, U.: Customer service chatbots: anthropomorphism and adoption. J. Bus. Res. **115**, 14–24 (2020)
58. Przegalinska, A., Ciechanowski, L., Stroz, A., Gloor, P., Mazurek, G.: In bot we trust: a new methodology of chatbot performance measures. Bus. Horiz. **62**(6), 785–797 (2019)
59. Davenport, T., Guha, A., Grewal, D., Bressgott, T.: How artificial intelligence will change the future of marketing. J. Acad. Mark. Sci. **48**, 24–42 (2020)
60. Wirtz, J., et al.: Brave new world: service robots in the frontline. J. Serv. Manag. **29**(5), 907–931 (2018)

61. Wirtz, J.: Organizational ambidexterity: cost-effective service excellence, service robots, and artificial intelligence. Organ. Dyn. **49**(3), 1–9 (2020)
62. Radford, A., Narasimhan, K., Salimans, T., Sutskever, I.: Improving language understanding by generative pre-training (2018)
63. Dwivedi, Y.K., et al.: Opinion paper: 'so what if ChatGPT wrote it?' Multidisciplinary perspectives on opportunities, challenges and implications of generative conversational AI for research, practice and policy. Int. J. Inf. Manag. **71**, 102642 (2023)
64. Eapen, T., Finkenstadt, D.J., Folk, J., Venkataswamy, L.: How generative AI can augment human creativity. Harv. Bus. Rev. **101**(4) (2023)
65. Fui-Hoon Nah, F., Zheng, R., Cai, J., Siau, K., Chen, L.: Generative AI and ChatGPT: applications, challenges, and AI-human collaboration **25**(3), 277–304. Taylor & Francis (2023)

Dynamic Capabilities in AI-powered Business Ecosystems

An Outline of an Emerging Research Agenda

Annika Steiber and Swapan Ghosh

Abstract The rise of artificial intelligence (AI) is transforming how firms innovate and compete, giving birth to AI-powered business ecosystems that span multiple organizations. This paper reviews highly cited research from the last five years on AI in business ecosystems through the lens of dynamic capabilities. Dynamic capabilities are a firm's ability to integrate, build, and reconfigure competencies in rapidly changing environments. They also provide a valuable perspective for understanding how organizations sense and seize AI opportunities and reconfigure their resources in ecosystem contexts. Key findings from recent studies were synthesized, highlighting that AI can amplify value creation through data-driven learning and network effects while introducing new value capture and governance challenges. Empirical evidence shows that leading firms leverage dynamic capabilities to orchestrate ecosystems and co-create value with partners and that an optimal level of AI utilization is needed for sustainable ecosystem performance. The paper will then discuss how companies develop new sensing, seizing, and transforming capabilities to drive AI-based business model innovation. The paper concludes with implications for theory and practice, suggesting that firms must strategically develop AI-related dynamic capabilities to thrive in increasingly AI-centric, inter-organizational business environments.

Keywords Ecosystem · Dynamic capability · AI · SDG · Industry innovation infrastructure

A. Steiber (✉)
Management Insights, Menlo Park, CA, USA
e-mail: drsteiber.managementinsights@gmail.com

S. Ghosh
California State University, Northridge, CA 91330, USA
e-mail: swapan.csun@gmail.com

A. Visvizi et al. (eds.), *Research and Innovation Forum 2025 Volume 1*,
Springer Proceedings in Complexity, https://doi.org/10.1007/978-3-032-09066-9_11

1 Introduction

Artificial intelligence (AI) is reshaping industry competition by enabling new business models and forming AI-powered business ecosystems—networks of firms, startups, partners, and users that co-create value through AI integration [1]. These ecosystems transcend traditional value chains, shifting competition from firm-level rivalry to collaborative, data-driven innovation. In such ecosystems, AI does not merely automate tasks but generates strategic differentiation by embedding intelligence into products, services, and decision-making processes. As data, algorithms, and digital infrastructure become shared resources, ecosystem boundaries blur. This evolution raises critical questions: How do firms sense and seize AI-driven opportunities alongside partners? What capabilities enable ongoing adaptation to fast-evolving AI landscapes?

This study adopts a dynamic capabilities lens to address these questions. Dynamic capabilities, defined as the firm's ability to purposefully create, extend, or reconfigure its resource base in response to environmental changes [2, 3], are critical for navigating AI-powered ecosystems. Unlike operational capabilities that support efficiency, dynamic capabilities foster innovation, agility, and strategic renewal. Recent scholarship emphasizes the importance of dynamic capabilities in leveraging AI for innovation and ecosystem orchestration [4]. AI acts as both a catalyst and a disruptor, transforming value creation, capture, and coordination mechanisms. Firms must thus reconfigure internal processes, collaborate across organizational boundaries, and align with shifting technological frontiers to sustain competitive advantage [5].

Despite growing interest, a structured synthesis of recent research at the intersection of AI, dynamic capabilities, and business ecosystems is lacking. This paper addresses this gap by reviewing high-impact studies from 2019 onward, offering insights across three focal areas: (1) *Theoretical foundations*, linking AI, dynamic capabilities, and ecosystems through the lens of sensing, seizing, and transforming; (2) *Empirical findings*, examining how firms develop capabilities to innovate and adapt within AI-powered ecosystems; (3) *Strategic implications*, identifying managerial practices and research opportunities to guide ecosystem engagement and innovation.

Subsequent sections detail the theoretical background, methodology for literature selection, and thematic synthesis of insights, culminating in a framework for future research and managerial action.

2 Methodology

This study employs a literature-based conceptual review methodology to examine how dynamic capabilities intersect with AI in business ecosystems. The review focuses on peer-reviewed academic publications from 2019 to 2024.

2.1 Source Selection and Search Strategy

High-impact journals such as the *Academy of Management Review*, the *Strategic Management Journal*, and *Business Research* were prioritized. Searches were conducted via Google Scholar, Scopus, and Web of Science using keywords including: "AI business ecosystem," "dynamic capabilities," "AI adoption capabilities," and "ecosystem orchestration AI." Citation snowballing was applied to capture foundational and derivative contributions.

2.2 Inclusion Criteria

Both conceptual and empirical studies were included, spanning qualitative case studies, quantitative surveys, and mixed-methods research. Select grey literature (e.g., *Harvard Business Review*) was integrated for practical context. All sources explicitly referenced AI-enabled transformation and dynamic capability constructs.

2.3 Analytical Techniques

A hybrid thematic analysis was adopted:

- **Based on Teece's dynamic capabilities framework, deductive coding** utilized pre-defined categories (sensing, seizing, reconfiguring).
- **Inductive coding** surfaced emergent themes (e.g., algorithmic governance, data-driven agility, AI-human co-evolution).

Cross-case comparison and pattern matching were employed to enhance analytical robustness, enabling evaluation of how theoretical expectations aligned with observed practices across industries and geographies with varying AI maturity.

2.4 Synthesis Strategy

A synthesizing logic integrates dispersed findings into a coherent conceptual structure. The review identified thematic convergence (e.g., criticality of data capabilities), contradictions (e.g., agility vs. control), and gaps (e.g., AI's effect on organizational learning routines).

2.5 Outcome

The result is a thematic synthesis that advances understanding of how AI reshapes the development and orchestration of dynamic capabilities in evolving business ecosystems.

3 Theoretical Background and Findings

3.1 Dynamic Capabilities and Business Ecosystems

Dynamic capabilities, defined as a firm's ability to integrate, build, and reconfigure internal and external competencies in dynamic environments [2, 3], are foundational for strategic adaptation. Teece's framework emphasizes three core processes: sensing opportunities, seizing them, and transforming organizational resources.

As firms transition from linear value chains to interconnected business ecosystems, dynamic capabilities must extend beyond firm boundaries. Ecosystem orchestration capabilities—such as aligning partner goals, resources, and governance—become critical for sustained value co-creation [7].

However, theoretical ambiguity persists. Critics argue that dynamic capabilities are inherently firm-centric and lose conceptual clarity in inter-organizational contexts [1]. Others advocate a distributed view, where capabilities are co-constructed with partners across the ecosystem [8]. This raises key challenges around agency, control, and coordination.

Further, while ecosystem leadership facilitates coherence, over-centralization can induce power asymmetries and stifle innovation among complementors [9]. Firms must balance centralized orchestration and decentralized experimentation to preserve ecosystem vitality.

3.2 AI as a Catalyst for Dynamic Capability Development

AI technologies, machine learning, intelligent automation, and predictive analytics enhance dynamic capabilities by augmenting sensing (pattern detection), seizing (data-driven decision-making), and transforming (process reconfiguration) [10]. AI thus acts as both an enabler and a disruptor of capability development.

AI integration, however, demands new learning routines, organizational redesign, and agile governance. Firms often confront internal capability gaps when embedding AI [11]. Scholars propose that dynamic capabilities must evolve into second-order capabilities to accommodate these shifts [12].

Platform-based ecosystems further reveal AI's duality. Data-driven network effects enhance personalization and scalability, yet concentrate value capture with platform leaders [13]. This can undermine ecosystem equity, creating tension between innovation gains and stakeholder disenfranchisement [14].

The literature also debates causality: do dynamic capabilities enable AI adoption, or does AI drive capability evolution? Some posit that strong dynamic capabilities are foundational for AI deployment [15]; others contend AI alters routines and stimulates capability emergence [16]. This bidirectional view remains theoretically unresolved.

3.3 Summary

AI expands dynamic capability functions but introduces governance, ethical, and distributional challenges. Effective orchestration of AI-powered ecosystems requires balancing innovation acceleration with collaborative equity and adaptive governance structures.

4 Discussion

Recent research confirms that dynamic capabilities are central to leveraging AI in business ecosystems. Several key themes emerge across studies.

4.1 AI Amplifies Value Creation and Demands New Capabilities

AI fosters value creation by enabling real-time learning, personalization, and automation. Data network effects, where AI systems improve through user interactions, generate a feedback loop that enhances platform utility and ecosystem engagement [6]. For example, AI in e-commerce ecosystems personalizes experiences, driving engagement and competitive advantage.

Firms need robust sensing (e.g., detecting algorithmic advances) and seizing (e.g., resource reallocation, AI investment) capabilities to harness AI. Empirical studies affirm that digital maturity and AI integration strengthen dynamic capabilities, enabling rapid business model transformation, e.g., from product-centric to AI-enabled services [4, 17, 18].

4.2 Orchestrating AI Ecosystems via Dynamic Capabilities

Ecosystem orchestrators platform leaders offering core AI functionalities leverage dynamic capabilities to align partners, scale offerings, and adapt configurations. Case studies (e.g., Amazon, iFLYTEK) show these leaders excel in sensing market signals, seizing technological shifts, and transforming ecosystem architectures [2].

Effective orchestration requires alliance management, regulatory navigation, and data governance. Strategic actions such as forming AI marketplaces or enabling third-party innovation depend on sensing external knowledge, seizing collaborations, and integrating outcomes—core tenets of dynamic capability theory [19].

4.3 Balancing Value Creation and Capture

AI ecosystems pose value distribution challenges. Overemphasis on value capture (e.g., monetizing data via opaque algorithms) may alienate users or partners [5]. Dynamic capabilities help orchestrators detect misalignments (sensing discontent), implement responsive changes (transforming governance), and preserve ecosystem health.

Capabilities such as user-centric design, legitimation, and building credibility and trust are critical in AI contexts. These are interpreted as higher-order sensing and seizing capabilities responsive to stakeholder expectations [6].

4.4 Stage-Dependent Capability Needs and Spillovers

Dynamic capabilities evolve with the ecosystem lifecycle. Sensing and legitimacy-building are critical; in maturity, transformation and renewal dominate [2]. Empirical work shows an inverted U-shaped relationship between AI intensity and ecosystem success, indicating that beyond a point, over-automation may reduce adaptability [20, 21].

Knowledge spillovers from AI-rich environments moderate this effect [22]. Access to external AI expertise and networks enhances sensing and integration capabilities, reinforcing the need for outward-facing dynamic capabilities.

4.5 Evolving Dynamic Capabilities for AI Adoption

Firms adapt to AI by extending existing capabilities rather than developing entirely new ones. For example, companies experienced in supply chain reconfiguration repurpose these capabilities to reengineer workflows around AI [3].

AI-specific capabilities, such as data governance, experimentation, and cross-functional integration, are layered atop existing routines. Organizations with prior agility transition more effectively, using learning, integration, and reconfiguration processes to drive innovation.

4.6 Sustaining Competitive Advantage in AI Ecosystems

Even as AI becomes ubiquitous, dynamic capabilities remain the foundation for long-term advantage. While AI tools are increasingly commoditized, the ability to align, deploy, and evolve them strategically is not.

Top-performing firms (e.g., Amazon, Google, Haier) distinguish themselves by embedding dynamic capabilities, ecosystem orchestration, ethical AI governance, and adaptive culture into their core strategy [23]. Responsible AI practices (e.g., transparency, ethical oversight) reflect reconfigured governance structures aligned with societal expectations.

4.7 Synthesis

AI and dynamic capabilities form a mutually reinforcing loop. Firms that leverage AI to enhance sensing, seizing, and transforming while managing ecosystem dynamics are better positioned to thrive in AI-driven environments (Table 1).

5 Conclusions and Implications

AI-powered business ecosystems represent a frontier where technology and strategy intersect. This review, grounded in recent high-impact research, underscores that dynamic capabilities provide a crucial strategic lens for navigating this frontier. In conclusion, the key takeaways and implications are:

- **Dynamic Capabilities as a Strategic Imperative**: Firms must cultivate dynamic capabilities (sensing, seizing, transforming) to successfully implement AI and adapt their business models. This involves investing in learning and agility at all organizational levels. Companies proficient in continuously reconfiguring resources can better capitalize on AI innovations and pivot when the ecosystem shifts.
- **Ecosystem Orchestration and Collaboration**: Leading or orchestrating an AI-centric ecosystem requires more than adopting new technology; it demands skillful coordination of partners and resources. Firms should develop capabilities in ecosystem governance, such as setting fair rules, facilitating knowledge sharing,

Table 1 AI-powered dynamic capability and business ecosystem

Theme	Key insights	Supporting references
AI amplifies value creation	AI drives personalization, automation, and innovation through data network effects. Firms need strong sensing and seizing capabilities to leverage AI	Almheiri et al. [4], Chen et al. [6], Verhoef et al. [17], Bughin et al. [18]
Ecosystem orchestration	Orchestrators use dynamic capabilities to manage partnerships, align goals, and adapt services. Capabilities include alliance management, regulatory alignment, and knowledge integration	Teece et al. [2], Ghosh [19]
Value creation versus value capture	Excessive focus on AI-based value capture may harm ecosystem cohesion. Dynamic capabilities help rebalance strategies through sensing user discontent and transforming governance	Clough and Wu [5], Chen et al. [6]
Stage-dependent capability needs	Ecosystem maturity influences capability emphasis, early stages need sensing; later stages require transformation. AI benefits peak at moderate use; spillovers moderate outcomes	Teece et al. [2], Mikalef et al. [20], Linde et al. [21]
Adapting dynamic capabilities for AI	Firms extend existing dynamic capabilities (e.g., supply chain agility) to support AI. AI-specific capabilities include data management, experimentation, and integration	Teece [3]
Sustaining competitive advantage	Dynamic capabilities differentiate AI adopters. Leading firms embed ecosystem thinking, agile governance, and ethical AI practices into strategic routines	Teece [3], Chen et al. [6], Linde et al. [21]

Source Authors

and aligning incentives. These capabilities enable the focal firm to co-create value with stakeholders and maintain ecosystem health.

- **Balanced Value Creation and Capture**: Managers must be mindful of AI's dual-edged nature in ecosystems. While AI can drive superior value for users (through personalization, efficiency, etc.), it can also tempt firms to overemphasize value capture (e.g., monetizing data aggressively). The research suggests that long-term success comes from balancing these forces. Practical implication: Incorporate feedback loops (customer and partner input) to recalibrate AI strategies to benefit all participants, thereby sustaining engagement and trust.
- **Context and External Knowledge Matter**: A firm's external environment—including regulatory context and availability of AI talent/knowledge—can amplify or constrain its success with AI. Policymakers and industry groups can facilitate ecosystems by supporting data-sharing standards and innovation sandboxes. For firms, an implication is to engage in alliances, consortiums, or open innovation

platforms to tap into knowledge spillovers. An active node in the broader AI knowledge network enhances a firm's dynamic capabilities.
- **Organizational Change and Leadership:** Embracing AI at scale often entails significant organizational change. Leaders should champion a vision of integrating AI into the company's strategy and culture. This includes retraining employees, redesigning workflows, and possibly restructuring units to be more cross-functional and data-driven. Dynamic managerial capabilities—the ability of leaders to adapt and steer the organization—are crucial in this regard. Companies should thus invest in leadership development focusing on digital savvy and change management.

For scholars, this synthesis reveals fertile areas for future research. Longitudinal studies could examine how dynamic capability development for AI unfolds over time or how small firms and startups (not just tech giants) build such capabilities in resource-constrained settings. There is also room to explore the dark side of AI in ecosystems—for instance, when does AI-driven automation erode a partner's willingness to collaborate, and how can dynamic capabilities mitigate such risks?

In conclusion, dynamic capabilities offer a robust framework to make sense of AI's transformative impact on business ecosystems. The studies reviewed consistently indicate that it is not just the possession of AI technology but the *orchestration and adaptive reconfiguration* of that technology within an ecosystem that differentiates winners from laggards. Firms that internalize this lesson will be better prepared to thrive in an era where ecosystems and the capabilities to navigate them determine strategic success.

6 Limitations

This study presents a literature-based conceptual review exploring how dynamic capabilities enable firms to navigate and orchestrate AI-powered business ecosystems. While offering integrative insights, the review is subject to several limitations. First, despite prioritizing high-impact and recent academic sources, the study may overlook relevant contributions from less prominent journals or non-English publications, potentially limiting comprehensiveness. Second, while rigorous, the interpretive thematic analysis introduces subjectivity inherent in qualitative synthesis. Third, given the rapid pace of AI and ecosystem evolution, recent findings may become temporally constrained as technologies and strategic practices evolve. Lastly, the paper is conceptual and does not present primary empirical evidence; thus, conclusions should be viewed as theoretical generalizations rather than empirically validated outcomes.

Future research should empirically test and extend the conceptual propositions identified here. Comparative, cross-industry, and longitudinal studies can assess how dynamic capabilities evolve in different AI maturity stages and regulatory environments. In particular, empirical investigations into second-order dynamic capabilities, data governance routines, and ecosystem trust mechanisms can deepen understanding of AI-driven transformation. Additionally, exploring AI adoption's ethical and social dimensions within ecosystem contexts may reveal new facets of strategic capability development. Such an inquiry will ensure AI-powered business ecosystems' sustainable and inclusive orchestration.

References

1. Jacobides, M.G., Brusoni, S., Candelon, F.: The evolutionary dynamics of the artificial intelligence ecosystem. Strateg. Sci. **6**(4), 412–435 (2021). https://doi.org/10.1287/stsc.2021.0148
2. Teece, D.J., Pisano, G., Shuen, A.: Dynamic capabilities and strategic management. Strateg. Manag. J. **18**(7), 509–533 (1997). https://doi.org/10.1002/(SICI)1097-0266(199708)18:7%3c509::AID-SMJ882%3e3.0.CO;2-Z
3. Teece, D.J.: Explicating dynamic capabilities: the nature and microfoundations of(sustainable) enterprise performance. Strateg. Manag. J. **28**(13), 1319–1350 (2007). https://doi.org/10.1002/smj.640
4. Almheiri, H.M., Ahmad, S.Z., Khalid, K., Ngah, A.H.: Examining the impact of artificial intelligence capability on dynamic capabilities, organizational creativity and organization performance in public organizations. J. Syst. Inf. Technol. **26**(1), 1–20 (2024). https://doi.org/10.1108/JSIT-10-2022-0239
5. Clough, D.R., Wu, A.: Artificial intelligence, data-driven learning, and the decentralized structure of platform ecosystems. Acad. Manag. Rev. **47**(1), 184–189 (2022). https://doi.org/10.5465/amr.2020.0222
6. Chen, Y., Chen, Y., Guo, Y.: Dynamic capabilities for value creation from AI innovation ecosystem: a comparison between iFLYTEK and Amazon. Int. J. Knowl. Manage. Stud. **14**(2), 212–236 (2023). https://doi.org/10.1504/IJKMS.2023.130031
7. Liu, Y., Sun, J., Zhang, Z., Wu, M., Sima, H., Ooi, Y.M.: How AI impacts companies' dynamic capabilities: lessons from six Chinese construction firms. Res. Technol. Manag. **67**(3), 64–76 (2024). https://doi.org/10.1080/08956308.2024.2324407
8. Gregory, R.W., Henfridsson, O., Kaganer, E., Kyriakou, H.: The role of artificial intelligence and data network effects for creating user value. Acad. Manag. Rev. **46**(3), 534–551 (2021). https://doi.org/10.5465/amr.2019.0178
9. Adner, R.: Ecosystem as structure: an actionable construct for strategy. J. Manag. **43**(1), 39–58 (2017). https://doi.org/10.1177/0149206316678451
10. Fjeldstad, Ø.D., Snow, C.C., Miles, R.E., Lettl, C.: The architecture of collaboration. Strateg. Manag. J. **33**(6), 734–750 (2012). https://doi.org/10.1002/smj.1968
11. Tiwana, A., Konsynski, B., Bush, A.A.: Research commentary—Platform evolution: coevolution of platform architecture, governance, and environmental dynamics. Inf. Syst. Res. **21**(4), 675–687 (2010). https://doi.org/10.1287/isre.1100.0323
12. Chakraborty, S., de Mauro, A., Scorrano, F.: Artificial intelligence and dynamic capabilities: a systematic literature review. Technol. Forecast. Soc. Chang. **178**, 121581 (2022). https://doi.org/10.1016/j.techfore.2022.121581
13. Wamba-Taguimdje, S.-L., Fosso Wamba, S., Kala Kamdjoug, J.R., Tchatchouang Wanko, C.E.: Influence of artificial intelligence (AI) on firm performance: the business value of AI-based transformation projects. Bus. Process. Manag. J. **26**(7), 1893–1924 (2020). https://doi.org/10.1108/BPMJ-10-2019-0411
14. Liu, H., Chen, J., Chou, T.C.: Resource orchestration for digital transformation: evidence from AI capabilities and strategic alignment. Information & Management **58**(4), 103447 (2021). https://doi.org/10.1016/j.im.2021.103447

15. Parker, G., Van Alstyne, M., Choudary, S.P.: Platform revolution: how networked markets are transforming the economy—and how to make them work for you. W. W. Norton & Company (2016)
16. Cusumano, M.A., Gawer, A., Yoffie, D.B.: The business of platforms: strategy in the age of digital competition, innovation, and power. Harper Business (2021)
17. Verhoef, P.C., Broekhuizen, T., Bart, Y., Bhattacharya, A., Qi Dong, J., Fabian, N., Haenlein, M.: Digital transformation: a multidisciplinary reflection and research agenda. J. Bus. Res. **122**, 889–901 (2021). https://doi.org/10.1016/j.jbusres.2019.09.022
18. Bughin, J., Seong, J., Manyika, J., Chui, M., Joshi, R. (2019). Notes from the AI frontier: modeling the impact of AI on the world economy. McKinsey Global Institute. https://www.mckinsey.com
19. Ghosh, S.: Developing artificial intelligence (AI) capabilities for data-driven business model innovation: roles of organizational adaptability and leadership. J. Eng. Tech. Manage. **75**, 101851 (2025). https://doi.org/10.1016/j.jengtecman.2024.101851
20. Mikalef, P., Boura, M., Lekakos, G., Krogstie, J.: Big data analytics capabilities and innovation: the mediating role of dynamic capabilities and moderating effect of the environment. Technol. Forecast. Soc. Chang. **168**, 120528 (2020). https://doi.org/10.1016/j.techfore.2021.120528
21. Linde, L., Sjödin, D., Parida, V., Wincent, J.: Dynamic capabilities for ecosystem orchestration: a capability-based framework for smart city innovation initiatives. Technol. Forecast. Soc. Chang. **166**, 120614 (2021). https://doi.org/10.1016/j.techfore.2021.120614
22. Chin, T., Ghouri, M.W.A., Jin, J., Deveci, M.: AI technologies affording the orchestration of ecosystem-based business models: the moderating role of AI knowledge spillover. Humanit. Soc. Sci. Commun. **11**(1), 1–13. Palgrave Communications. https://doi.org/10.1057/s41599-024-03003-7
23. Steiber, A., Alvarez, D.: AI-driven digital business ecosystems: a study of Haier's EMCs. Eur. J. Innov. Manage. (2024). https://doi.org/10.1108/EJIM-01-2024-0076

Acceptance of Generative AI in the Creative Industry: The Role of Brand Recognition, Trust, and Experience Moderation in AI Adoption

Dominika Weglarz, Cintia Pla-Garcia, and Ana Isabel Jiménez Zarco

Abstract This study investigates the adoption of Generative AI text-to-image tools in the creative industry using an extended Unified Theory of Acceptance and Use of Technology (UTAUT) model. The objective is to assess how brand recognition and trust, alongside UTAUT factors such as performance expectancy, effort expectancy, facilitating conditions, and social influence, influence behavioral intentions to adopt Generative AI tools. Additionally, it evaluates how prior experience with such tools moderates these relationships. While prior research underscores the significance of UTAUT factors in technology adoption, the role of brand equity factors—crucial for evaluating quality in industries with limited technological expertise—remains underexplored. This study addresses this gap by incorporating brand recognition and trust as key variables. Data were collected via standardized questionnaires from 208 creative professionals in the US and Spain, distributed through online creative communities. Partial Least Squares Structural Equation Modeling was employed to validate the hypotheses. Results indicate that performance expectancy, facilitating conditions, and brand trust positively influence behavioral intention, while brand recognition negatively impacts it. Social influence and effort expectancy showed no significant effects. Importantly, prior experience with Generative AI tools significantly moderates the relationship between brand trust and behavioral intention, with a stronger influence among experienced users, though it has limited effects on other variables. The findings enhance understanding of Generative AI adoption and offer practical insights for creative professionals and marketers, emphasizing the importance of brand trust and experience in driving adoption within the creative industry.

D. Weglarz (✉) · C. Pla-Garcia · A. I. J. Zarco
Universitat Oberta de Catalunya, Barcelona, Spain
e-mail: dweglarz@uoc.edu

C. Pla-Garcia
e-mail: cplag@uoc.edu

A. I. J. Zarco
e-mail: ajimenezz@uoc.edu

A. Visvizi et al. (eds.), *Research and Innovation Forum 2025 Volume 1*,
Springer Proceedings in Complexity, https://doi.org/10.1007/978-3-032-09066-9_12

Keywords Generative artificial intelligence (SDG-9) · UTAUT (SDG-9) · Brand recognition (SDG-8) · Brand trust (SDG 12) · Creative industries (SDG-8 · SDG-11)

1 Introduction

Creative industries are undergoing significant transformation driven by the Artificial Intelligence revolution. Understanding the adoption criteria among professionals is crucial for the success of this innovation. Generative AI is being integrated into creative workflows, offering potential benefits in productivity [32]. The expanding role of it prompted a research agenda to explore its impact on the creative workforce. While Generative AI surged in 2024, AI's early creative deployment began in 2016 with The Next Rembrandt, a 3D-printed painting generated from Rembrandt's portfolio.

The Next Rembrandt was the first attempt to use AI in the creative industries, but its adoption has increased exponentially since then. As AI-generated arts booms, galleries now showcase art with their prompt writers as authors. World prominent galleries have showcased such art. The Whitney Museum featured a collection of Harold Cohen's AARON, one of the first artificial intelligence programs for art [29]. Refik Anadol created an installation 'Unsupervised' showcased in MoMA, using the Museum's data set to create new forms of art based on 200 years of museum's collections [2].

These exhibitions demonstrate the integration of AI into the artistic process, pushing the boundaries of creativity. The adoption of Generative AI in the creative industry is growing because professionals use them to streamline tasks and enhance their productivity [26]. Research shows that using Generative AI could automate up to 26% of tasks in the arts, and design [12]. The launch of Chat GPT made the use of AI simple using natural language prompts and gained rapid adoption due to user-friendly interface [3, 8]. As the usage of Generative AI in the creative sector increases, the need for understanding its adoption grows, with the The Unified Theory of Acceptance and Use of Technology being widely used for adoption prediction [5, 21, 35]. UTAUT has been used to examine Generative Artificial Intelligence adoption in various industries such as health and safety [10], financial services [17], and lastly creative industries [21, 35]. UTAUT takes into consideration four key factors, performance expectancy, effort expectancy, social influence, and facilitating conditions [30]. The UTAUT model defines Performance Expectancy as the perceived benefits of new technology, Effort Expectancy as the ease of use, Social Influence as the perception of others' expectations to use it and Facilitating Conditions as the belief in the supporting infrastructure for its use [30]. Figure 1 presents the original UTAUT model.

Performance expectancy, effort expectancy, and social influence drive Generative AI adoption, while facilitating conditions has less impact [5, 21, 35]. According to Yin

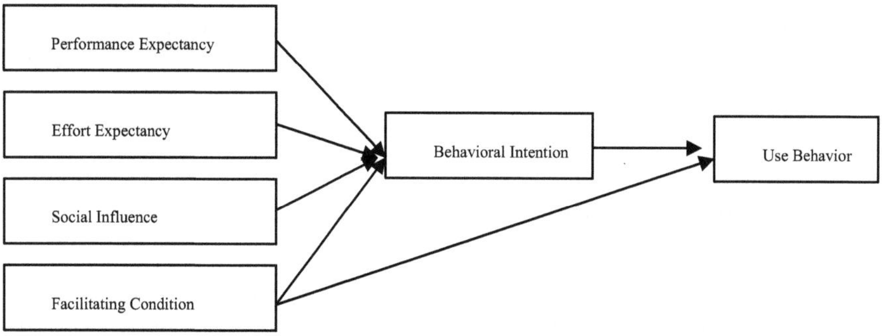

Fig. 1 Model of unified theory of acceptance and use of technology. *Source* Venkatesh et al. [30]

et al. [35] most creatives perceive this technology as a tool to complement their performance rather than replace their work completely [35]. Previous studies implementing UTAUT showed that Behavioral Intention to use AI, can be attributed strongly to Performance Expectancy [5, 19, 21], Effort Expectancy [21, 36], Facilitating Condition [21], and Social Influence [19, 21]. Additionally, studies showed how prior experience with generative AI tools positively influences adoption by enhancing UTAUT variables such as performance expectancy, effort expectancy, and social influence, while also moderating concerns around risk and trust. While these studies enhance the UTAUT framework by incorporating variables such as user trust and emotional responses, they overlook additional factors that could support users in industries characterized by limited technological expertise. One of the barriers of adoption of technologies in creative industries is lack of formal technical training, understanding of innovation, and lack of technical skills [1, 15]. This gap necessitates reliance on different ways to find reassurance of quality. Unfortunately, there is sparse research that investigates the role of brand recognition, and trust in acceptance and use of new technologies. In industries with limited technological knowledge, those factors are substitutes for direct quality evaluation. The confidence provided by brand factors supports decision-making and reduces uncertainty [1]. Brand recognition refers to the ability of consumers to identify a brand within the category. Brand recognition plays a key role in the initial stages of the consumer journey. It ensures that the brand is considered by the consumer [20]. High recognition enhances decision-making, and builds loyalty [16, 24]. Brand trust builds upon recognition, representing confidence in a brand's ability to deliver on promises [20]. Brand trust can lessen perceived risk, and foster customer loyalty [14]. Both recognition and trust impact brand use. Recognition facilitates decisions by creating familiarity and positive associations [22]. Trust enhances loyalty by aligning brand values with consumers' values reducing perceived risks [23]. This study seeks to address the research gap by investigating the role of brand-related factors in shaping the acceptance of Generative AI among individuals in creative industries, particularly to enhance quality assurance in sectors with limited technological expertise. This research is grounded in the UTAUT model because its validity has been proven in professional, business settings [7, 36]. It aims to address this gap by exploring how brand equity factors interact with established UTAUT constructs to shape user behavior and help creative professionals to adapt Generative AI effectively.

2 Materials and Method

This study adopts a design science research approach, which is well-suited for developing models that address practical problems. Given the emerging role of Generative AI in the creative industry, there is a need to construct a theoretical framework that not only explains adoption behavior but also offers guidance to practitioners for enhancing user engagement. Based on the objectives of this study, the following theoretical model and hypotheses were formulated.

2.1 Performance Expectancy

Performance Expectancy refers to the belief that using a Generative AI text-to-image tool will improve a creative professional's job performance. Recent studies showed that behavioral intention to use AI is highly determined by PE [5, 19, 21]. Individuals anticipate that AI technology will improve their work efficiency and effectiveness [35] Therefore, we structured the hypothesis as follows:

H1: Performance Expectancy positively influences the increase of Behavioral Intention of use of the Generative Artificial Intelligence text to image service-based brand.

2.2 Effort Expectancy

Effort Expectancy is the ease of using Generative AI text-to-image tools by creative professionals. To gain widespread adoption among new users, generative AI tools need to have user-friendly and intuitive interfaces. Prior research proved that EE positively influences behavioral intention, especially in fields where complex or non-intuitive interfaces may deter regular use [9]. In dynamic sectors such as the creative industry, ease of use is a crucial factor in encouraging the adoption of new tools and reducing barriers. This makes EE an important predictor in such industries [21, 36]. Therefore, we structured the hypothesis as follows:

H2: Effort Expectancy positively influences the increase of Behavioral Intention of using the Generative Artificial Intelligence text to image service-based brand.

2.3 Social Influence

Social Influence is the perception that important others believe professionals should use Generative AI text to image tools. Research from the design industry showed that individuals are likely to adopt AI tools if their use is supported by influential peers or leaders [9]. Furthermore, other research described that individuals from the music industry often adopt AI tools based on industry-wide trends, reflecting the impact of SI on AI adoption [19]. Therefore, we structured the hypothesis as follows:

H3: Social Influence positively influences the increase of Behavioral Intention of use of the Generative Artificial Intelligence text to image service-based brand.

2.4 Facilitating Condition

Facilitating Condition is the belief that the organizations and technical infrastructure exists to support the use of Generative AI text to image tools. It has been proven that Facilitating Conditions enhance the adoption of Generative AI [21]. Access to training, and educational resources positively influence users' confidence and facilitates AI adoption, indicating that FC is a predictor of Generative AI use [9]. Therefore, we structured the hypothesis as follows:

H4: Facilitating Condition positively influences the increase of Behavioral Intention of use of the Generative Artificial Intelligence text to image service-based brand.

2.5 Brand Recognition

Brand Recognition is the ability to recall the brand in a Generative AI text-to-image services category. Brand recognition serves as the initial step in the consumer journey and influences the decision to adopt a product [27]. Additionally, consumers tend to choose recognizable brands with a good performance history [18]. Therefore, we structured the hypothesis as follows:

H5: The more recognized the brand of Generative Artificial Intelligence text to image is, the more positive influence it has on Behavioral Intention of use.

2.6 Brand Trust

Brand Trust is a confidence in the brand's ability to deliver on its promises. Previous research showed that trust in an AI tool's ability to deliver the best task performance along with confidence in its promise delivery, are predictors of AI adoption [5]. Additionally, satisfaction and service quality impacts positively on AI service adoption [7]. Therefore, we structured the hypothesis as follows:

H6: The more trusted the brand of Generative Artificial Intelligence text to image is, the more positive influence it has on behavioral intention to use it.

2.7 Moderation Effect

Prior experience with technology moderates Performance Expectancy, Effort Expectancy, Facilitating Condition, as it reduces complexity, enhances utility and lowers the need of external support and resources [25]. When it comes to AI adoption, prior experience with similar technologies significantly influences the perceived ease of use and performance expectancy of that technology [25]. Other studies showed that the familiarity with specific AI tools enhances adoption. Experienced users may have formed trust because of the repeated use, strengthening the relationship between brand trust and use intention [19]. Therefore, we structured the hypothesis as follows:

H7: Prior experience with Generative AI tools moderates the relationship between Effort Expectancy, Social Influence, Facilitating Conditions, Brand Recognition, and Brand Trust, and their influence on Behavioral Intention to use Generative AI tools.

As a resume, the theoretical model is presented in Fig. 2.

2.8 Data Collection

We used standardized Qualtrics questionnaires to test the hypotheses, divided into Sociodemographic and factors influencing use intention. The second section followed [31], with added brand-related variables (Recognition, Trust) from [33], focusing on Adobe Firefly. The study targeted creative professionals in Spain and the US. Based on the criteria provided by Soper [28] and Westland [34] for conducting PLS-SEM, 224 responses were collected, of which 208 were complete and valid, meeting the PLS-SEM sample size requirements with a margin of error under 5%.

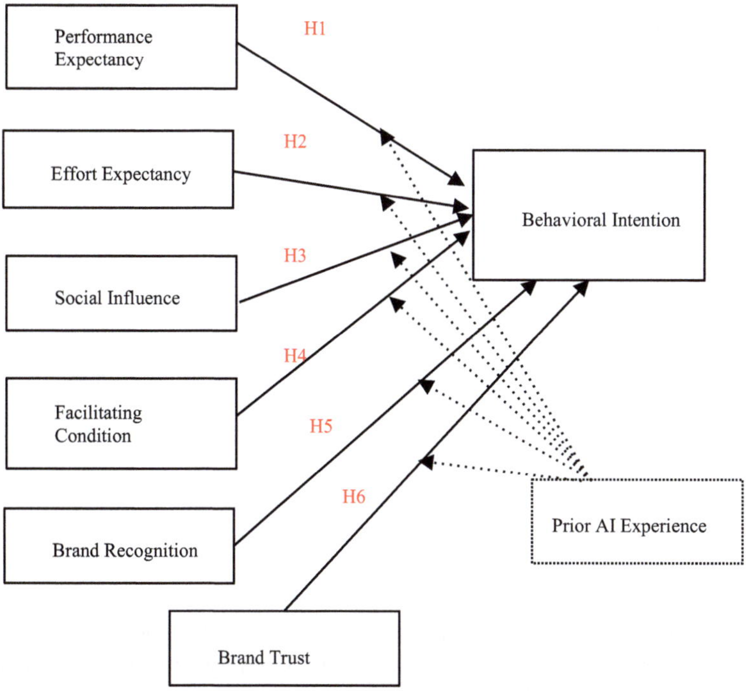

Fig. 2 Theoretical model and hypotheses

3 Result and Discussion

3.1 Data Analysis

Sample details are in Table 1.

To validate the proposed model, the data was subjected to analysis using the Partial Least Squares (PLS) method. Data were analyzed with SmartPLS 4.0, in two phases: measuring the model and analyzing the structural model.

3.2 Measurement Model Evaluation

First, we assessed items' validity following the 0.708 threshold for outer loadings [4, 11], ensuring indicators explained at least 50% of the variance in the construct. Using PLS-SEM, we evaluated the measurement model's reliability and validity, then modified it, as shown in Fig. 3.

Table 2 confirms item validity, with outer loadings for reflective constructs (PE, EE, SI, BR, BT, BI) exceeding 0.708, ensuring indicator reliability. Facilitating

Table 1 Description of the sample

	Sample	%
Age		
Less than 35	103	46
More than 35	121	54
Gender		
Female	93	41.5
Male	130	58
Prefer not to say	1	0.5
Place of residence		
Spain	104	46.5
US	104	46.5
Other	16	7
Experience with Gen AI		
Yes	138	62
No	86	38

Condition, being formative, doesn't meet this criterion. Composite reliability was acceptable, with all construct loadings above 0.7. Convergent validity, assessed with AVE, following the Fornell and Larcker criterion, showed all constructs exceeded the 0.5 threshold, confirming they explain at least 50% of the variance in their indicators.

Discriminant validity test was conducted using Fornell-Larcker criteria and Heterotrait-Monotrait Ratio. The results in Table 3, reveal that the correlations between constructs are lower than the square root of the AVE, suggesting that each construct shares more variance with its own indicators than with any other constructs in the model. All six reflective constructs demonstrate strong internal consistency and convergent validity.

We assessed multicollinearity for the FC variable using the Variance Inflation Factor (VIF), confirming no critical issues (values below 3, [11]). Table 4 shows that, while 4 out of 5 indicators had strong contributions (loadings > 0.5), FC1's low contribution was retained for conceptual relevance. Bootstrapping analysis confirmed statistical significance for 4 indicators ($p < 0.05$).

As shown in Table 5, discriminant validity was assessed using the HTMT ratio, which fell within acceptable thresholds.

3.3 Structural Model Evaluation

The structural model was evaluated by calculating path coefficients, p-values, T-values, Coefficient of determination (R2), and predictive relevance (Q2). As

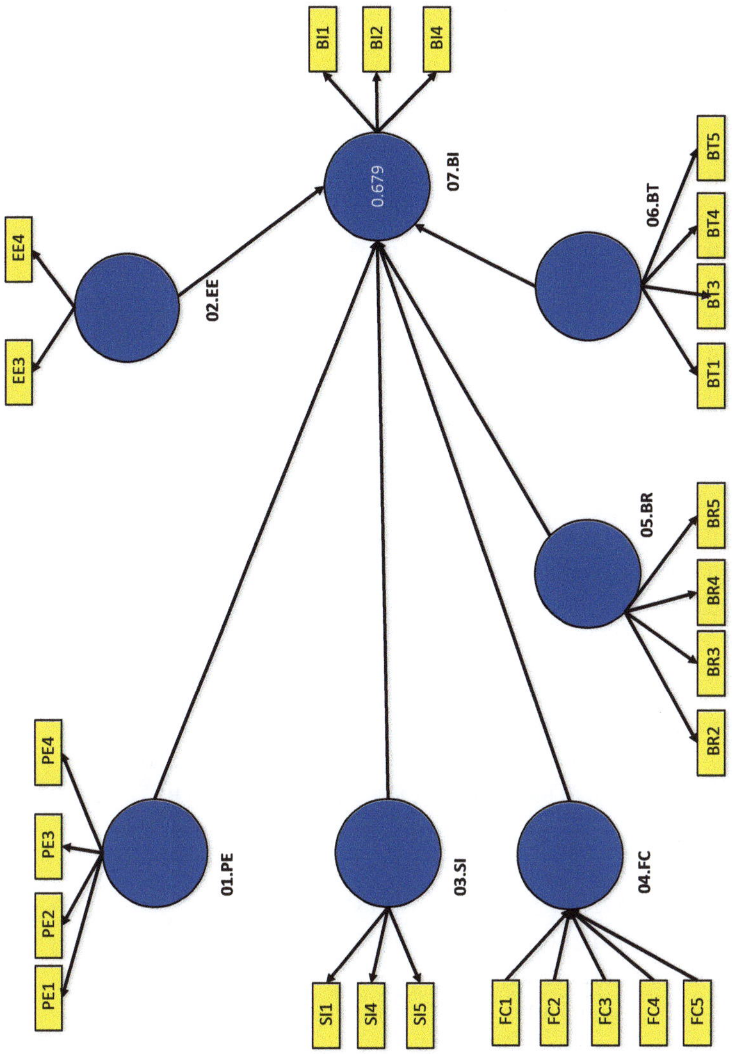

Fig. 3 Measurement model. *Source* Own elaboration

Table 2 Construct reliability and validity overview. *Source* Own elaboration

		Outer loading	Cronbach's alpha	Composite reliability (rho_a)	Composite reliability (rho_c)	AVE
PE	PE1	0.908	0.905	0.917	0.906	0.709
	PE2	0.842				
	PE3	0.900				
	PE5	0.877				
EE	EE3	0.899	0.775	0.776	0.775	0.633
	EE4	0.908				
SI	SI1	0.820	0.768	0.786	0.760	0.520
	SI4	0.827				
	SI5	0.824				
BR	BR2	0.863	0.919	0.933	0.917	0.738
	BR3	0.909				
	BR4	0.892				
	BR5	0.920				
BT	BT1	0.880	0.906	0.914	0.902	0.700
	BT3	0.874				
	BT4	0.873				
	BT5	0.902				
BI	BI1	0.909	0.872	0.876	0.874	0.698
	BI2	0.914				
	BI4	0.854				

Table 3 Discriminant validity—Fornell-Larcker criterion and Heterotrait-Monotrait ratio (HTMT). *Source* Own elaboration

	Fornell-Larcker criterion					Heterotrait-Monotrait ratio (HTMT)					
	PE	EE	SI	BR	BT	PE	EE	SI	BR	BT	BI
PE	0.842										
EE	0.637	0.795				0.639					
SI	0.711	0.517	0.721			0.713	0.515				
BR	0.324	0.340	0.474	0.859		0.332	0.332	0.494			
BT	0.337	0.287	0.512	0.809	0.837	0.347	0.285	0.532	0.827		
BI	0.892	0.613	0.679	0.290	0.384	0.890	0.612	0.669	0.288	0.382	

displayed in Table 6, the path coefficients for Performance Expectancy and Behavioral Intention (H1), Facilitating Condition and Behavioral Intention (H4), Brand Recognition and Behavioral Intention (H5), and Brand Trust and Behavioral Intention (H6) were 0.638, 0.137, −0.134, 0.147 with p-values 0.000, 0.011, 0.019, 0.017 respectively. Therefore, four out of six hypotheses were supported. The model

Table 4 Multicollinearity evaluation. VIF. *Source* Own elaboration

	VIF
FC1	1.022
FC2	1.314
FC3	1.280
FC4	1.223
FC5	1.127

Table 5 Heterotrait-Monotrait ratio (HTMT). *Source* Own elaboration

	01.PE	02.EE	03.SI	04.FC	05.BR	06.BT	08.BI
01.PE							
02.EE	0.639						
03.SI	0.713	0.515					
04.FC	0.577	0.620	0.509				
05.BR	0.332	0.332	0.494	0.333			
06.BT	0.347	0.285	0.532	0.377	0.827		
08.BI	0.890	0.612	0.669	0.608	0.288	0.382	

accounts for 67% of explained variance, indicating that the model offers a significant explanation for the variation in users' intentions to adopt the technology. The explained variance is shaped by the factors of Performance Expectancy, Facilitating Condition, Brand Recognition, and Brand Trust. The strongest predictors is Performance Expectancy ($\beta = 0.638$, $p < 0.001$).

The Q^2 value of 0.649 presents that the model has substantial predictive relevance for Behavioral Intention, demonstrating the model's strong ability to predict BI based on selected predictors (Fig. 4).

Table 6 Path coefficients. *Source* Own elaboration

	Model $R^2 = 0.679$ (67%) Model $Q^2 = 0.649$ (64%)		
	Path coefficient	P values	Support
H1: PE -> BI	0.638	0.000	H1 accepted
H2.EE -> BI	0.057	0.139	H2 rejected
H3 SI -> BI	0.084	0.088	H3 rejected
H4 FC -> BI	0.137	0.010	H4 accepted
H5. BR -> BI	−0.134	0.020	H5 partially accepted
H6.BT -> BI	0.147	0.017	H6 accepted

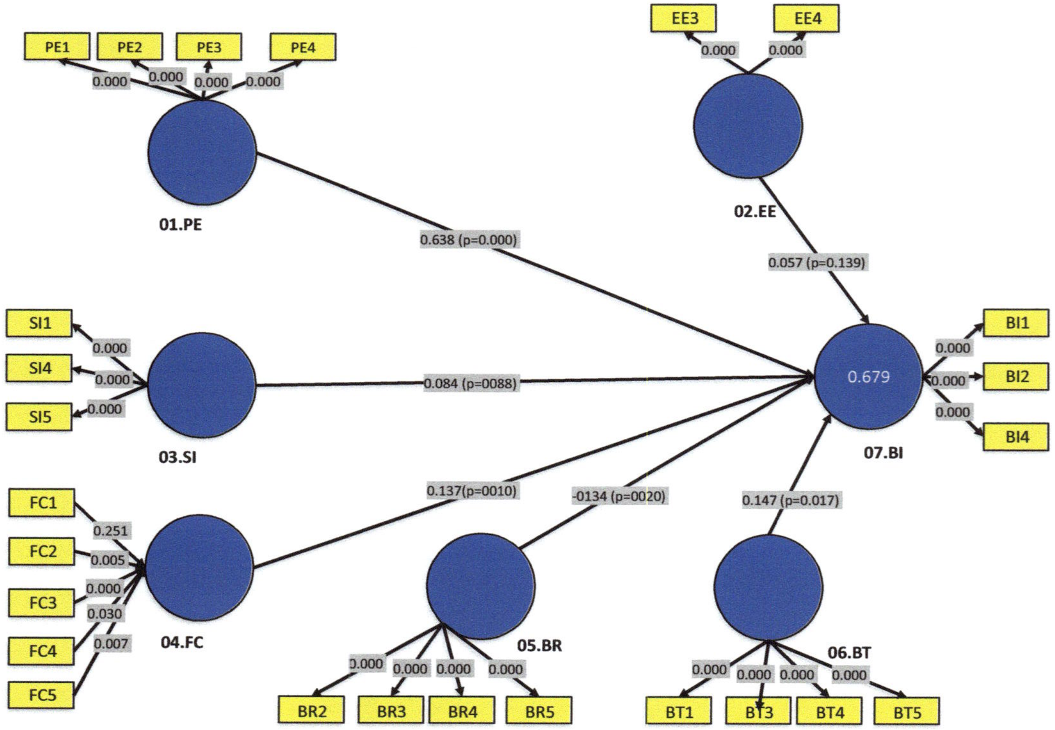

Fig. 4 Final structural model with path coefficients, p-values and R square. *Source* Own elaboration

3.4 Moderation

To establish if the strength of the relationship between the predictors, and BI depends on the third- moderator variable of prior experience, a moderation test was run. Details results in Table 7 demonstrate one statistically significant moderation. The interaction between Generative AI Experience and Brand Trust (BT) significantly predicts BI (0.499, p = 0.005p), meaning that for users with more experience with generative AI, the impact of BT on BI is stronger. In contrast, interactions involving other variables do not exhibit significant moderating effects, indicating their limited influence on BI within this model, only partially confirming H7.

The simple slope chart revealed that for individuals with prior generative AI experience (1), higher Brand Trust positively influences Behavioral Intention to use AI tools. For those without experience (0), Brand Trust has a slightly negative effect. This suggests that prior experience makes users more receptive to Brand Trust, while it has little impact on those without experience (Fig. 5).

Table 7 Moderation. *Source* Own elaboration

Gen AI experience × 01.PE -> 07.BI	−0.086	−0.094	0.226	0.378	0.705
Gen AI experience × 02.EE -> 07.BI	−0.115	−0.100	0.137	0.834	0.404
Gen AI experience × 03.SI -> 07.BI	0.132	0.128	0.138	0.959	0.338
Gen AI experience × 04.FC -> 07.BI	−0.049	−0.046	0.147	0.334	0.738
Gen AI experience × 05.BR -> 07.BI	−0.085	−0.092	0.182	0.467	0.641
Gen AI experience × 06.BT -> 07.BI	0.499	0.504	0.178	2.803	0.005

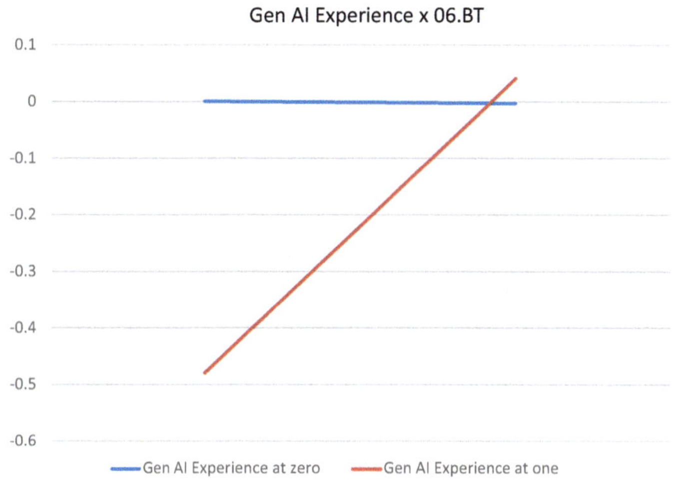

Fig. 5 Simple slope chart. *Source* Own elaboration

3.5 Discussion and Conclusion

This study aimed to identify key factors influencing generative AI adoption in the creative industry by extending the UTAUT with Brand Recognition and Brand Trust. Performance Expectancy emerged as the strongest predictor, followed by Facilitating Conditions and Brand Trust. Brand Recognition had a negative effect, likely due to Adobe's traditional brand associations. Effort Expectancy and Social Influence were not significant, supporting four of six hypotheses.

These results align with previous research [19, 21], which found that Performance Expectancy is a major driver of behavioral intention to adopt generative AI tools. Creative professionals seek tools that enhance productivity and streamline workflows, making Performance Expectancy the primary factor driving adoption. Facilitating Conditions, which include resources and support were also significant predictors of adoption [3]. Tools that offer comprehensive support and align with professional goals are more likely to be adopted. Brand Recognition traditionally increases the likelihood of engaging with new technologies [5], as it brings reassurance of quality when technical knowledge is limited. However, our study found a negative relationship between Brand Recognition and adoption of Adobe's generative AI tools. Professionals familiar with Adobe AI were less likely to adopt these tools, a result that contrasts with the usual positive correlation between brand recognition and intention to use [20]. This could be due to the misalignment between Adobe's established identity and new offering [13]. The familiarity with Adobe, paradoxically, might lead to heightened scrutiny and higher expectations, limiting the adoption. On the other hand, Brand Trust plays a crucial role in mitigating these barriers. Users tend to adopt tools from trusted brands, reducing perceived risks and increasing long-term adoption. Our study confirmed that Brand Trust positively influenced Behavioral Intention to use generative AI tools, especially for users with prior experience. For experienced users, higher Brand Trust was a strong predictor of adoption. However, Brand Trust had minimal impact on users without prior generative AI experience, suggesting that familiarity with the technology is essential for building trust. This implies that prior exposure to AI increases reliance on brand credibility in shaping adoption behavior. Generative AI may be seen as threatening human creativity, potentially deterring adoption [6]. Trusted brands can address this by reassuring users and promoting a positive view of AI's role.

From a managerial perspective brand strategies for AI tools should highlight performance benefits, streamline workflows, and offer strong support. Subbranding can differentiate new AI services from traditional offerings, as in Adobe's case. Building Brand Trust through ethical practices, transparency, and strategic partnerships will enhance adoption and ensure smooth integration into creative workflows.

The study faced some limitations. The study's sample size, though sufficient for PLS-SEM, was near the minimum threshold, which may limit generalizability. Focusing on Adobe may have influenced the negative impact of Brand Recognition, suggesting future research should explore other brands. Including additional consumer-based brand equity factors like Brand Loyalty and Perceived Quality could also yield deeper insights.

In conclusion, this study contributes to the growing body of research on AI adoption, specifically within the creative sector, and offers practical implications for both creative professionals and brand managers. By laying the groundwork for future research, it provides valuable insights that can inform theoretical and practical approaches for promoting AI tools in creative industries.

References

1. Abbasi, M., Vassilopoulou, P., Stergioulas, L.: Technology roadmap for the creative industries. Creative Ind. J. **10**, 40–58 (2017). https://doi.org/10.1080/17510694.2016.1247627
2. Anadol, R.: Refik Anadol on AI, Algorithms, and the Machine as Witness | Magazine. MoMA (2022). Retrieved January 20, 2025, from https://www.moma.org/magazine/articles/821
3. Anantrasirichai, N., Bull, D.: Artificial intelligence in the creative industries: a review. Artif. Intell. Rev. **55**, 589–656 (2022). https://doi.org/10.1007/s10462-021-10039-7
4. Bagozzi, R.P., Yi, Y.: On the evaluation of structural equation models. J. Acad. Mark. Sci. **16**(1), 74–94 (1988). https://doi.org/10.1007/BF02723327
5. Cabrera-Sánchez, J.P., Villarejo-Ramos, Á.F., Liébana-Cabanillas, F., Shaikh, A.A.: Identifying relevant segments of AI applications adopters: expanding the UTAUT2's variables. Telematics Inform. **58**, 101529 (2021). https://doi.org/10.1016/j.tele.2020.101529
6. Caporusso, N.: Generative artificial intelligence and the emergence of creative displacement anxiety. Res. Directs Psychol. Behav. **3**(1) (2023). https://doi.org/10.53520/rdpb2023.10795
7. Chatterjee, S., Chaudhuri, R., Vrontis, D., Thrassou, A., Ghosh, S.K.: Adoption of artificial intelligence-integrated CRM systems in agile organizations in India. Technol. Forecast. Soc. Chang. **168**, 120783 (2021). https://doi.org/10.1016/j.techfore.2021.120783
8. Chui, M., Hall, B., Mayhew, H., Singla, A.: The state of AI in 2022—and a half decade in review. McKinsey & Company (2022). https://bit.ly/40Lbo7o. Accessed 20 Jan 2025
9. Chuyen, N., Vinh, N.: An empirical analysis of predictors of AI-powered design tool adoption. TEM J. **12**(3), 1012–1021 (2023). https://doi.org/10.18421/tem123-28
10. De Almeida Pedro, E., Panizzon, M., & Weber, C.: OHS professionals AI adoption: A UTAUT research in Brazilian industry. In: 2023 15th IEEE international conference on industry applications (INDUSCON), pp. 850–857 (2023). https://doi.org/10.1109/INDUSCON58041.2023.10374850
11. Hair, J.F., Risher, J.J., Sarstedt, M., Ringle, C.M.: When to use and how to report the results of PLS-SEM. Eur. Bus. Rev. **31**(1), 2–24 (2019). https://doi.org/10.1108/EBR-11-2018-0203
12. Hatzius, J., Briggs, J., Kodnani, D., Pierdomenico, G., Goldman Sachs & Co. LLC.: The potentially large effects of artificial intelligence on economic growth. Goldman Sachs & Co. LLC (2023). https://bit.ly/3ElOOKX
13. Hem, L.E., De Chernatony, L., Iversen, N.M.: Factors influencing successful brand extensions. J. Mark. Manag. **19**(7–8), 781–806 (2003). https://doi.org/10.1362/026725703322498109
14. Hess, J., Story, J.: Trust-based commitment: multidimensional consumer-brand relationships. J. Consum. Mark. **22**(6), 313–322 (2005). https://doi.org/10.1108/07363760510623902
15. Irjayanti, M.: Implementing technology in creative industry (benchmarking study in developed countries). Asia Proc. Soc. Sci. **1**(1) (2023). https://doi.org/10.31580/apss.v1i1.155
16. Izza, A.M., Ardiansyah, M.N., Barkah, F., Romdonny, J.: Synergistic effects of content marketing and influencers marketing on the formation of brand awareness and purchase interest of TikTok Shop users (Cirebon City case study). Int. J. Soc. Serv. Res. **4**(5), 1339–1347 (2024). https://doi.org/10.46799/ijssr.v4i05.781

17. Jiang, J., Ma, J., Huang, X., Zhou, J., Chen, T.: Extend UTAUT2 model to analyze user behavior of China Construction Bank Mobile App. SAGE Open **14**(4) (2024). https://doi.org/10.1177/21582440241287070
18. Kathuria, S., Bansal, H., Balhara, S.: Impact of brand recognition on consumer attraction: a study of telecom sector. Res. World J. Arts Sci. Commer. **9**(1), 57–63 (2018). https://doi.org/10.18843/rwjasc/v9i1/07
19. Maican, C., Sumedrea, S., Tecău, A., Nichifor, E., Chițu, I., Lixăndroiu, R., Brătucu, G.: Factors influencing the behavioural intention to use AI-generated images in business. J. Organ. End User Comput. (2023). https://doi.org/10.4018/joeuc.330019
20. Medrano García, M.L., Yábar, D.: Impact of brand awareness on consumer loyalty. Sci. J. Appl. Soc. Clin. Sci. **3**(12), 2–9 (2023). https://doi.org/10.22533/at.ed.2163122307068
21. Menon, D., Shilpa, K.: Chatting with ChatGPT: analyzing the factors influencing users' intention to use OpenAI's ChatGPT using the UTAUT model. Heliyon **9**(11) (2023). https://doi.org/10.1016/j.heliyon.2023.e20962
22. Rios, R.E., Riquelme, H.: Sources of brand equity for online companies. J. Res. Interact. Mark. **4**(3), 214–240 (2010). https://doi.org/10.1108/17505931011070587
23. Roets, C.R.Q., Bevan-Dye, A.L., Viljoen, W.P.: Influence of social image and brand trust on mobile phone brand equity amongst African Generation Y students. Medit. J. Soc. Sci. **5**(21), 75–84 (2014). https://doi.org/10.5901/mjss.2014.v5n21p75
24. Rubio, N., Oubiña, J., Villaseñor, N.: Brand awareness–brand quality inference and consumers' risk perception in store brands of food products. Food Qual. Prefer. **32**, 289–298 (2014). https://doi.org/10.1016/j.foodqual.2013.09.006
25. Russo, D.: Navigating the complexity of generative AI adoption in software engineering. arXiv (Cornell University) (2024). https://doi.org/10.48550/arXiv.2307.06081
26. Sanchez, T.: Examining the text-to-image community of practice: why and how do people prompt generative AIs? In: Creativity and cognition (2023). https://doi.org/10.1145/3591196.359305
27. Sasmita, J., Mohd Suki, N.: Young consumers' insights on brand equity: Effects of brand association, brand loyalty, brand awareness, and brand image. Int. J. Retail Distrib. Manag. **43**(3), 276–292 (2015). https://doi.org/10.1108/IJRDM-02-2014-0024
28. Soper, D.S.: A-priori sample size calculator for structural equation models [Software] (2024). https://www.danielsoper.com/statcalc
29. The Whitney Museum of American Art.: Harold Cohen: AARON. Whitney Museum (2024). Retrieved January 20, 2025, from https://whitney.org/exhibitions/harold-cohen-aaron
30. Venkatesh, V., Morris, M., Davis, G.B., Davis, F.D.: User acceptance of information technology: toward a unified view. MIS Q. **27**(3), 425–478 (2003). https://doi.org/10.2307/30036540
31. Venkatesh, V., Thong, J.Y., Xu, X.: Consumer acceptance and use of information technology: extending the unified theory of acceptance and use of technology. MIS Q. **36**(1), 157–178 (2012). https://doi.org/10.2307/41410412
32. Vinchon, F., Lubart, T., Bartolotta, S., Gironnay, V., Botella, M., Bourgeois-Bougrine, S., Burkhardt, J., Bonnardel, N., Corazza, G.E., Glăveanu, V., Hanson, M.H., Ivcevic, Z., Karwowski, M., Kaufman, J.C., Okada, T., Reiter-Palmon, R., Gaggioli, A.: Artificial intelligence and creativity: a manifesto for collaboration. J. Creative Behav. **57**(4), 472–484 (2023). https://doi.org/10.1002/jocb.597
33. Wang, H., Wei, J., Yu, C.: Global brand equity model: combining customer-based with product-market outcome approaches. J. Prod. Brand. Manag. **17**(5), 305–316 (2008). https://doi.org/10.1108/10610420810896068
34. Westland, J.C.: Lower bounds on sample size in structural equation modeling. Electron. Commer. Res. Appl. **9**(6), 476–487 (2010). https://doi.org/10.1016/j.elerap.2010.07.003
35. Yin, M., Han, B., Ryu, S., Min, H.: Acceptance of generative AI in the creative industry: examining the role of AI anxiety in the UTAUT2 model. In: Lecture Notes in Computer Science, pp. 288–310 (2023). https://doi.org/10.1007/978-3-031-48057-7_18
36. Zhang, W.: A study on the user acceptance model of artificial intelligence music based on UTAUT. J. Korea Soc. Comput. Inf. **25**(1), 25–33 (2020). https://doi.org/10.9708/jksci.2020.25.06.025

Beyond Imagination: Design Thinking, Creativity and the Power of AI

Mario Testa, Maddalena Della Volpe, Adriana Apuzzo, and Stefania Manetti

Abstract The adoption of generative artificial intelligence and automation technologies is reshaping business dynamics, accelerating innovation, and transforming workforce skills. This study examines the role of creativity and Design Thinking in integrating AI into organizational processes, highlighting the synergy between artificial and human intelligence in both incremental and disruptive innovation. AI adoption generates a dual effect: on the one hand, it enhances productivity by automating routine tasks; on the other, it necessitates strengthening cognitive, social, and emotional skills to support adaptation in a rapidly evolving environment. The analysis underscores how DT, with its human-centered approach, can serve as a catalyst for the effective integration of AI, balancing analytical thinking with intuitive creativity. The proposed framework illustrates how AI can support each phase of the DT process, from data collection and interpretation to prototyping and solution validation. Furthermore, the study explores the impact of AI on knowledge-based work and the changing nature of decision-making, collaboration, and creative processes within enterprises. This research contributes to the academic debate on AI's impact in business, proposing an integrated approach that leverages the complementarity between human capabilities and artificial intelligence, outlining strategies for sustainable and inclusive innovation.

Keywords Generative AI · Design thinking · Innovation · Human-AI collaboration · Creativity · SDG-9

M. Testa · M. D. Volpe · A. Apuzzo (✉)
University of Salerno, Fisciano, Italy
e-mail: adapuzzo@unisa.it

M. Testa
e-mail: mtesta@unisa.it

M. D. Volpe
e-mail: mdellavolpe@unisa.it

S. Manetti
San Raffaele Roma University, Roma, Italy
e-mail: stefania.manetti@uniroma5.it

A. Visvizi et al. (eds.), *Research and Innovation Forum 2025 Volume 1*,
Springer Proceedings in Complexity, https://doi.org/10.1007/978-3-032-09066-9_13

1 Introduction

The introduction of generative artificial intelligence (GAI) and automation technologies is increasingly emerging as a strategic lever for innovation and productivity within business contexts. However, fully unlocking this potential depends on substantial investment in human capital and the swift, effective implementation of technological solutions [1, 2]. AI's transformative effects are already reshaping organizational structures, especially in cognitively intensive sectors such as marketing, customer service, software development, and R&D [3]. By 2030, the demand for highly skilled professionals—particularly in STEM fields—is projected to rise, while traditional administrative and operational roles will decline [2]. To respond to this shift in the labor market, organizations must rethink their training models, promoting reskilling and upskilling programs that not only develop technical capabilities but also enhance soft skills such as emotional intelligence, critical thinking, and problem-solving [4]. In an environment where skill life cycles are shortening, continuous learning becomes a strategic asset for maintaining business competitiveness [5, 6]. AI—especially through natural language processing applications—is redefining the boundaries of cognitive labor. It significantly boosts productivity when companies can reallocate tasks toward higher value-added activities. Some projections estimate annual productivity gains between 0.5 and 3.4% by 2040 [2]. Simultaneously, new opportunities are emerging in creative domains, where AI not only automates processes but also enhances decision-making and collaboration, enriching rather than replacing human capabilities [7, 8]. As a result, the skillsets in demand are evolving: analytical and creative thinking, alongside self-efficacy traits like resilience, adaptability, motivation, and self-awareness, are becoming essential [9, 10]. AI is increasingly viewed as a complement to human labor, augmenting qualities such as empathy, time management, and teamwork—alongside robust STEM knowledge [11, 12]. Current adoption trends confirm this outlook. Current adoption trends reinforce this view: by 2024, 72% of companies had already integrated AI into their operations, compared to 50% six years earlier, and 67% plan to invest further within the next three years [2]. The most significant outcomes have been observed in human resources—via analytical AI—and in marketing and sales, where generative AI has proven particularly effective [13]. However, the intersection of AI and creativity raises important questions about originality and human ingenuity. Today, AI is contributing to the creation of new markets, the transformation of business operations, and influencing consumer behavior. The so-called "creative economy"—driven by independent writers, artists, and musicians using AI tools—is now valued at an estimated $14 million annually. It exemplifies the growing interdependence between technology and creativity, but also poses new challenges in balancing automation and authenticity. [14, 15]. Three plausible future scenarios emerge [16]:

- co-creation—AI as a creative partner within a hybrid intelligence framework;
- content saturation—the risk that mass production of AI-generated content may marginalize human creativity;

- human valorization—the paradoxical outcome in which algorithmic overproduction increases the perceived value of rare, dynamic human expression.

Within this evolving landscape, the present study is guided by the following research question:

RQ_1: *To what extent can artificial intelligence support and assist human involvement in business innovation processes?*

To address this question, the study aims to explore and systematize the synergistic potential of creativity, Design Thinking (DT), and generative artificial intelligence (GAI) within business innovation processes. Through a selective review of academic literature and international reports, a theoretical framework is proposed that integrates GAI and DT. The findings suggest that collaboration between artificial intelligence and human creativity can strengthen incremental innovation by enhancing the effectiveness of the DT approach, without diminishing the unique value of human contribution.

2 Theoretical Insights on Innovation: Evolution, Drivers, and Implications

2.1 Innovation as a Driver of Global Growth and Resilience. An Overview of Key Trends

Innovation has become a core driver of economic and social development, especially amid intensifying global competition and rapid technological change [17, 18]. Across industries and firm sizes, innovation is increasingly adopted as a strategic tool to strengthen market positioning, enhance organizational resilience, and proactively respond to environmental volatility [19, 20]. No longer merely reactive, innovation is now recognized as a structured and deliberate process that enables firms to seize growth opportunities, solve problems creatively, and adapt to evolving economic, technological and geopolitical conditions—ultimately ensuring continuity and cultivating long-term competitive advantage [21, 22]. Empirical evidence confirms the strategic value of innovation: a recent global survey found that 97% of companies consider it essential for revenue growth and customer satisfaction, while also linking it to brand reinforcement and resilience in turbulent markets [23, 24]. This effect is especially significant in high value-added sectors, where R&D investments correlate with export performance and international competitiveness [1, 25, 26]. Leading firms adopt proactive innovation strategies—diversifying supply chains, adjusting business models, and responding agilely to macroeconomic disruptions. In this light, innovation emerges not as a siloed function but as a systemic organizational capability, increasingly synonymous with agility and adaptability [27]. At the global level, the innovation landscape is undergoing geopolitical realignment. Asian countries—most notably China, South Korea and Japan—have rapidly strengthened their positions through coordinated public policies, investment in innovation ecosystems, and strategic public–private partnerships [28–30]. China now leads in global patent filings, driven by a national strategy for technological autonomy [31], while South Korea exemplifies effective R&D translation into scalable, market-ready innovations. Western economies, though still influential, face rising competition. The United States remains a global leader in AI and biotechnology but must contend with emerging innovation hubs. Europe presents a more fragmented picture: Nordic countries (Denmark, Sweden, Finland) lead thanks to strong educational systems and effective academia-industry collaboration [32, 33], while Southern and Eastern European nations lag behind, underscoring the need for tailored policy support [27, 34]. Switzerland, despite being outside the EU, tops European innovation rankings due to its world-class research institutions and industrial strength. Nonetheless, Europe grapples with persistent challenges: R&D investment levels remain lower than in China and the U.S., and a fragmented single market hinders the scalability of innovative firms [35]. Italy shows moderate progress: between 2020 and 2022, 58.6% of firms with over ten employees engaged in innovation, with adoption rates surpassing 80% in sectors like pharmaceuticals, electronics, and automotive [36]. However, Italy's innovation ecosystem remains constrained by structural challenges—such as bureaucratic inefficiencies, inadequate digital infrastructure, limited access to capital, and continued brain drain—which hinder its global competitiveness [24, 37].

Global innovation trends reveal sharp regional disparities in adoption and strategic focus (see Table 1). Asia leads, with South Korea (85%) and China (80%) driven by robust R&D and targeted policies. The U.S. and Japan follow, while Europe (68%) shows internal divides—Nordic countries excel, but Southern and Eastern regions, including Italy (58.6%), lag behind [38, 39]. Worldwide, over 70% of innovation is incremental (see Table 2), offering stable but limited returns. Disruptive innovation—under 10%—is concentrated in countries with long-term strategies like China, South Korea, and the U.S., exemplified by firms like Apple and Tesla [38, 39]. Italy shows strength in pharma, electronics and traditional industries, but its SME-dominated system limits disruptive capacity. Overcoming barriers such as limited funding and weak digital infrastructure is key to expanding innovation beyond large firms like ENEL and Leonardo [37].

Incremental innovation prevails among European firms, yet disruptive advances—like blockchain and green hydrogen—are gaining traction, led by major players such as Siemens and BASF through collaborative initiatives [34].

2.2 *The Creative Foundations of Disruptive and Incremental Innovation*

In current innovation discourse, creativity is widely seen as essential to how organizations adapt and evolve within an increasingly complex global environment

Table 1 Innovation propensity

Region/Country	Percentage of innovative enterprises (%)	Leading sectors	Main challenges
South Korea	85	Robotics, semiconductors	Maintaining technological leadership
China	80	AI, advanced manufacturing	Long-term sustainability, Global patents
Japan	78	Electronics, automation	Ageing workforce
Nordic countries (Denmark, Sweden, Finland)	75–80	Green energy, biotechnology, ICT	Gap with Southern and Eastern Europe
United States	75	Technology, biotechnology	Competition with China and Asia
World	72	Technology, pharmaceuticals, energy	Access to funding, governance
European Union	68	Technology, manufacturing, R&D	Market fragmentation, digital infrastructure
Italy	58.6	Pharmaceuticals, electronics, automotive	Access to funds, bureaucracy, brain drain

Source Edited by the authors considering OECD, UNESCO and European Commission Reports [7, 28, 33]

[40, 41]. Rather than a linear process, the link between creativity and innovation unfolds as a layered dynamic—where intuition and insight, when paired with strategic thinking and structured execution, generate scalable and impactful solutions [42–44]. Creativity is not merely individual brilliance; it is a collective, cultural phenomenon fostered by interdisciplinary exchange and environments that promote exploration [45]. This creative potential expresses itself in different forms: sustainable (refining existing systems), incremental (improving current processes), and disruptive (breaking established paradigms) [46]. However, transforming ideas into innovation demands a balance—between freedom and discipline, flexibility and structure. Organizations that master this balance strengthen their resilience and capacity to shape alternative futures [47]. Despite its importance, creativity is often underdeveloped due to a lack of supportive environments [45, 48, 49]. Here, AI functions both as an enabler and a challenge. Its ability to process large data sets and generate novel combinations enhances creative capacity—provided organizations possess the absorptive capability to recognize and apply new knowledge [14, 50, 51]. Yet, amidst these developments, a fundamental question persists: *Can AI replace human creativity?*

Table 2 Comparison of incremental and disruptive innovation

Typo of innovation	Global	Europe	Italy
Incremental	70–80% of companies focus on gradual improvements. This approach is predominant across all sectors due to its more predictable return on investment	78% of European companies focus on incremental innovations, with a strong emphasis on sustainability and the improvement of products and processes	84% of Italian companies focus on incremental innovations, prioritizing quality, design, and process optimization
Disruptive	20–30% of global companies develop radical innovations, driven by major players like Apple, Tesla, and Amazon, capable of transforming markets and industries	22% of European companies are dedicated to disruptive innovations, with a higher concentration in Nordic countries and in sectors like energy and ICT	Only 16% of Italian companies introduce disruptive innovations, primarily in the automotive and pharmaceutical sectors. These are mostly limited to large enterprises

Source Edited by the authors considering Community Innovation Survey [25], Deloitte Insights for the Innovation in Europe [24] and PwC Innovation Report [23]

AI complements but does not replace human creativity. While algorithms can analyze patterns and generate suggestions, they lack emotional depth, contextual awareness, and intuitive judgment. Examples in music, art, and advertising (e.g., GumGum) show how AI can assist human creativity, but not replicate it [52].

The future of innovation lies in hybrid intelligence: a synergistic collaboration between human imagination and technological precision—where creativity remains the uniquely human force that gives meaning and direction to data-driven possibilities.

2.3 *The Algorithmic Artist: Exploring the Limits of AI in Creative Expression*

While AI demonstrates exceptional computational power, its creative capabilities remain inherently constrained by statistical models and existing datasets. This raises critical ethical and cultural concerns, including questions of intellectual authorship and the growing "algorithmization" of creativity [40, 53, 54]. Although AI can produce content that appears innovative, it often lacks the narrative depth, symbolic coherence, and emotional resonance that characterize authentic human creativity. Human creativity is grounded not only in novelty, but also in lived experience, cultural context, and meaning-making. As Cropley et al. [55] AI can address the "how" of creation but not the "why." Similarly, Rhodes' four-part framework—person, process, product, and press—highlights the complex interplay of individual and environmental factors that machines cannot replicate [56]. Where AI excels at data processing and generating plausible outputs, human creativity is rooted in intuition, reinterpretation, and visionary thinking. History demonstrates that paradigm-shifting innovations—such as Brunelleschi's linear perspective, Schönberg's twelve-tone system, Le Corbusier's modernist architecture, or Descartes' Cogito—were not the result of pattern recognition, but of subjective reflection and cultural dialogue. Even in science and literature, breakthroughs like Einstein's theory of relativity, Joyce's Ulysses, or Freud's psychoanalytic theories stem from experiential insight and conceptual risk—dimensions inaccessible to AI. These cases (see Table 3) affirm that while AI may enhance and accelerate creative processes, it cannot replace the intentionality, symbolic depth, or unprogrammed originality that define human creativity.

The future of innovation thus lies in hybrid intelligence—a synergy between algorithmic power and human imagination, balancing data with vision, and computation with meaning.

AI is highly effective in processing data and supporting incremental innovation, helping firms optimize existing processes with minimal risk. Over 70% of leading companies have adopted or plan to adopt AI for this purpose [32]. In Europe, AI adoption is growing: by 2020, 42% of firms had implemented at least one solution, and 58% planned further adoption [32]. However, disruptive innovation—which redefines paradigms—often stems from intuition, emotion, and cultural insight, aspects still beyond AI's reach (see Table 3). Human creativity, with its non-linear and counterintuitive nature, remains essential for breakthrough innovation.

2.4 *Balancing the Equation Through the of Integrating AI and Human Creativity for Innovation*

A 2024 survey by Aspen Institute Italia reports that 58% of companies have adopted AI, with 56% planning further expansion. However, adoption remains uneven—30.4% among large firms versus just 8% among businesses—highlighting disparities in resources and innovation readiness [57]. AI tools such as recommendation systems and predictive analytics are proving valuable, particularly in early innovation phases like ideation and prototyping. Over 60% of manufacturing firms using AI report improved performance [39], while 65% of businesses see enhanced customer satisfaction [28, 39]. AI is increasingly viewed not as a tool of automation, but as a creative collaborator that amplifies human potential. It supports incremental innovation by combining efficiency with adaptability. Yet, its rise raises important questions about the future of work. In a world of automation, human contributions—empathy, creativity, storytelling—gain strategic importance. A vivid example is a driver in Rome who transformed a routine trip into a rich cultural narrative, demonstrating how emotional resonance and meaning remain uniquely human. These capacities can't be replicated by machines, underscoring the need for a human-AI partnership.

Table 3 Human creativity versus AI: limitations in disruptive innovations across disciplines

Discipline	Innovation	AI limitations	Proponents
Art	The invention of linear perspective during the Renaissance by Filippo Brunelleschi transformed the way space was represented	AI would struggle to conceive such a revolutionary concept without a profound symbolic and cultural understanding of the need to represent space realistically, tied to humanistic evolution	
Music	Arnold Schönberg's break with traditional harmony through twelve-tone composition redefined musical language	AI, based on existing data, could replicate or vary known models but would find it challenging to conceive a completely new and deliberately unconventional system without preexisting harmonic rules	
Architecture	The introduction of modernism by Le Corbusier, with the idea of the house as a "machine for living," challenged decorative traditions	AI might optimize designs based on functional needs, but Le Corbusier's intuition stemmed from a cultural dialogue with industrial movements and philosophical interpretations of human space-elements hard to replicate without cultural awareness	
Philosophy	René Descartes' cogito ("Cogito, ergo sum") redefined the understanding of being and human thought	AI, incapable of experiencing subjective consciousness, could not have formulated an idea so deeply rooted in reflection on the nature of thought and individual awareness	
Sciences	Albert Einstein's theory of relativity revolutionized physics by breaking with established Newtonian laws	While AI can process vast amounts of data, formulating the theory of relativity required speculative and counterintuitive thinking that went beyond the empirical evidence of the time	

(continued)

Table 3 (continued)

Discipline	Innovation	AI limitations	Proponents
Literature	James Joyce's Ulysses overturned traditional narrative by introducing the stream-of-consciousness technique	AI could mimic a style like the stream of consciousness but would struggle to originate it, as this type of innovation arises from deep reflection on human interiority and the limits of conventional narrative forms	
Psychology	Sigmund Freud's introduction of psychoanalysis opened new perspectives on the unconscious mind	AI could analyze behavioral data, but Freud's insights were rooted in observation, self-reflection, and profoundly human interaction-capabilities beyond computational models	

Source Created by the authors

As Creely [58] emphasizes, AI follows two paths: non-intrusive (enhancing human tasks) and disruptive (redefining creativity and knowledge). Both demand human guidance. The future of innovation lies in this hybrid model, where technology enables and humans give meaning—placing creativity at the core of augmented intelligence.

3 Methodology

3.1 Abductive Methodology and Narrative Review for the Construction of an Emerging Conceptual Framework

From a methodological standpoint, this study adopts a theoretical-conceptual approach, grounded in a selective and narrative review of academic literature and international reports in the fields of management, innovation, creativity, artificial intelligence, and design. The research follows an abductive-conceptual logic, aiming to construct an exploratory theoretical framework that synthesizes fragmented knowledge and contributes to the structuring of an emerging research domain.

The analysis focuses on key theoretical contributions and emblematic case studies related to the evolution of creativity in business contexts and the potential of generative artificial intelligence to produce original content. The methodological core of the study lies in the development of a conceptual framework describing the integration of AI within the five phases of DT (Empathize, Define, Ideate, Prototype, Test). This model is supported by conceptual evidence and practical cases drawn from institutional sources and industry studies.

4 Results

4.1 *Design Thinking Meets AI: The Human-Centered Solutions*

In today's innovation discourse, any reflection on creativity naturally leads to the concept of DT—a user-centered methodology that integrates empathy, ideation, and experimentation to creatively address complex problems. Within this framework, AI emerges as a strategic partner, capable of enhancing design phases through data analysis, insight generation, and optimization of decision-making processes. The intersection of AI and DT thus opens promising scenarios for the development of increasingly personalized, sustainable, and user-oriented solutions. Originally developed in the field of industrial design, DT has expanded into business management due to its ability to tackle so-called wicked problems—ambiguous and multidimensional challenges. The model integrates three essential dimensions: desirability for the user, technological feasibility, and economic viability [59]. Introduced into the business world by David Kelley and later reinforced by institutions such as Stanford and the Hasso Plattner Institute [60], DT is built on multidisciplinary teams, collaborative environments, continuous iteration, and a radically human-centered approach [61]. The synergy between AI and DT is particularly effective due to their shared emphasis on personalization and user centrality. AI algorithms—especially those based on machine learning and generative models—not only support predictive analytics and behavioral pattern recognition but also assist designers in idea generation, prototype validation, and decision-making optimization [62]. The integration of AI's quantitative analysis with the qualitative intuition of DT produces tangible added value, especially in uncertain, high-variability contexts (see Table 4) [63]. Each phase of the DT process can be strengthened by the contribution of AI:

- Empathy: AI enables the analysis of large volumes of data (behavioral, textual, biometric), uncovering latent needs, sentiments, and user preferences with depth otherwise unattainable. Natural language processing, sentiment analysis, and opinion mining transform raw data into strategic insights.

Table 4 The role of AI in enhancing design thinking

Design thinking phases	Positive impact of AI for incremental innovation	Examples of incremental innovation
Empathize	*Predict and interpret*: AI analyses user data to reveal patterns, preferences, and hidden needs, aiding design teams with data-driven insights	Sogei leverages AI to analyze user behavioral data, enhancing the understanding of their needs and optimizing the design of public services [65]
Define	*Extract, synthesize and prioritize*: AI aggregates and processes complex data, clarifying design problems and uncovering hidden priorities for strategic structuring	Assist digital utilizes AI to identify patterns in customer interaction data, allowing for precise problem definition and improved solutions [65]
Ideate	*Suggest and augment*: AI enhances design capabilities by generating solutions, anticipating scenarios, and providing innovative insights through data and predictive models	The Design Thinking Observatory at Politecnico di Milano highlights how generative AI supports the ideation phase, offering new perspectives based on data and emerging trends [66]
Prototype	*Simulate and optimize*: AI accelerates prototyping by reducing costs and time, using simulations and adaptive algorithms to optimize design performance	BMW employs machine learning to analyze prototype test data, optimizing vehicle design and reducing development time [67]
Test	*Evaluate, adapt and learn*: AI converts user feedback into insights, enabling adaptive systems to refine solutions and enhance user alignment	BBVA enhanced its mobile banking app through Design Thinking and AI, earning international recognition for improving user experience [68]

Source Created by the authors

- Problem definition: AI's ability to aggregate and synthesize heterogeneous information helps refine the design challenge, prioritizing issues with greater precision and relevance.
- Ideation: Here, AI becomes a true co-creator. Generative algorithms can propose alternative scenarios, suggest innovative solutions, and inspire new design directions based on historical data, emerging trends, and predictive models. Rather than replacing human creativity, AI amplifies its divergent potential.
- Prototyping: Predictive simulations and adaptive algorithms help reduce development time and cost by anticipating usability or functionality issues. AI also enables fast iteration of prototypes based on specific design parameters.
- Testing: Leveraging machine learning, AI can rapidly analyze user feedback, dynamically adapt solutions, and anticipate future challenges through predictive analytics—fostering a process of continuous improvement.

During The convergence between DT and AI goes beyond a purely instrumental relationship. As Poleac [64] observes, tools like MidJourney or ChatGPT provide designers with opportunities for creative exploration, enriching divergent thinking and freeing up time for more conceptual work. This methodological alliance opens new avenues not only for product and service innovation but also for entrepreneurial education: integrating AI and DT into training programs means equipping future innovators with a hybrid mindset, capable of combining analytical rigor with design sensibility. The collaboration between AI and DT shapes a dynamic ecosystem in which technology enhances—but does not replace—human creativity. On the contrary, it expands its boundaries, supporting the development of more empathetic, inclusive, and effective solutions.

Considering, therefore, these aspects, we present the following conceptual framework (Fig. 1), in which AI can be integrated into the DT process to foster incremental innovation. In this perspective, AI does not replace human creative thinking but enhances its effectiveness through advanced analytical capabilities, predictive simulations, and decision-support tools. The goal is to make DT more structured, adaptive, and data-driven, overcoming the traditional limitations associated with subjectivity and the complexity of information management.

At the core of this model lies the assumption that incremental innovation, understood as the progressive improvement of products, services and processes, can be significantly accelerated through the interaction between human creativity and computational analysis. Specifically, the framework is structured around the five canonical phases of DT highlighting how each stage can be enhanced through AI integration.

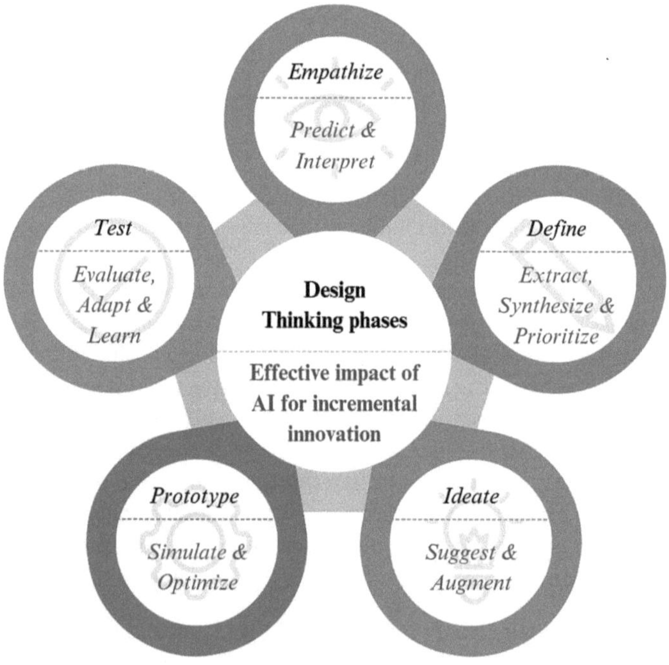

Fig. 1 Conceptual framework. *Source* Edited by the authors

The proposed framework redefines DT as a hybrid process, where artificial intelligence acts as a facilitator of incremental innovation, enhancing the effectiveness and precision of each methodological phase. Adopting this model allows organizations to develop faster, data-driven, and adaptive solutions while maintaining the centrality of human experience in value creation. The emergence of DT, based on the interaction between AI and creative intelligence, thus represents a new frontier for innovation, offering a competitive advantage to companies that strategically and consciously integrate these tools into their processes.

5 Discussion

Our study develops a conceptual framework integrating three key pillars of contemporary innovation: creativity as a transformative impulse, DT as a human-centered methodology, and artificial intelligence as a disruptive driver reshaping business strategies. AI accelerates digital transformation and, when combined with DT, enhances the ability to translate creative insights into value. Rather than replacing human creativity, AI complements it—expanding its operational scope while remaining limited in emotional depth, relational understanding, and complex cognition.

This synergy calls for a new educational paradigm—Education 5.0—that unites technological precision with empathic design. The integration of AI and DT fosters personalized, adaptive, and collaborative learning, shifting focus from technical proficiency to broader skills such as creativity, critical thinking and problem solving.

In this way, this co-creative approach repositions technology not as a substitute for human ingenuity, but as a catalyst for augmented humanity.

6 Conclusions

The integration of AI into creative processes is not merely a technical innovation, but a transformation in how organizations generate, interpret, and harness creative thinking. Far from being just a functional extension, AI introduces a shift in perspective that redefines the very logics of innovation, urging a reconsideration of the balance between rationality, intuition, and human interaction.

One of the most significant insights to emerge concerns the paradoxical nature of AI as an enabling technology. While it does not entirely replace existing paradigms, it acts as a systemic discontinuity: disrupting established practices, reshaping operational priorities, and altering the ways in which solutions are explored. AI takes on the characteristics of a disruptive innovation that, rather than aiming for total rupture, accelerates pathways of incremental innovation. This hybrid dynamic is particularly relevant in organizational contexts seeking agile and progressive transformation.

Regardless of the application domain—be it educational, productive or creative—the proposed approach is rooted in a collaborative logic, focused on co-creation and the shared generation of value. The synergy between AI and DT functions as both a cognitive and operational framework that merges advanced data analysis with human-centered design sensitivity. This convergence enhances not only the effectiveness of decision-making processes, but also their ability to dynamically respond to changing, complex, and context-specific needs. However, the success of this integration is contingent upon non-trivial enabling conditions. Technological limitations, organizational rigidity, and cultural resistance may still hinder widespread and informed adoption of these solutions. For the collaboration between AI and human creativity to become a truly generative asset, it is essential to foster ecosystems that support interdisciplinarity, design responsibility, and lifelong learning.

What makes this scenario promising is not the replacement of human ingenuity, but its amplification. AI can multiply the trajectories of the possible, but it remains the task of human intelligence to choose, interpret, and create meaning. In an environment where standardization can be automated, the true competitive advantage will increasingly lie in the ability to generate difference—through vision, empathy, and imagination.

This contribution is conceptual and theoretical in nature. As such, it does not provide empirical validation of the proposed framework, nor does it directly explore the relationship between artificial intelligence and disruptive creativity. In particular, there remains a need for deeper reflection on how the absence of embodied experience and emotional sensitivity in AI limits its capacity to generate radical breaks from existing paradigms.

Future research should thus move beyond the analysis of operational synergies between AI and creativity, and investigate the cultural, symbolic and political implications that the adoption of artificial intelligence brings to managerial contexts.

References

1. OECD.: OECD Employment Outlook (2023). https://doi.org/10.1787/08785bba-en
2. McKinsey.: The state of AI in early 2024: Gen AI adoption spikes and starts to generate value (2024)
3. Visvizi, A.: Artificial intelligence (AI) and sustainable development goals (SDGs): exploring the impact of AI on politics and society. Sustainability **14**(3), 1730 (2022). https://doi.org/10.3390/su14031730
4. Troisi, O., Visvizi, A., Grimaldi, M.: Rethinking innovation through industry and society 5.0 paradigms: a multileveled approach for management and policy-making. Eur. J. Innov. Manag. **27**(9), 22–51 (2024). https://doi.org/10.1108/EJIM-08-2023-0659
5. Barra, C., Grimaldi, M., Muazzam, A., Troisi, O., Visvizi, A.: Digital divide, gender gap, and entrepreneurial orientation: how to foster technology adoption among Pakistani higher education students? Socioecon. Plann. Sci. **93**, 101904 (2024). https://doi.org/10.1016/j.seps.2024.101904
6. Visvizi, A., Lytras, M.D., Aljohani, N.R.: Research and innovation forum 2020: disruptive technologies in times of change. Springer Nature (2021)

7. UNESCO.: Reimagining our futures together. A new social contract for together education (2021)
8. Visvizi, A.: Artificial Intelligence (AI): Explaining, Querying, Demystifying, pp. 13–26 (2021). https://doi.org/10.1007/978-3-030-88972-2_2
9. LinkedIn.: Future of work report (2023)
10. Visvizi, A.: Artificial Intelligence and Its Contexts. Springer International Publishing, Cham (2021). https://doi.org/10.1007/978-3-030-88972-2
11. World Economic Forum.: Future of jobs report (2023)
12. Visvizi, A., Troisi, O.: Effective management of the smart city: an outline of a conversation. In: Managing Smart Cities. Springer International Publishing, Cham, pp. 1–10 (2022). https://doi.org/10.1007/978-3-030-93585-6_1
13. Lytras, M.D., Visvizi, A.: Artificial intelligence and cognitive computing: methods, technologies, systems, applications and policy making. Sustainability **13**(7), 3598 (2021). https://doi.org/10.3390/su13073598
14. Bouschery, S.G., Blazevic, V., Piller, F.T.: Augmenting human innovation teams with artificial intelligence: exploring transformer-based language models. J. Prod. Innov. Manag. **40**(2), 139–153 (2023). https://doi.org/10.1111/jpim.12656
15. Wingström, R., Hautala, J., Lundman, R.: Redefining creativity in the era of AI? Perspectives of computer scientists and new media artists. Creat. Res. J. **36**(2), 177–193 (2024). https://doi.org/10.1080/10400419.2022.2107850
16. De Cremer, D., Bianzino, N.M., Falk, B.: AI and machine learning how generative AI could disrupt creative work. Harv. Bus. Rev. (13) (2023)
17. Rodríguez-Espíndola, O., et al.: The role of circular economy principles and sustainable-oriented innovation to enhance social, economic and environmental performance: evidence from Mexican SMEs. Int. J. Prod. Econ. **248**, 108495 (2022). https://doi.org/10.1016/J.IJPE.2022.108495
18. Visvizi, A.: Computers and human behavior in the smart city: issues, topics, and new research directions. Comput. Human Behav. **140**, 107596 (2023). https://doi.org/10.1016/j.chb.2022.107596
19. Huang, Y., Li, K., Li, P.: Innovation ecosystems and national talent competitiveness: a country-based comparison using fsQCA. Technol. Forecast. Soc. Change **194**, 122733 (2023). https://doi.org/10.1016/J.TECHFORE.2023.122733
20. Zhu, Z.Y., Xie, H.M., Chen, L.: ICT industry innovation: knowledge structure and research agenda. Technol. Forecast. Soc. Change **189**, 122361 (2023). https://doi.org/10.1016/J.TECHFORE.2023.122361
21. Abbas, J.: Does the nexus of corporate social responsibility and green dynamic capabilities drive firms toward green technological innovation? The moderating role of green transformational leadership. Technol. Forecast. Soc. Change **208**, 123698 (2024). https://doi.org/10.1016/J.TECHFORE.2024.123698
22. Visvizi, A.: Special issue on the Covid-19 inflicted digital transformation in ambient and more distant environments. J. Ambient. Intell. Humaniz. Comput. **14**(11), 14629–14630 (2023). https://doi.org/10.1007/s12652-023-04697-4
23. PwC Italia.: PwC innovation report (2018). Accessed 12 Dec 2024. https://www.pwc.com/it/it/publications/assets/docs/survey_innovation_report_eng.pdf
24. Deloitte.: Innovation in Europe (2019). Accessed 13 Jan 2025. https://www2.deloitte.com/content/dam/insights/us/articles/DE_897_Innovation-in-Europe/DI_Innovation-In-Europe.pdf
25. European Commission.: Community innovation survey 2022-key indicators statistics explained (2024). Accessed 15 Dec 2024. https://ec.europa.eu/eurostat/statisticsexplained/
26. ISTAT.: L'innovazione nelle imprese (2024)

27. European Commission.: I risultati dell'Europa in materia di innovazione, Brussels (2024). Accessed 15 Dec 2024. https://www.google.com/search?q=I+risultati+dell%27Europa+in+materia+di+innovazione+sono+in+costante+miglioramento%2C+ma+a+velocit%C3%A0+diverse+da&rlz=1C1GCEA_enIT1046IT1046&oq=I+risultati+dell%27Europa+in+materia+di+innovazione+sono+in+costante+miglioramento%2C+ma+a+velocit%C3%A0+diverse+da&gs_lcrp=EgZjaHJvbWUyBggAEEUYOTIGCAEQRRhAMgYIAhBFGEDSAQcyNzZqMGo0qAIAsAIB&sourceid=chrome&ie=UTF-8
28. OECD.: The impact of AI on the workplace: main findings from the OECD AI surveys of employers and workers (2023). https://doi.org/10.1787/ea0a0fe1-en
29. United Nations.: Transforming our world: the 2030 agenda for sustainable development (2021). https://sustainabledevelopment.un.org/post2015/transformingourworld/publication
30. UNESCO.: Global education monitoring report 2023: technology in education: a tool on whose terms?, GEM report UNESCO (2023). https://doi.org/10.54676/UZQV8501
31. UNESCO.: World commission on the ethics of scientific knowledge and technology. Preliminary study on the ethics of artificial intelligence (2019)
32. Eric, H., et al.: A new future of work: the race to deploy AI and raise skills in Europe and beyond (2024). www.mckinsey.com/mgi
33. European Commission.: European enterprise survey on the use of technologies based on artificial intelligence final report (2020). https://doi.org/10.2759/759368
34. European Commission.: EU study calls for strategic AI adoption. Executive summary (2024)
35. European Commission (2021) Community innovation survey (2021). https://ec.europa.eu/eurostat/statisticsexplained/
36. Ufficio Rapporti con l'Unione Europea.: Il rapporto sul futuro della competitività europea di Mario Draghi (2024)
37. Confidustria.: Rapporto Innovazione Italia 2024—La tenuta del Made in Italy globale e il ruolo dell'innovazione per la competitività futura (2024). Accessed 27 Nov 2024. https://www.confindustria.it/home/appuntamenti/eventi-associazioni/dettaglio-evento/ASSOCONSULT+CONFINDUSTRIA
38. IBM.: AI revolution: productivity boom and beyond (2024)
39. Deloitte.: Now decides next: getting real about Generative AI (2024). http://deloitte.com/us/state-of-generative-ai
40. Gajdzik, B., Wolniak, R.: Smart production workers in terms of creativity and innovation: the implication for open innovation. J. Open Innov. Technol. Market Complex. **8**(2), 68 (2022). https://doi.org/10.3390/JOITMC8020068
41. Lam, L., Nguyen, P., Le, N., Tran, K.: The relation among organizational culture, knowledge management, and innovation capability: its implication for open innovation. J. Open Innov. Technol. Market Complex. **7**(1), 66 (2021). https://doi.org/10.3390/JOITMC7010066
42. Peng, Y., Tao, C.: Can digital transformation promote enterprise performance? —From the perspective of public policy and innovation. J. Innov. Knowl. **7**(3), 100198 (2022). https://doi.org/10.1016/J.JIK.2022.100198
43. Troisi, O., Visvizi, A., Grimaldi, M.: Rethinking innovation through industry and society 5.0 paradigms: a multileveled approach for management and policy-making. Emerald Publishing (2023). https://doi.org/10.1108/EJIM-08-2023-0659
44. Visvizi, A., Troisi, O., Grimaldi, M.: Exploring the drivers for digital transformation in smart education: an ecosystems approach. ITM Web Conf. **51**, 01003 (2023). https://doi.org/10.1051/itmconf/20235101003
45. Brandt, A.: Blueprints for a creativity curriculum. In: Creative Provocations: Speculations on the Future of Creativity, Technology & Learning. Springer (2022)
46. Grilli, L., Pedota, M.: Creativity and artificial intelligence: a multilevel perspective. Creativity Innov. Manage. **33**(2), 234–247 (2024). https://doi.org/10.1111/caim.12580
47. Mazzone, M., Elgammal, A.: Art, creativity, and the potential of artificial intelligence. Arts **8**(1), 26 (2019). https://doi.org/10.3390/arts8010026
48. Shen, S., Xu, K., Sotiriadis, M., Wang, Y.: Exploring the factors influencing the adoption and usage of augmented reality and virtual reality applications in tourism education within the

context of COVID-19 pandemic. J. Hosp. Leis. Sport Tour. Educ. **30** (2022). https://doi.org/10.1016/j.jhlste.2022.100373

49. Maital, S., Seshadri, D.V.R.: Innovation Management: Strategies, Concepts and Tools for Growth and Profit. SAGE publications (2012)
50. Haefner, N., Wincent, J., Parida, V., Gassmann, O.: Artificial intelligence and innovation management: a review, framework, and research agenda☆. Technol. Forecast. Soc. Change **162** (2021). https://doi.org/10.1016/j.techfore.2020.120392
51. Obschonka, M., Audretsch, D.B.: Artificial intelligence and big data in entrepreneurship: a new era has begun. Small Bus. Econ. **55**(3), 529–539 (2020). https://doi.org/10.1007/s11187-019-00202-4
52. Schneider, J., Richner, R., Riser, M.: Towards trustworthy autograding of short, multi-lingual, multi-type answers. Int. J. Artif. Intell. Educ. **33**(1), 88–118 (2023). https://doi.org/10.1007/s40593-022-00289-z
53. Amabile, T.M., Hagtvedt, L., Harrison, S., Holzle, K., Runco, M.: Creativity, artificial intelligence, and a world of surprises. Acad. Manage. Discov. **6**(3), 351–354 (2020)
54. Yan, L., Greiff, S., Teuber, Z., Gašević, D.: Promises and challenges of generative artificial intelligence for human learning. Nat. Hum. Behav. (2024)
55. Cropley, D.H., Medeiros, K.E., Damadzic, A.: The intersection of human and artificial creativity. In: *Creative Provocation: Speculations on the Future of Creativity, Technology & Learning*. Springer (2022)
56. Rhodes, M.: Analysis of creativity. The Phi Delta Kappan, vol. 42, no. 7, pp. 305–310 (1961). https://about.jstor.org/terms
57. Aspen Istitute Italia.: Osservatorio Permanente sull'Adozione e l'Integrazione della Intelligenza Artificiale (IA2) (2024)
58. Creely, E.: Conceiving creativity and Learning in a world of artificial intelligence: a thinking model. In: Creative Provocation: Speculations on the Future of Creativity, Technology & Learning. Springer (2022)
59. Brown, T.: Design thinking. Harv. Bus. Rev. **86**(6) (2008)
60. Plattner, H., Meinel, C., Weinberg, U.Y.: Design Thinking: Innovation lernen -- Ideenwelten öffnen. Mi-Wirtschaftsbuch, München (2009)
61. Carlgren, L., Rauth, I., Elmquist, M.: Framing design thinking: the concept in idea and enactment. Creativity Innov. Manage. **25**(1), 38–57 (2016). https://doi.org/10.1111/caim.12153
62. Jarrahi, M.H.: Artificial intelligence and the future of work: human-AI symbiosis in organizational decision making. Bus. Horiz. **61**(4), 577–586 (2018). https://doi.org/10.1016/j.bushor.2018.03.007
63. Sreenivasan, A., Suresh, M.: Design thinking and artificial intelligence: a systematic literature review exploring synergies. KeAi Publishing Communications Ltd (2024). https://doi.org/10.1016/j.ijis.2024.05.001
64. Poleac, D.: Design thinking with AI. Proc. Int. Conf. Bus. Excellence **18**(1), 2891–2900 (2024). https://doi.org/10.2478/picbe-2024-0240
65. NetworkDigital.: Analisi o intuizione? Come l'AI può aiutare il design thinking. Accessed 29 Jan 2025. https://www.zerounoweb.it/cio-innovation/analisi-o-intuizione-come-lai-puo-aiutare-il-design-thinking/?utm_source=chatgpt.com
66. EconomyUp.: La Generative AI nell'Innovazione e nel Design Thinking: 3 aspetti chiave. Accessed 29 Jan 2025. https://www.economyup.it/innovazione/la-generative-ai-nellinnovazione-e-nel-design-thinking-3-aspetti-chiave/?utm_source=chatgpt.com
67. BMWGROUP.: Intelligenza artificiale Tecnologia chiave per la trasformazione digitale. Accessed 29 Jan 2025. https://www.bmwgroup.com/en/innovation/artificial-intelligence.html
68. Yeeply.: Approccio Design Thinking per decisioni strategiche e innovazione digitale d'impresa. Accessed 29 Jan 2025. https://yeeply.com/it/blog/digitalizzazione/design-thinking-decisioni-strategiche-impresa/?utm_source=chatgpt.com

Education and Higher Education Institutions in the GenAI Era

Redesigning Skills for the Twin Transition: A Path Toward Sustainable Innovation

Adriana Apuzzo and Vincenzo Piluso

Abstract The transition to a circular economy has reshaped organizational processes, demanding a workforce equipped with skills capable of managing and driving change. The twin transition, which integrates ecological and digital imperatives, introduces an additional layer of complexity, requiring strategic adaptations at both individual and organizational levels. Building on recent literature, this study develops a framework to redefine skills in the digital era, emphasizing the interplay between hard and soft skills. European Economic and Social Committee has recognized the urgency of addressing the skills gap within the green labor market, advocating for targeted training programs to align workforce capabilities with evolving industrial and regulatory demands. Similarly, the International Labor Organization highlights the potential positive impact of circular business models on employment, stressing the need for structured skill development initiatives. Despite these policy efforts, there remains a lack of an integrated approach to defining, identifying and measuring the skills necessary for navigating the twin transition. This study addresses this gap by: (a) *identifying key skills required for ecological and digital transitions*; (b) *examining the mechanisms through which these skills are embedded in organizational practices*; and (c) *assessing whether these skills emerge as entirely new constructs or as transformations of pre-existing skills.* Methodologically, the research employs Content Analysis to examine job postings on major online platforms, offering an empirical perspective on labor market trends. By bridging theoretical insights, this study contributes to the academic discourse by proposing a systematic approach to competence development in the context of sustainable and digital transitions, providing valuable guidance for policies aimed at narrowing the gap between traditional and emerging professional roles.

Keywords Twin transition · Circular economy · Workforce development · Sustainable innovation

A. Apuzzo (✉) · V. Piluso
University of Salerno, Fisciano, Italy
e-mail: adapuzzo@unisa.it

V. Piluso
e-mail: vpiluso@unisa.it

A. Visvizi et al. (eds.), *Research and Innovation Forum 2025 Volume 1*,
Springer Proceedings in Complexity, https://doi.org/10.1007/978-3-032-09066-9_14

1 Introduction

The shift toward circular economies and digitalization is driving deep structural changes in both society and business, affecting production models, organizational structures, and workforce composition. Known as the Twin Transition, this process presents both opportunities and adaptation challenges, particularly in redefining the concept of work. Key issues include a growing skills mismatch in the green labor market, as highlighted by the EESC and ILO [1–3] and of fragmented efforts without a unified framework. There is a pressing need to restructure business processes in line with ecological, social, and economic changes, and to clearly define the workforce skills required. While the EU and UN are promoting common sustainability standards, a shared interpretative framework for labor skills is still lacking. For this reason, our study identifies three research question:

RQ_1 What skills do companies consider essential for driving the ecological and digital transition?

RQ_2 How are these skills integrated into existing organizational processes?

RQ_3 To what extent do these skills emerge in response to corporate changes, and to what extent do they represent a transformation of existing skills?

To explore how companies seek talent for the Twin Transition, a qualitative analysis was conducted on 84,136 LinkedIn job postings from 2024 that explicitly reference digital and ecological transformation challenges.

The study identified five emerging professional profiles, linked to the reconfiguration of organizational processes. Job postings were analyzed across ten skill macro-areas, based on frequency, semantic relevance, and explanatory value.

Findings reveal a redefinition of in-demand skills, with a strong focus on digital competencies and strategic sustainability management.

The remainder of this paper is structured as follows (See Fig. 1).

The paper is structured into six sections: Sect. 1 (Introduction) analyzes the context of the Twin Transition and identifies the research questions. Section 2 (Literature Review) explores the role of the Fourth Industrial Revolution and the key skills required for the sustainable and digital transition. Section 3 (Methodology) describes the PRISMA approach for the systematic literature review and content analysis. Section 4 (Results) presents the identification of five key professional profiles and a classification of skills into ten macro areas. Section 5 (Discussion) delves into the transformation of existing skills and the implications for training and corporate strategies. Section 6 (Conclusion) summarizes the study's contributions, discusses its limitations, and suggests future research directions.

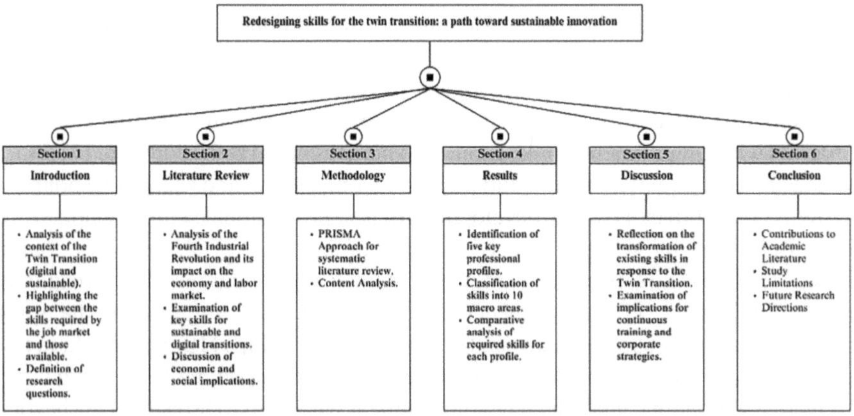

Fig. 1 Workflow structure

2 Literature Review

The debate on the economic and social implications of the twin transition, which intertwines ecological and digital challenges, is assuming a central role in current economic systems. The Fourth Industrial Revolution represents the driving force behind this transformation, bringing about changes whose overall scope has only been partially explored within the scientific debate [4, 5]. Fueled by the simultaneous integration of advanced technologies, the twin transition, on the one hand, contributes to increased productivity and efficiency but, on the other, reshapes the industrial landscape, fostering the emergence of new production chains and the reconfiguration of traditional sectors [6, 7]. At the core of this process is the need to face external changes, such as the climate crisis, which push companies toward adopting an organizational model that considers both economic efficiency and social and environmental responsibility [8]. A crucial aspect of this process lies in the management and processing of information, which serve as strategic levers for the dissemination of sustainable practices and the acquisition of the skills necessary to tackle the new challenges of the labor market [9–11].

2.1 Work and Society in the Era of Digital and Green Transformation

The twin transition unfolds within a complex global context marked by interconnected challenges such as population aging, demographic growth, migration, and climate change, all of which underscore the need for systemic socio-economic restructuring [4, 12, 13]. This transformation resembles an industrial revolution in scale but is broader in scope, affecting not just production but also consumption, lifestyles, and cultural values [14]. It entails a shift toward sustainability that combines digital and green technologies to foster circular economies and decarbonization strategies [15]. However, lasting impact depends on balancing economic progress with environmental protection, with human capital at the core [16, 17]. Innovation should serve as a tool to build inclusive, competitive, and socially responsible systems. This shift calls for the development of new professional skills—adaptability, interdisciplinary knowledge, and continuous learning—as key elements for organizations to adjust to evolving demands and sustainability regulations [18]. Companies must rethink their operations to meet the rising demand for low-impact products and services, reshaping supply chains and economic relationships [18, 19]. Work, in this context, becomes more than an economic activity—it is essential to individual and collective well-being, enabling human fulfillment when tied to broader goals beyond productivity [20, 21]. The twin transition thus triggers a triple transformation across structural (new organizational forms), cultural (changing perceptions of work), and agency (emerging worker identities) dimensions [15, 22]. This leads to a flexible, decentralized system fueled by intelligent technologies, where "relational work" regains importance as a driver of innovation, community, and value creation [20].

The integration of digital technologies is reshaping collaboration and labor dynamics, but a significant skills gap remains between what the market demands and what is currently available [23, 24]. Companies must leverage data-driven decision-making to assess environmental impacts and guide strategic interventions [25]. The UN Sustainable Development Goals (SDGs) offer a framework to align business strategies with responsible growth, emphasizing the need for a systemic shift in production and consumption patterns [26].

The recombinant knowledge approach, rooted in Schumpeterian theory, frames innovation as the result of combining diverse ideas, skills, and technologies, leading to new and sustainable industrial models [27]. This gives rise to recombinant capabilities—the ability to merge external and internal knowledge for innovation [28] Green technologies exemplify this, requiring interdisciplinary knowledge and posing development challenges due to their complexity [29], yet offering high innovation potential [30, 31]. The integration of emerging technologies also transforms the labor market, creating new opportunities while deepening inequality and making hybrid, digital, and sustainable skills essential [32]. The Just Transition framework addresses these challenges by promoting inclusivity and fairness across three dimensions: digital, green, and skills [4]. It emphasizes reskilling and continuous learning to ensure employability in a rapidly evolving job market [14, 33]. Human skills remain crucial; interdisciplinary competence—blending technical, digital, and critical thinking—is increasingly necessary. The UN has embedded this vision in the SDGs, highlighting human capital investment as key to facing the digital and ecological transitions [34]. The European Commission similarly urges investment in green and digital skills to prevent labor market disruptions [35]. National responses vary: Sweden and Denmark emphasize digital skills and lifelong learning; Finland sees transversal competencies as vital, while Poland faces urgent needs due to a widespread digital skills deficit [34]. Across all cases, targeted interventions are essential to bridge disparities.

Finally, sustainability is now integral to business strategies, requiring advanced measurement tools. Regulatory pressures, ESG compliance, and shifting market expectations are driving the shift toward sustainable models and accelerating the Twin Transition [36].

2.2 *Dynamic Skills and Digitalization: The Role of Companies in the Twin Transition*

The acceleration of the Twin Transition has led to the emergence of hybrid skills, which combine technological knowledge with a solid understanding of environmental and social challenges[37]. To fully harness this potential, a joint effort is essential to redefine the role of skills, enhancing human capital. In order for companies to adapt swiftly to rapid market transformations, it is necessary to develop dynamic capabilities, meaning the ability to continuously reconfigure resources and strategies in response to changes in the competitive landscape. These capabilities are structured into three main dimensions [4, 35]:

1. Identification of opportunities, i.e., the ability to recognize emerging trends and strategic innovations;
2. Capitalization on opportunities, by translating strategic insights into innovative solutions and concrete actions;
3. Continuous adaptation, which involves constant revision of business processes to maintain and strengthen competitiveness.

In addition to these, digital skills are also essential to cope with ongoing evolution, going beyond mere technological mastery to include adaptability, problem-solving, and strategic data interpretation [8]. A digital mindset not only enhances operational efficiency but also fosters innovation and sustainability, making companies more resilient to market changes [25]. At an individual level, digital skills develop along a continuum ranging from basic literacy to advanced skills, such as programming, cybersecurity, and big data analysis [8]; however, digitalization does not end with the acquisition of technical skills: it requires a cultural shift, demanding cognitive flexibility and a propensity for continuous innovation [8].

The importance of digital skills is particularly evident within the resource-based view (RBV), according to which these skills influence three key aspects of organizational performance [33]:

- Development of individual capabilities, providing workers with tools to handle the complexity of digital transformation;
- Increase in motivation and engagement, creating a stimulating and innovative work environment;
- Creation of an ecosystem favorable to innovation, where continuous learning becomes the driving force of business progress.

The Twin Transformation Capability Maturity Model (TTCMM) offers a scalable framework to guide organizations in developing the skills needed for the twin transition, updating the Resource-Based View (RBV) [18, 38]. It emphasizes the importance of strategic leadership and a unified approach to digital and environmental goals [33], supported by a culture of innovation that embraces change [8]. A data-driven approach is essential, leveraging perceptive, predictive, and prescriptive capabilities to optimize performance [8]. Success depends on hybrid skills—blending tech expertise with environmental understanding—and the development of dynamic capabilities to adapt to change [39]. Despite growing interest, there is still no clear framework to define or measure the skills required, creating a skills mismatch that may hinder the adoption of sustainable models and limit adaptability.

Based on these reflections, this study aims to explore the relationship between theory and practice, contributing to the literature through a systematic analysis of the skills required to face the twin transition. From this perspective, the research is developed around the following questions:

RQ_1: What skills do companies consider essential to drive the ecological and digital transition?

RQ_2: How are these skills integrated into pre-existing organizational processes?

RQ_3: To what extent do these skills emerge in response to business changes, and to what extent do they represent a rethinking of existing ones?

This study adopts an approach aimed at providing an analytical framework to support companies and institutions in adapting to the challenges of the twin transition. Through an empirical investigation of the skills required by the labor market, the research focuses on identifying the key capabilities needed to address ongoing changes, with particular attention to the integration between digitalization and sustainability.

3 Methodology

The study adopted a methodological approach based on the PRISMA framework (Preferred Reporting Items for Systematic Reviews and Meta-Analyses), which represents a standard for conducting systematic literature reviews. PRISMA was chosen for its ability to structure in a clear and replicable manner the identification, selection, evaluation, and inclusion of studies relevant to the research topic. This approach is particularly useful in interdisciplinary fields such as the skills required to support the Twin Transition, which demands a broad and detailed view of the available sources. The methodological process was structured into four main phases: Identification, Screening, Eligibility Assessment, and Inclusion. The PRISMA flow diagram, presented in Fig. 2, provides a visual representation of the decisions made and the steps followed during the systematic review.

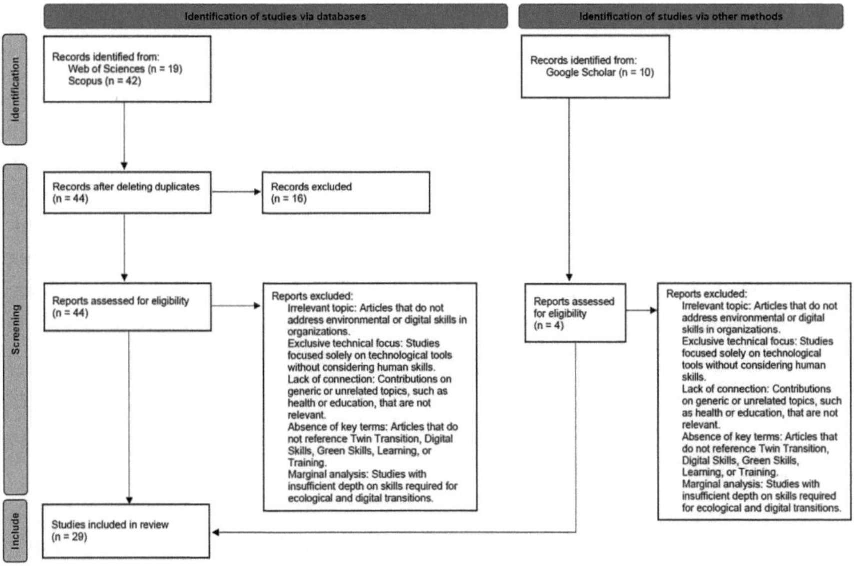

Fig. 2 Flowchart of the steps of a systematic review

3.1 Data Collection

In the initial identification phase, the Scopus and Web of Science (WoS) databases were selected. The search strings were designed to include key terms related to the Twin Transition and green and digital skills. Specifically, the following filters were applied:

- Document type: Peer-reviewed scientific articles (Article);
- Language: English, to ensure access to a widely disseminated international literature.

Table 1 summarizes the combinations used and the results obtained.

This phase led to the preliminary identification of 61 articles, representing the initial corpus for subsequent analyses.

Table 1 Query for systematic review

Query	SCOPUS	WOS
"Twin transition" AND "Skill*"	4	4
"Twin transition" AND "Competence*"	1	1
"Twin transition" AND "Capabil*"	34	13
"Twin transition" AND "Labour Market"	2	1
Totals by database	42 results	19 results

3.2 Screening

In the screening phase, RStudio was used to compare the results obtained from the two main databases, allowing for the precise identification of duplicates within the initial corpus. This ensured a transparent, reproducible analysis free from manual errors. Through this cross-analysis, 16 duplicate articles were removed, reducing the preliminary sample from 61 to 44 articles. Subsequently, each article underwent a thorough screening of titles and abstracts to assess its relevance to the specific objectives of the study. For this phase, inclusion criteria (see Sect. 3.3) were established to identify contributions directly relevant to understanding the skills required for the Twin Transition. Articles that did not meet these criteria were excluded.

3.3 Eligibility Assessment

The articles selected during the screening phase underwent a full and thorough reading to ensure compliance with the following inclusion criteria, defined based on the research objectives:

- Thematic relevance: The articles had to explicitly address the topic of skills required to support environmental and digital sustainability in organizations. This criterion was essential to ensure that the selected sources directly contributed to understanding the mechanisms and strategies related to the Twin Transition.
- Perspective on skills: Contributions were required to analytically and concretely discuss aspects related to the development, implementation, or integration of skills connected to the Twin Transition. The focus was placed on studies providing practical examples or theoretical models useful for the topic.
- Breadth and relevance of contributions: Even when the term "Twin Transition" was not explicitly mentioned, articles were included if they addressed related topics, such as green skills, digital skills, or forms of integration between these dimensions, which represent the essence of the investigated phenomenon.

  Articles that did not fully meet the above criteria were excluded.

3.4 Inclusion

The final sample consists of 25 articles, and to complement the systematic review, four contributions from the grey literature were included, offering a complementary and interdisciplinary perspective.

3.5 Content Analysis

Alongside the systematic literature review, a Content Analysis was conducted following the approach proposed by Krippendorff [40]. The analysis focused on job advertisements within Online Job Advertising (OJA) to identify the skills required by companies engaged in the Twin Transition [41]. Osgood introduced a methodological shift in content analysis by developing sophisticated analytical procedures that allow for preserving and analyzing the structural complexity of texts [42]. This approach, applied within the analysis, goes beyond simple frequency analysis by introducing two complementary methodologies that enhance interpretative tools.

The first, the analysis of evaluative assertions, enables a better understanding of the evaluative nuances present in texts. This approach focuses on measuring the intensity of both positive and negative assessments expressed toward specific textual objects. As highlighted by Krippendorff [40], this technique captures not only the direction of an attitude (positive or negative) but also its strength and nuances, offering a rich and detailed understanding of the positions expressed in the text. The second, contingency analysis, introduces a relational dimension into content analysis. This technique follows a two-step process: initially, the units of analysis are classified into distinct categories; subsequently, co-occurrences of elements belonging to different categories within the same text or corpus of texts are identified. This approach formally surpasses a purely frequency-based analysis [43].

The use of these analytical techniques allowed for mapping a segment of labor market needs, highlighting the most in-demand skills and comparing them with those identified in the literature review. This exploratory approach, which integrates academic sources with data from digital platforms, contributes to the academic debate by providing an analytical method and an interpretative framework aimed at supporting policies to reduce the gap between traditional and emerging professional roles. The job advertisement collection covers the entire period of 2024 (January 1, 2024–December 31, 2024) through a direct API call on the LinkedIn platform via a Python script. The Data Collection process generated a textual corpus of approximately 84,136 job postings. From the data analysis, five predominant professional profiles emerge: [a] Sustainability and Digital Transformation Manager; [b] Data Analyst; [c] Circular Economy Specialist; [d] Energy Manager 4.0; [e] Supply Chain Coordinator.

4 Results

The results of the analysis define areas of required skills within the five identified professional profiles, based on a sample of 84,136 job advertisements published in 2024. The dataset was organized into five specific tables for each professional profile, allowing both a comparative analysis and an in-depth examination of individual roles. The skills were classified into ten main areas, defined based on the intersubjective

Table 2 Supply chain coordinator

Skill area	V.A	%
Industry knowledge	14.130	90.0
Digital technologies	13.345	85.0
Project management	12.560	80.0
Data analytics	11.775	75.0
Sustainability strategy	10.990	70.0
Risk management	9.420	60.0
ESG reporting	8.635	55.0
Strategic management	7.850	50.0
Stakeholder management	7.065	45.0
Change management	6.280	40.0

Sample size (n) = 15.700

agreement of the research group. The aggregation of skills into macro-areas was based on three fundamental elements: frequency weight, semantic affinity, and explanatory capacity. Frequency weight represents the number of times certain skills appear within a group of job postings; semantic affinity serves as a grouping principle aimed at connecting semantically related skills; explanatory capacity determines the suitability of a set of skills to meaningfully enrich the group to which they belong (Table 2).

The conducted analysis indicates a configuration of skills strongly oriented toward the integration of sectoral knowledge and sustainable supply chain management capabilities. The predominance of Industry Knowledge (90.0%), followed by Digital Technologies (85.0%) and Project Management (80.0%), outlines a professional profile characterized by a foundation of technical and managerial skills. Notably, Data Analytics ranks fourth (75.0%), highlighting the growing importance of data analysis in supply chain management (See Table 3).

The analysis of job postings related to this professional role reveals an evolution compared to the traditional Energy Manager profile. The dominance of Digital Technologies (90.0%), followed by Industry Knowledge (85.0%) and Data Analytics (80.0%), indicates a profile focused on the digital transformation of energy management processes. Technological skills surpass sustainability-related ones in terms of weight and relevance, with Sustainability Strategy ranking only fifth (60.0%). The relatively lower presence of transversal skills, such as Stakeholder Management (30.0%), suggests a predominantly technical-operational characterization of the role, with a primary focus on implementing advanced technological solutions for energy efficiency (See Table 4).

Sustainability Strategy leads the ranking (90.0%), followed closely by Industry Knowledge (85.0%) and Digital Technologies (75.0%). This configuration defines a professional profile where a strategic vision of sustainability is necessarily combined with a deep understanding of industrial processes. The significant presence of Project Management (70.0%) and Data Analytics (60.0%) in intermediate positions suggests

Table 3 Energy manager 4.0

Skill area	V.A	%
Digital technologies	15.683	90.0
Industry knowledge	14.811	85.0
Data analytics	13.940	80.0
Project management	12.198	70.0
Sustainability strategy	10.455	60.0
ESG reporting	9.584	55.0
Strategic management	8.713	50.0
Change management	6.970	40.0
Risk management	6.099	35.0
Stakeholder management	5.228	30.0

Sample size (n) = 17.425

Table 4 Circular economy specialist

Skill area	V.A	%
Sustainability strategy	15.201	90.0
Industry knowledge	14.357	85.0
Digital technologies	12.668	75.0
Project management	11.823	70.0
Data analytics	10.134	60.0
ESG reporting	9.290	55.0
Strategic management	8.445	50.0
Change management	7.601	45.0
Stakeholder management	6.756	40.0
Risk management	5.067	30.0

Sample size (n) = 16.890

an active role in the operational implementation of circular economy principles. The gradual distribution of percentages, extending to Risk Management (30.0%), indicates a versatile profile, characterized by the ability to integrate strategic, technical, and operational skills (See Table 5).

Job postings for this profile highlight a prevalence of Data Analytics (90.0%), followed by Digital Technologies (85.0%), outlining a role highly specialized in the quantitative analysis of environmental performance. It is significant to observe the substantial gap between technical and managerial skills, with Change Management (30.0%) and Risk Management (25.0%) ranking at the bottom. This distribution depicts a highly specialized professional figure, whose primary focus is the processing and interpretation of environmental data through advanced digital tools. The presence of ESG Reporting in fourth place (75.0%) indicates a role in environmental reporting and compliance processes (See Table 6).

Table 5 Data analyst

Skill area	V.A	%
Data analytics	14.288	90.0
Digital technologies	13.495	85.0
Sustainability metrics	12.701	80.0
ESG reporting	11.907	75.0
Industry knowledge	11.113	70.0
Project management	9.526	60.0
Strategic management	7.938	50.0
Stakeholder management	6.350	40.0
Change management	4.763	30.0
Risk management	3.969	25.0

Sample size (n) = 15.876

Table 6 Sustainability and digital transformation manager

Skill area	V.A	%
Strategic management	16,420	90.0
Digital technologies	15,508	85.0
Sustainability strategy	14,596	80.0
Change management	13,684	75.0
ESG reporting	12,772	70.0
Project management	11,859	65.0
Data analytics	10,947	60.0
Stakeholder management	10,035	55.0
Risk management	9123	50.0
Industry knowledge	8210	45.0

Sample size (n) = 18.245

The job posting analysis reveals a hierarchy of required skills, with Strategic Management in a dominant position (90.0%), followed by Digital Technologies (85.0%) and Sustainability Strategy (80.0%). This distribution outlines a leadership profile strongly oriented toward the strategic management of corporate transformation. Change Management ranks fourth (75.0%), emphasizing the importance of the ability to lead organizational change processes. The progressive decrease in the incidence of skills, extending to Industry Knowledge (45.0%), suggests a role characterized more by managerial and transformational capabilities than by specific technical expertise.

5 Discussion

The analysis addressed the first research question, highlighting that the key skills required for the twin transition are not limited to technical knowledge or the ability to adopt new technologies but require a systemic vision and the ability to manage change.

The second research question, related to the mechanisms through which these skills are developed and integrated into business practices, is answered by the emergence of three main transformation drivers:

1. Continuous learning, made necessary by the rapid evolution of technologies and sustainability regulations;
2. Integration of skills into business processes, through governance tools based on data and ESG metrics that allow measuring the impact of adopted strategies;
3. Adoption of a collaborative approach, involving companies, institutions, and universities in the creation of innovation ecosystems, promoting knowledge sharing and best practices.

The third research question, concerning the emergence of new skills or the transformation of existing ones, revealed that the twin transition has not led to the creation of entirely new skills but rather to their reorganization and hybridization. Strategic management and project management skills, for example, have long existed but must now be integrated with advanced digital tools to ensure more effective and sustainable management. Similarly, industry knowledge, once sufficient to operate effectively in a specific production sector, must now be combined with data analytics skills to support evidence-based decision-making.

A particularly relevant element that emerged from the analysis concerns the gap between currently available skills and those demanded by the market. The growing demand for hybrid skills suggests that educational systems and traditional career paths are not yet fully aligned with new requirements. The concept of just transition, proposed by Bianchi et al. [4], plays a crucial role in this context, emphasizing the need to ensure continuous upskilling and reskilling to prevent new forms of labor exclusion. The collected evidence indicates that advanced companies are already investing in upskilling and reskilling programs, while public institutions will need to strengthen training policies to mitigate the risk of labor market polarization between highly skilled workers and those with obsolete skills.

The overall picture that emerges highlights that the balance between digitalization and sustainability is the central axis around which the redefinition of professional roles revolves, with implications for business strategies and labor policies.

6 Conclusion

This study has highlighted how the convergence between digitalization and sustainability necessitates the development of hybrid skills, which combine advanced technical skills with managerial and strategic abilities, enabling organizations to address environmental and technological challenges in an integrated manner.

The labor market already reflects this need, with a growing demand for professionals capable of operating in contexts characterized by the continuous evolution of technologies and the need to ensure increasingly sustainable production processes. The analyzed data demonstrate that the very concept of competency is undergoing transformation; rather than solely the emergence of new skills, there is a recombination and restructuring of existing skills. The ability to integrate advanced digital tools with sustainability strategies is becoming a key factor for corporate competitiveness.

The literature has extensively discussed how dynamic capabilities represent a critical success factor in the twin transition [18], confirming that companies investing in training and innovation are more resilient and better able to adapt to new business models. However, numerous challenges remain, including the uneven adoption of twin transition strategies across companies and sectors [23]. While large enterprises have already initiated transformation processes supported by investments in technologies and skills, many SMEs struggle to keep pace due to limited resources and a lower capacity to attract talent with advanced skills.

This imbalance could exacerbate disparities within the labor market, generating new forms of exclusion for workers who lack access to upskilling opportunities or operate in less innovation-driven sectors. The literature has emphasized that just transition represents an essential tool to prevent the twin transition from producing asymmetric effects on the labor market, penalizing specific segments of the workforce and increasing inequalities.

Looking ahead, the effectiveness of the twin transition will depend on the ability of businesses and institutions to develop integrated strategies that enhance human capital and promote inclusive innovation. The collected evidence suggests that the twin transition is not a linear process but rather a complex phenomenon that requires coordination among various stakeholders, including companies, governments, educational institutions, and regulatory bodies.

Without joint efforts, there is a risk of a fragmented transition, where only certain entities manage to seize the opportunities offered by digitalization and sustainability, while others remain excluded from this transformation.

References

1. International Labour Organization.: Decent Work in the Circular Economy: An Overview of the Existing Evidence Base (2023)
2. International Labour Organization.: Social Finance. Annual Report 2023 (2024)

3. European Economic and Social Committee Skills.: Skills Mismatches. An Impediment to the Competitiveness of EU Businesses (2018). https://doi.org/10.2864/448258
4. Bianchi, P., De Propris, L., Labory, S.: People-centred policies for a just transition (digital, green and skills). Contemp. Soc. Sci. **19**(1–3), 262–282 (2024). https://doi.org/10.1080/21582041.2024.2351479
5. Hofmann Trevisan, A., Acerbi, F., Dukovska-Popovska, I., Terzi, S., Sassanelli, C.: Skills for the twin transition in manufacturing: a systematic literature review. Elsevier Ltd. (2024). https://doi.org/10.1016/j.jclepro.2024.143603
6. Skellern, K., Markey, R., Thornthwaite, L.: Identifying attributes of sustainable transitions for traditional regional manufacturing industry sectors—a conceptual framework. J. Clean. Prod. **140**, 1782–1793 (2017). https://doi.org/10.1016/J.JCLEPRO.2016.07.183
7. Alimam, H., Mazzuto, G., Tozzi, N., Emanuele Ciarapica, F., Bevilacqua, M.: The resurrection of digital triplet: a cognitive pillar of human-machine integration at the dawn of industry 5.0. J. King Saud Univ. Comput. Inf. Sci. **35**(10), 101846 (2023). https://doi.org/10.1016/J.JKSUCI.2023.101846
8. Sjödin, D., Parida, V., Kohtamäki, M.: Artificial intelligence enabling circular business model innovation in digital servitization: conceptualizing dynamic capabilities, AI capacities, business models and effects. Technol. Forecast. Soc. Change **197**, 122903 (2023). https://doi.org/10.1016/J.TECHFORE.2023.122903
9. Bailey, D., de Ruyter, A.: Plant closures, precariousness and policy responses: revisiting MG Rover 10 years on. Policy Stud. **36**(4), 363–383 (2015). https://doi.org/10.1080/01442872.2015.1073248
10. Garay, I.S., Quintana, M.G.B.: 21st Century Skills. An Analysis of Theoretical Frameworks to Guide Educational Innovation Processes in Chilean Context, pp. 37–46 (2019). https://doi.org/10.1007/978-3-030-30809-4_4
11. Testa, M.: The Managerial Skills in Complex Systems: An Integrated Approach to Learning Process, pp. 435–447 (2024). https://doi.org/10.1007/978-3-031-44721-1_33
12. Troisi, O., Grimaldi, M., Visvizi, A.: Digital Transformation in Tourism Ecosystems: What Impact on Sustainability and Innovation?, pp. 73–81 (2023). https://doi.org/10.1007/978-3-031-19560-0_5
13. Simpson, N., Mulhaney, A.: The Death of Business as Usual: The Impact of VUCA on the Sustainability of SMEs, pp. 65–83 (2024). https://doi.org/10.1007/978-3-031-44721-1_6
14. Rollnik-Sadowska, E.: Labor market in sustainability transitions: a systematic literature review. Econ. Environ. **87**(4), 1–31 (2023). https://doi.org/10.34659/eis.2023.87.4.681
15. Mehmood, K., Kiani, A., Rehman, H., Alshibani, S.M., Piccardi, P.: Can platform leadership drive twin transitions in greening SMEs? Exploring the nexus between AI infrastructure flexibility, information effects, and green sustainable practices. In: Business Ethics, The Environment and Responsibility. Wiley and Sons Inc. (2024). https://doi.org/10.1111/beer.12761
16. Helmi, A., Bastidas, V., Oti-Sarpong, K., Schooling, J.: Sustainable urban digital innovation: a socio-technical competency-based approach to evaluation. Sustain. Cities Soc. **117**, 105946 (2024). https://doi.org/10.1016/J.SCS.2024.105946
17. Visvizi, A.: Artificial intelligence (AI) and sustainable development goals (SDGs): exploring the impact of AI on politics and society. Sustainability **14**(3), 1730 (2022). https://doi.org/10.3390/su14031730
18. Breiter, K., Crome, C., Oberländer, A.M., Schnaak, F.: Dynamic capabilities for the twin transformation climb: a capability maturity model. Inf. Syst. Front. (2024). https://doi.org/10.1007/s10796-024-10520-y
19. Kumar, M., Kumar, P.: Valuation of the ecosystem services: a psycho-cultural perspective. Ecol. Econ. **64**(4), 808–819 (2008). https://doi.org/10.1016/J.ECOLECON.2007.05.008
20. Rodriguez-Lluesma, C., García-Ruiz, P., Pinto-Garay, J.: The digital transformation of work: a relational view. Bus. Ethics **30**(1), 157–167 (2021). https://doi.org/10.1111/beer.12323
21. Visvizi. A., Troisi, O., Grimaldi, M.: Digital Divide and Entrepreneurial Orientation in the Global South: Quantifying and Explaining the Nexus, pp. 723–734 (2023). https://doi.org/10.1007/978-3-031-19560-0_62

22. Visvizi, A., Lytras, M.D., Aljohani, N.R.: Research and innovation forum 2020: disruptive technologies in times of change. Springer Nature (2021)
23. Mas, F.D., Valentina, N., Massaro, M., Bagnoli, C.: Combining deep and digital technologies as a path towards twin transition: the "future farming" case study. IEEE Trans. Eng. Manage. (2024). https://doi.org/10.1109/TEM.2024.3444592
24. Fusillo, F., Quatraro, F., Santhià, C.: Leveraging on circular economy technologies for recombinant dynamics: do localised knowledge and digital complementarities matter? Reg. Stud. (2024). https://doi.org/10.1080/00343404.2024.2329255
25. Martínez-Peláez, R., et al.: Role of digital transformation for achieving sustainability: mediated role of stakeholders, key capabilities, and technology. Sustainability (Switzerland) 15(14) (2023). https://doi.org/10.3390/su151411221
26. Ferreira, J.J., Fernandes, C.I., Veiga, P.M., Caputo, A.: The interactions of entrepreneurial attitudes, abilities and aspirations in the (twin) environmental and digital transitions? A dynamic panel data approach. Technol. Soc. **71**, 102121 (2022). https://doi.org/10.1016/J.TECHSOC.2022.102121
27. Wang, T., Yu, C., Shi, X., Su, H.N.: From recombination to diffusion: exploring the impact of knowledge recombination characteristics on geographical knowledge diffusion. J. Eng. Tech. Manage. **72**, 101818 (2024). https://doi.org/10.1016/J.JENGTECMAN.2024.101818
28. Meng, Y., Wu, H., Wang, Y., Duan, Y.: International trade diversification, green innovation, and consumption-based carbon emissions: the role of renewable energy for sustainable development in BRICST countries. Renew. Energy **198**, 1243–1253 (2022). https://doi.org/10.1016/J.RENENE.2022.08.045
29. Lee, J.M., Narula, R., Hillemann, J.: Unraveling asset recombination through the lens of firm-specific advantages: a dynamic capabilities perspective. J. World Bus. **56**(2), 101193 (2021). https://doi.org/10.1016/J.JWB.2021.101193
30. Xu, Y., Zhao, X., Wang, J., Xie, P.: Clarifying the dispute of corporate social responsibility: evidence from green technological innovation. Technol. Soc. **75**, 102392 (2023). https://doi.org/10.1016/J.TECHSOC.2023.102392
31. Pan, C.L., Liao, H.T., Zhang, Y.: Knowledge mapping of resilience and human rights in supply chains: a roadmapping taxonomy for twin green and digital transition design. Front. Environ. Sci. **11** (2023). https://doi.org/10.3389/fenvs.2023.1152345
32. Christmann, A.S., Crome, C., Graf-Drasch, V., Oberländer, A.M., Schmidt, L.: The twin transformation butterfly: capabilities for an integrated digital and sustainability transformation. Bus. Inf. Syst. Eng. **66**(4), 489–505 (2024). https://doi.org/10.1007/s12599-023-00847-2
33. Singh, B., Srivastava, A.P., Chatterjee, S., Durana, P., Kliestik, T.: Assessing digital capability for twin transition and profitability: from firm and people perspectives with leadership support as moderator. In: Business Ethics, the Environment and Responsibility. Wiley and Sons Inc. (2024). https://doi.org/10.1111/beer.12698
34. Rehman, S.U., Giordino, D., Zhang, Q., Alam, G.M.: Twin transitions & industry 4.0: unpacking the relationship between digital and green factors to determine green competitive advantage. Technol. Soc. **73**, 102227 (2023). https://doi.org/10.1016/J.TECHSOC.2023.102227
35. Aljohani, N.R., Aslam, A., Khadidos, A.O., Hassan, S.U.: Bridging the skill gap between the acquired university curriculum and the requirements of the job market: a data-driven analysis of scientific literature. J. Innov. Knowl. **7**(3), 100190 (2022). https://doi.org/10.1016/J.JIK.2022.100190
36. Jiang, Y., Ni, H., Guo, X., Ni, Y.: Integrating ESG practices and natural resources management for sustainable economic development in SMEs under the double-carbon target of China. Resour. Policy **87**, 104348 (2023). https://doi.org/10.1016/J.RESOURPOL.2023.104348
37. Boeren, E.: Post-COVID-19 European country-specific recommendations on adult education, training, and skills: content, focus, shifts, and patterns. J. Adult Continuing Educ. (2024). https://doi.org/10.1177/14779714241276465
38. Harfouche, A., Merhi, M.I., Albizri, A., Dennehy, D., Thatcher, J.B.: Sustainable development through technological innovations and data analytics. Inf. Syst. Front. (2025). https://doi.org/10.1007/s10796-024-10570-2

39. Peças, P., et al.: Holistic framework to data-driven sustainability assessment. Sustainability (Switzerland) **15**(4) (2023). https://doi.org/10.3390/su15043562
40. Krippendorff, K.: Content Analysis: An Introduction to its Methodology. Sage publications (2018)
41. Palazzo, M., Notari, F.: Big data per l'open innovation. Trends, opportunità e nuove sfide per i decision makers nel mercato del lavoro. G. Giappichelli Editore srl (2024)
42. Osgood, C.E.: The Measurement of Meaning. Urban III/Universit of Illinois (1957)
43. Stone, P.J., Hunt, E.B.: A computer approach to content analysis: studies using the general inquirer system. In: Proceedings of the May 21–23, 1963, Spring Joint Computer Conference, pp. 241–256 (1963)

Building the Next Generation of Entrepreneurs: The Role of Group Projects and Experiential Education in Nurturing Entrepreneurial Mindset and Skills

Hajar Makry

Abstract This study investigates the effectiveness of experiential learning in fostering an entrepreneurial mindset and soft skills among students from science and technology disciplines (computer science and biology) as well as management disciplines (finance and marketing), focusing on cohorts graduating in 2022 and 2024. Throughout a semester, students engaged in real entrepreneurial projects while working in interdisciplinary teams. A follow-up assessment was conducted one year after graduation to evaluate the longer-term impact of this educational approach. A quantitative methodology was employed, using a structured questionnaire to collect data on students' experiences with group-based experiential learning, their perceived development of entrepreneurial competencies, and their post-graduation outcomes. The results indicate that experiential education significantly enhances the development of an entrepreneurial mindset and transversal competencies, with notable disciplinary differences. The study contributes to theory by reinforcing experiential learning models such as those of Dewey and ET-7 within the context of entrepreneurial education. Practically, it offers insights for educators and policymakers aiming to improve employability, innovation readiness, and curriculum design across disciplines.

Keywords Experiential learning · Entrepreneurship · Entrepreneurial skills · Education

1 Introduction

In a constantly evolving economy, developing transversal skills has become a key priority in higher education. Recent reforms in Morocco's higher education system, particularly the PACTE ESRI, clearly underscore this shift. However, traditional

H. Makry (✉)
Polydisciplinary Faculty Sidi Bennour, Chouaib Doukkali University, El Jadida, Morocco
e-mail: Makry.h@ucd.ac.ma

A. Visvizi et al. (eds.), *Research and Innovation Forum 2025 Volume 1*,
Springer Proceedings in Complexity, https://doi.org/10.1007/978-3-032-09066-9_15

academic structures often struggle to promote these competencies, especially in disciplines outside the field of management. While entrepreneurship education has gained momentum, it remains largely compartmentalized and theoretical, limiting its broader academic impact. This study addresses a critical gap in the literature by examining how experiential learning—through interdisciplinary group projects—can effectively foster entrepreneurial thinking and soft skills among students in both science/technology and management fields.

The paper is structured into five main sections. Following this introduction, the literature review explores the evolution of entrepreneurship education and outlines key theoretical frameworks, including Dewey's experiential learning and the ET-7 model. The methodology section presents the research design, sample, data collection tools, and analytical methods. The results and discussion sections examine how interdisciplinary, project-based learning shapes entrepreneurial competencies and post-graduation outcomes. Finally, the conclusion highlights the study's implications for curriculum development and education policy, while acknowledging its limitations and proposing directions for future research.

2 Literature Review

2.1 Experiential Learning and Entrepreneurial Education

Entrepreneurship education has seen significant development since the twentieth century. This dynamic began with Shigeru Fijii's initiative in 1938 at Kobe University in Japan and was later reinforced by the introduction of an entrepreneurship course by Myles Mace at Harvard in 1947. These early experiences paved the way for the significant expansion of entrepreneurship education worldwide, confirming its growing role in addressing the challenges of the modern economic landscape.

Several studies [1] have highlighted the positive effects of these programs within universities: improved attractiveness and feasibility of entrepreneurial projects, and even increased business startups. However, they note that traditional higher education institutions continue to structure entrepreneurship according to a model inherited from business schools—often confined to a specific department, with little cross-disciplinarity or interdisciplinary collaboration. This demonstrates the need for an evolution towards more integrated and multidisciplinary approaches.

It is in this perspective that the entrepreneurial pedagogy model presented by Klucznik-Törő [2] fits, based on strong theoretical foundations. This approach aims to integrate entrepreneurship into university educational programs not just as a transfer of knowledge, but as a lever for developing transversal skills useful in many professions. It encourages institutions to adopt a more integrated approach, incorporating entrepreneurial teaching methods in all BA and MA programs where appropriate.

This model thus broadens the concept of the "entrepreneurial university": it is no longer only about fostering the transfer of knowledge to external actors (industrials, partners, funders), but also about cultivating entrepreneurial skills and attitudes among students and teachers—as internal actors.

The educational objectives [2] of this pedagogy focus on value creation in different forms, going beyond financial outcomes to include economic, social, and ecological values. The emphasis is on experiential learning, in which students are encouraged to step outside their usual assumptions to face uncertain contexts comparable to those encountered by entrepreneurs in the creation of their projects. Entrepreneurial skills are developed through iterative learning processes and rely on several theoretical frameworks: the theory of competencies, the theory of learning processes, and the theory of entrepreneurial intentions. These approaches link theoretical knowledge to practical skills, preparing students to tackle real challenges and seize entrepreneurial opportunities in various fields.

Recent research highlights the importance of an organizational culture based on learning, data, and human resources for fostering innovation. Indeed, a data-driven orientation, when coupled with strategies activating human skills—notably through distributed teams, knowledge sharing, and co-creation—enables the generation of various forms of innovation: technological, social, educational, and organizational [3].

The synergistic integration of technical skills (hard skills) and relational abilities (soft skills), grounded in experiential learning practices, thus represents a central lever to stimulate agility, creativity, and an entrepreneurial mindset.

The study's findings show that innovation does not emerge solely from technology but primarily from activating human skills and strategic knowledge management. In this regard, experiential learning based on concrete, collaborative, and interdisciplinary projects fosters the adoption of a proactive and reflective entrepreneurial posture. This type of learning helps develop co-creation capabilities, which are essential in complex digital environments where innovation relies on the convergence of data, expertise, and diverse perspectives [3].

Other recent research investigates the integration of metaverse technologies into organizational learning. It aligns with experiential education models by offering immersive, project-based environments that enhance both hard and soft skill development. Like the interdisciplinary entrepreneurial projects explored in the present study, the metaverse fosters teamwork, creativity, and applied knowledge acquisition through real-world simulations. Recent findings underline its potential to support T-shaped knowledge—combining deep expertise with cross-functional capabilities—while also highlighting challenges related to implementation. This reinforces the role of experiential learning not only in academic contexts but also in digital workspaces, bridging education and professional innovation [4].

2.2 Theoretical Framework: Dewey, ET-7, and Entrepreneurial Competence

John Dewey (1938/1997) asserts that all authentic education is based on experience but immediately clarifies that this does not mean "all experiences are genuinely or equally educative." He distinguishes between genuinely formative experiences and those that, if poorly structured, can even hinder the learner's development. For Dewey, an educational experience must meet two essential criteria: **continuity** and **interaction**. Continuity means that "every experience lives on in further experiences" (p. 27), highlighting that each experience shapes future ones in a cumulative learning process. Interaction, on the other hand, refers to the dynamic between the learner and their environment, as "an experience is always what it is because of a transaction taking place between an individual and what, at the time, constitutes his environment" (p. 43). These principles laid down by Dewey provide a solid theoretical framework for designing pedagogical devices centered on active student engagement, particularly in entrepreneurship education.

Experiential learning refers to a process where knowledge is constructed through the transformation of experience [5]. It is a holistic and integrative approach to learning that combines experience, perception, cognition, and behavior. Specifically, an experiential learning activity includes several stages: students live a concrete experience, observe and reflect on this experience, extract abstract concepts from it, and then test these concepts through active experimentation. This type of learning makes the learner a central actor and fosters deep learning—meaning a personal and meaningful understanding of the content, beyond mere memorization.

Experiential learning is based on student engagement in practical and reflective experiences in uncertain contexts, similar to those encountered by entrepreneurs [2]. This iterative process encourages students to step outside of assumed truths, promoting learning by action and value creation. The inclusion of **effectuation** in education is crucial for developing an entrepreneurial mindset, especially in higher education. Furthermore, social and collaborative learning—which supports the development of social capital—plays a key role in the effectiveness of entrepreneurship education, as highlighted by Klucznik-Tîrô [2] and Mansoori and Lackéus [6].

Experiential education, particularly through group projects, serves as a powerful tool in entrepreneurship education, offering practical, hands-on experience that allows students to apply theoretical knowledge in real-world scenarios, thereby developing a deeper understanding of entrepreneurial challenges and opportunities. These activities, simulating the environments entrepreneurs face, ensure entrepreneurial learning through action [5].

In this study, the identification of entrepreneurial competencies relies on the analysis provided by Houston Peschl, Deng, and Larson, which is a significant contribution to the pedagogical structuring of entrepreneurial thinking. These authors conducted a rigorous synthesis of the main existing conceptualizations of entrepreneurial competencies, resulting in the formulation of an integrated framework they named **ET-7** [5]. This framework is based on an in-depth comparative

Table 1 Entrepreneurial competencies: the ET-7 framework and theoretical correspondences

Entrepreneurial thinking compe-tencies (ET-7)	Imaginativeness [9]	Entrepreneurial mindset profile [8]	EntreComp frame-work [7]
Problem solving	Practical imagination	Idea generation	Seizing opportunities, taking initiative, mobilizing resources
Tolerance for ambiguity		Optimism	Facing uncertainty, ambiguity, and risk
Learning from failure (failing forward) empathy	Practical imagination	Self-confidence, perseverance	Learning from experience
Empathy	Social imagination	Interpersonal sensitivity	
Creativity with limited resources	Creative imagination	Perseverance, self-confidence	Creativity
Response to critical feedback	Practical imagination	Long-term vision, execution, motivation, perseverance	Valuing ideas
Teamwork			Mobilizing others, working with others

Source Adapted from Houston Peschl et al. [5]

analysis of three major references: the **EntreComp Framework** developed by Bacigalupo et al. [7], the **Entrepreneurial Mindset Profile** by Davis et al. [8], and the model of **entrepreneurial imagination** proposed by Kier and McMullen [9]. The table they constructed highlights this synthesis, cross-referencing the competencies identified in each model with the seven key dimensions of the ET-7 framework. This approach provides a coherent model—both rooted in scientific literature and adapted to educational purposes—emphasizing the development of transversal competencies, particularly relevant for preparing students for the uncertainty and complexity inherent in contemporary entrepreneurship (Table 1).

3 Materials and Methods

3.1 Sample and Data Collection

The study employed a self-selection non-probability sampling method, using an online questionnaire distributed to students from the 2022 and 2024 cohorts enrolled in business-related programs (Finance and Marketing) and science and technology fields (Biotechnology and Computer Science). A total of 87 students voluntarily participated in the survey.

Table 2 Distribution of respondents by program and gender (in %)

Program	% of total (%)	% Women (%)	% Men (%)
Management—finance	34.5	83.3	16.7
Management—marketing	55.2	66.7	33.3
Biotechnology	10.3	55.6	44.4
Computer science (GGT)	3.4	0.0	100.0
Total	100	71.3	28.7

Given the nature of the sampling strategy, the study may be subject to self-selection bias, as individuals with a particular interest in entrepreneurship or stronger opinions about group projects may have been more likely to respond. This potential bias is taken into account when considering the generalizability of the results.

All participants had previously been involved in group projects integrated into their academic curricula, including real-life entrepreneurial experiences. The gender and program distribution of the sample is detailed in Table 2.

The responses were anonymized to ensure the confidentiality of the data collected.

The questionnaire was designed specifically for this study, drawing on relevant academic literature and prior validated instruments assessing both hard and soft skill development in higher education contexts. It was adapted to reflect the learning context of group-based and experiential learning projects. A preliminary version of the questionnaire was reviewed by educational experts for content validity and clarity.

The survey collected data on:

- The development of technical (hard) and behavioral (soft) skills;
- Students' perceptions of the impact of group projects and experiential learning on their entrepreneurial mindset;
- Early professional outcomes post-graduation, including employment status, geographic location, and industry sector.

All responses were anonymized to ensure confidentiality and ethical compliance in data handling.

3.2 *Analysis Methods*

The analysis of the collected data is based on a quantitative approach. The results were processed using software such as Excel for data visualization through cross-tabulations and charts. Correlation analyses were performed to identify significant links between different variables, allowing the measurement of the impact of group projects on the development of entrepreneurial competencies and soft skills.

3.3 Intervention Research

Intervention research has emerged as a major methodological approach in various disciplinary fields—from health to education, social policies, and the environment—due to its ability to combine knowledge production and practical action [10]. It refers to deliberate involvement in a process or system, aiming to influence events and/or their outcomes. The term can refer to isolated actions, but more frequently describes a structured set of activities conducted within a project, program, or initiative [11].

Organizational interventions, for example, aim to improve individual, collective, or institutional outcomes by modifying the way work is organized, designed, or managed. These interventions are often implemented to prevent or correct dysfunctions, or to promote sustainable positive effects [12].

In this context, intervention research requires a balance between scientific rigor and practical relevance to ensure that the interventions implemented are based on interdisciplinary expertise and respond to real-world challenges [12].

Our study fits into this perspective, having been conducted over four years. The students were divided into workgroups to participate in real entrepreneurial projects. These projects aimed to develop both the entrepreneurial mindset and technical and behavioral competencies. The main goal was to test the effectiveness of this immersive pedagogical method on the concrete development of entrepreneurial thinking by exposing students to an environment similar to that of real entrepreneurship. In addition to the immediate assessment of learning outcomes, an impact measurement was conducted one year after graduation to examine the long-term effects of this educational experience.

3.4 Study Limitations

The study presents several limitations. Firstly, the data collection relies on self-assessments by the students, which introduces a reporting bias. Additionally, the sample is limited to the 2022 and 2024 cohorts, which restricts the scope of the results and their generalizability to other cohorts or different educational contexts.

Moreover, while the descriptive statistics provided offer useful insights, the absence of basic inferential analyses (e.g., chi-square tests, t-tests) limits the analytical depth and the ability to draw statistically supported conclusions. Future studies would benefit from including such analyses to assess the significance of observed patterns and enhance the validity of the findings.

3.5 Central Hypothesis

HC: Experiential learning plays a critical role in the development of the entrepreneurial mindset and the key competencies necessary for entrepreneurial success, particularly within the context of entrepreneurial education.

The three most relevant sub-hypotheses for this study, which specifically address the development of entrepreneurial mindset and competencies through experiential learning in entrepreneurial education, are as follows:

H1: Students engaged in experiential learning activities, especially those incorporated within entrepreneurship education programs, develop a more pronounced entrepreneurial mindset.

H2: Experiential learning enhances entrepreneurial competencies, with a specific focus on leadership and decision-making, within the context of entrepreneurship education.

H3: Students who have participated in entrepreneurial projects through experiential learning are more likely to start their own business after graduation.

These three hypotheses form a strong foundation for analyzing the role of experiential learning in shaping the entrepreneurial capabilities of future entrepreneurs. Would you like me to propose a methodology to test these hypotheses within the context of entrepreneurial education?

4 Implementation

As outlined previously in this paper, students participated in an entrepreneurship training program over the course of one semester, lasting 13 weeks with 4-h sessions held in person. This comprehensive program covered a wide range of topics, including positive thinking, entrepreneurial mindset, business models, and business plans. Some of these sessions were delivered as practical workshops by professionals, experts, and business leaders with extensive industry experience.

The students attended these training sessions in teams of five, competing against one another. Each team represented a project idea and used these sessions to refine and develop their business concepts, progressing from ideation to detailed business plans, including financial, marketing, and legal components. Throughout the program, teams had access to expert consultations to receive personalized advice tailored to their projects. The goal was to equip the teams with the tools and knowledge necessary to optimize their entrepreneurial ventures.

At the end of the semester, an entrepreneurial competition in the form of a two-day hackathon was organized. This event featured three qualification phases culminating in a final pitch, where teams presented their projects to a jury composed entirely of external professionals—business leaders and investors. This setup ensured a neutral and industry-focused evaluation of the students' projects.

Additionally, a questionnaire was distributed to the students at the conclusion of each training cycle to gather feedback and evaluate their learning experience.

5 Discussion

Hypothesis 1: Students involved in experiential learning activities, particularly entrepreneurship training programs, develop a more pronounced entrepreneurial mindset.

The data suggest that students engaged in experiential learning activities exhibit entrepreneurial mindset characteristics such as innovation, proactive problem-solving, and opportunity recognition. These are consistent with Klucznik-Törő's [2] proposition that experiential learning environments cultivate not only knowledge but also the transversal competencies critical to entrepreneurship.

Tables 3, 4 and 5 show inclinations toward innovation (especially among women), motivation to improve existing products and services, and interest in environmental scanning for new opportunities. These traits resonate with the EntreComp Framework [7], particularly the dimensions of "spotting opportunities," "mobilizing resources," and "taking the initiative."

Table 3 Perception of innovation and addressing unmet market needs by gender

Innovation and addressing unmet market needs	Female (%)	Male (%)
Agree	65	35
Somewhat agree	78	22
Strongly agree	76	24

Table 4 Motivation to improve existing products/services by gender

Motivation in seeking solutions to improve existing products/services	Female (%)	Male (%)
Agree	28	32
Somewhat agree	6	8
Strongly agree	66	60

Table 5 Interest in analyzing the environment for new opportunities by gender

Analyzing the environment for new opportunities	Female (%)	Male (%)
Agree	50	36
Disagree	2	4
Somewhat agree	6	12
Strongly agree	42	48

The findings also echo Dewey's (1938/1997) theory of experiential learning, which emphasizes learning through active engagement with real-world challenges and the development of reflective judgment. Students' ability to identify unmet needs and opportunities suggests that they are not just absorbing content but engaging in the kind of deep, situated learning that builds authentic entrepreneurial awareness.

Moreover, the emphasis on market analysis and innovation ties into the ET-7 model's dimensions of Problem Solving and Tolerance for Ambiguity, further supporting the claim that experiential activities foster real entrepreneurial thinking. These findings suggest that the experiential pedagogy used in the program activates entrepreneurial intention through identity formation and cognitive engagement—key processes outlined in recent entrepreneurial learning literature (Peschl et al., 2021; Loia et al., 2025). Therefore, Hypothesis 1 is supported.

Hypothesis 2: Experiential learning strengthens entrepreneurial competencies, particularly leadership and decision-making in an entrepreneurial context.

The data from Tables 6 and 7 suggest for Hypothesis H2. It suggests that students significantly developed entrepreneurial soft skills, particularly those related to leadership, collaboration, and creative thinking. These findings are consistent with the idea that project-based experiential learning enhances not just knowledge but also personal and interpersonal competencies.

The most frequently cited skills—team spirit (22.47%), creativity (20.22%), and self-confidence (19.10%)—are foundational elements of effective entrepreneurial leadership. These are consistent with the ET-7 framework, where "mobilizing others," "resilience," and "decision-making" are considered core entrepreneurial capacities. The presence of stress management and adaptability further highlights students' readiness to function in complex, high-uncertainty environments—a hallmark of entrepreneurship.

Table 4 adds further support, showing high agreement with the statement, "Bringing the right people together for my project is truly exciting." Over 89% of respondents agreed or strongly agreed, suggesting high levels of engagement in leadership and coordination tasks. This reinforces the notion that students are not just participating in tasks, but developing leadership through action and reflection, which mirrors Dewey's notion of "learning-by-doing" and the reflective cycle of Kolb.

Table 6 Exploring leadership practices in the selection and coordination of project teams

Bringing the right people together for my project is truly exciting						
Gender	DAgree (%)	Disagree (%)	Somewhat agree (%)	Somewhat disagree (%)	Strongly agree (%)	Total général (%)
Female	40.63		9.38	1.56	48.44	100.00
Male	48.00	4.00	4.00		44.00	100.00
Total général	42.70	1.12	7.87	1.12	47.19	100.00

Table 7 Distribution of self-reported entrepreneurship-related entrepreneurial soft skills acquired during the experience

Entrepreneurial soft skill	Percentage (%)
Adaptability	8.99
Communication skills	7.87
Self-confidence	19.10
Creativity	20.22
Empathy	1.12
Team spirit	22.47
Stress management	10.11
Motivation	5.62
Optimism	3.37
Rigor	1.12

These results also are consistent with findings in recent work on entrepreneurial education (e.g., Klucznik-Törő [2]; Loia et al. [4]), which advocate for learning environments that simulate real entrepreneurial ecosystems—spaces where students take initiative, work across disciplines, and learn from failure.

Hypothesis 3: Students who have participated in entrepreneurial projects are more likely to start their own business after graduation.

The data provided in the table suggest valuable insight into the motivations of students across different fields of study to pursue entrepreneurship after graduation. According to the table, the majority of students across all fields indicate a high likelihood of starting their own business, with a general percentage of 90.59% saying "Yes" to the question of entrepreneurial intent after graduation.

Table 8 shows particularly strong entrepreneurial intentions among students in both Management—Marketing (GC) and Management—Finance (GF). For instance:

Table 8 Analysis of entrepreneurial intentions: the motivation to launch a new business

Field of study	No (%)	Yes (%)
Biotechnology (Biotech)	0.00	100.00
Geoinformatics (GGT)	0.00	100.00
Management—marketing (GC) promotion 2023	7.69	92.31
Management—marketing (GC) promotion 2024	10.00	90.00
Management—finance (GF) promotion 2024	7.14	92.86
Management—marketing (GC) promotion 2022	0.00	85.71
Management—finance (GF) promotion 2023	0.00	100.00
Management—finance (GF) promotion 2022	0.00	100.00
Management—marketing (GC) 2025	0.00	100.00
Management—finance (GF) promotion 2023	18.75	81.25

- Management—Marketing (GC) Promotion 2023: 92.31% of students are inclined to start their own business after graduation.
- Management—Finance (GF) Promotion 2023: 100% of the students expressed a desire to become entrepreneurs, suggesting a link between participation in entrepreneurial projects and business start-up intentions.
- Management—Marketing (GC) Promotion 2024: 90% of students also expressed interest in entrepreneurship.

Additionally, Biotechnology (Biotech) and Geoinformatics (GGT) students, despite not being directly involved in business-oriented fields, still show a strong inclination toward entrepreneurship, with 100% of Biotechnology students and 100% of Geoinformatics students indicating their interest in creating their own business after graduation. This suggests that experiential learning in entrepreneurial projects may have a significant impact on students across diverse academic backgrounds, even those in more technical or scientific fields. This is consistent with Klucznik-Törő's argument for the transversal integration of entrepreneurial learning across curricula, and reinforces the effectiveness of cross-disciplinary project-based learning environments in activating entrepreneurial agency.

The consistently high percentages of "Yes" responses in all the field categories further support Hypothesis 3. It indicates that students who engage in entrepreneurial projects, regardless of their field of study, are more likely to pursue entrepreneurship upon graduation. This trend is especially prominent in the Management fields, which emphasize skills like marketing, finance, and strategy—key competencies for successful entrepreneurship.

6 Conclusion

This study highlights the significant impact of experiential learning on the development of students' entrepreneurial mindset. The results confirm that immersive pedagogical activities enhance both transversal skills, such as innovation, initiative, and opportunity recognition, and key behavioral and managerial competencies, including leadership, decision-making, and collaboration. Participation in entrepreneurial projects is also strongly linked to students' intention to start their own business after graduation, regardless of their academic background. These findings validate the relevance of the theoretical frameworks employed (Dewey, Kolb, EntreComp, ET-7) and reinforce the importance of project-based approaches in higher education, particularly in a context of economic uncertainty and social transformation.

The study makes a valuable contribution to the field of entrepreneurial learning by providing empirical evidence of the structuring role that real-world experience plays in the development of cognitive, emotional, and social skills.

Beyond its academic contribution, the study also offers important implications for public policy and educational governance, especially in light of recent reforms in Moroccan higher education, which emphasize the importance of transversal skills and entrepreneurship education. The demonstrated link between experiential learning and entrepreneurial intent suggests that educational reforms aimed at fostering entrepreneurship should prioritize practice-based pedagogical methodologies. Governments and policymakers can draw on these findings to design national programs and curricula that incorporate entrepreneurial projects, interdisciplinary collaboration, and real-world problem-solving, particularly at higher education levels.

For educators and academic institutions, the results of this study advocate for the integration of entrepreneurial learning across all disciplines, including technical and scientific fields. Promoting autonomy, creativity, and interdisciplinary collaboration appears essential for shaping the next generation of leaders and innovators.

However, this study is based on a small, self-selected sample and self-reported data, which may introduce bias. It also measures entrepreneurial intent rather than actual business creation outcomes. Our future research should adopt longitudinal approaches to assess whether these intentions are indeed translated into business ventures and to evaluate the long-term impact of experiential learning on students' professional trajectories.

References

1. Mathushan, P.: Instilling entrepreneurial mind-set through entrepreneurship education: a systematic review of literature. Sabaragamuwa Univ. J. **18**(1), 43–54 (2020). https://doi.org/10.4038/suslj.v18i1.7753
2. Klucznik-Tîrô, A.: The new progression model of entrepreneurial education—guideline for the development of an entrepreneurial university with a sustainability approach. Sustainability **13**(20), 11243 (2021). https://doi.org/10.3390/su132011243
3. Visvizi, A., Troisi, O., Grimaldi, M., Loia, F.: Think human, act digital: activating data-driven orientation in innovative start-ups. Eur. J. Innov. Manag. **25**(6), 452–478 (2022). https://doi.org/10.1108/EJIM-04-2021-0206
4. Loia, F., Capolupo, N., Adinolfi, P.: Training me softly: metaverse-based immersive and gamified learning through a knowledge management perspective. J. Knowl. Manag. **29**(5), 1465–1489 (2025). https://doi.org/10.1108/JKM-08-2024-0929
5. Houston Peschl, C., Deng, C., Larson, N.: Entrepreneurial thinking: a signature pedagogy for an uncertain 21st century. Int. J. Manage. Educ. **19**(1), 100427 (2021). https://doi.org/10.1016/j.ijme.2020.100427
6. Mansoori, Y., Lackéus, M.: Comparing effectuation to discovery-driven planning, prescriptive entrepreneurship, business planning, lean startup, and design thinking. Small Bus. Econ. **54**, 791–818 (2020). https://doi.org/10.1007/s11187-019-00233-3
7. Bacigalupo, M., Kampylis, P., Punie, Y., Van den Brande, G.: EntreComp: The entrepreneurship competence framework (JRC Science for Policy Report). Publications Office of the European Union (2016). https://doi.org/10.2791/593884
8. Davis, M.H., Hall, J.A., Mayer, P.S.: Developing a new measure of entrepreneurial mindset: reliability, validity, and implications for practitioners. Consult. Psychol. J. Pract. Res. **68**(1), 21–48 (2016). https://doi.org/10.1037/cpb0000045
9. Kier, A.S., McMullen, J.S.: Entrepreneurial imaginativeness in new venture ideation. Acad. Manag. J. **61**(6), 2265–2295 (2018). https://doi.org/10.5465/amj.2017.0395
10. Kilbourne, A.M., Garrido, M.M., Brown, A.F.: Translating research into policy and action. Health Serv. Res. **57**(Suppl. 1), 5–8 (2022). https://doi.org/10.1111/1475-6773.13980

11. Belcher, B., Palenberg, M.: Outcomes and impacts of development interventions: toward conceptual clarity. Am. J. Eval. **39**(4), 478–495 (2018). https://doi.org/10.1177/1098214018765698
12. Von Thiele Schwarz, U., Nielsen, K., Edwards, K., Hasson, H., Ipsen, C., Savage, C., Reed, J.E.: How to design, implement and evaluate organizational interventions for maximum impact: the Sigtuna principles. Eur. J. Work Organ. Psy. **30**(3), 415–427 (2020). https://doi.org/10.1080/1359432X.2020.1803960

University Education and Entrepreneurial Intention: The Role of University Climate and Entrepreneurship Education

Raquel Ferreras-Garcia, Jordi Sales-Zaguirre, and Enric Serradell-López

Abstract Entrepreneurship is a key driver of economic growth, innovation, and societal well-being. Higher education institutions play a crucial role in fostering entrepreneurial intention among students through their learning environments, curricula, and pedagogical approaches. This study, conducted within the framework of the Global University Entrepreneurial Spirit Students' Survey (GUESSS) Spain, aims to identify the university-related factors that significantly predict entrepreneurial intention levels among students. Using data from the GUESSS Spain 2023 survey and a sample of more than 36,000 students, we examine how different dimensions of university education influence students' entrepreneurial intentions. To analyze these effects, a multiple linear regression analysis is conducted in order to assess how these predictors contribute to the variance in entrepreneurial intention indicators. The independent variables include measures of university climate, entrepreneurial education received, and entrepreneurial learning experiences. Preliminary findings suggest that a positive and supportive university climate, along with experiential and well-structured entrepreneurial education, significantly enhances students' entrepreneurial intentions. The results of this study provide valuable insights for policymakers, educators and university administrators to design more effective entrepreneurship education programs that contribute to the competitiveness and sustainable economic development of regions. By strengthening the entrepreneurial ecosystem within universities, this research contributes to fostering innovation and growth through education and technology.

R. Ferreras-Garcia · E. Serradell-López
Universitat Oberta de Catalunya, Barcelona, Spain
e-mail: rferreras@uoc.edu

E. Serradell-López
e-mail: eserradell@uoc.edu

J. Sales-Zaguirre (✉)
IQS, Universitat Ramon Llull, Barcelona, Spain
e-mail: jordi.sales@iqs.url.edu

A. Visvizi et al. (eds.), *Research and Innovation Forum 2025 Volume 1*,
Springer Proceedings in Complexity, https://doi.org/10.1007/978-3-032-09066-9_16

Keywords Entrepreneurial intention · Entrepreneurship education · Higher education · GUESSS Spain · SDG-4,8

1 Introduction

Entrepreneurship is widely recognized as a critical driver of economic growth, innovation, and societal progress. Higher education institutions play an essential role in cultivating entrepreneurial intentions among students by providing structured learning environments, experiential opportunities, and targeted pedagogical approaches [1, 2]. Within this context, universities are increasingly integrating entrepreneurship education into their curricula, thereby creating ecosystems that inspire innovation, mitigate risk aversion, and empower students to perceive entrepreneurship as a viable career option [2, 3].

University ecosystems are designed to go beyond traditional instruction, incorporating essential components such as mentorship programs, incubators, and partnerships with industry and policymakers. These initiatives provide students with the tools and confidence to transition from intention to action, as they foster entrepreneurial self-efficacy and reduce barriers to venture creation [4]. Research underscores the importance of inspirational elements in entrepreneurship education, delivered through exposure to entrepreneurial role models, experiential learning opportunities, and creative problem-solving, as these aspects significantly enhance students' attitudes and intentions toward self-employment [2].

The broader sociocultural and institutional context also exerts significant influence on entrepreneurial behavior. Factors such as gender, cultural norms, and perceived societal expectations shape entrepreneurial intentions, often requiring tailored interventions to address disparities in access and engagement [5, 6]. Collaborative networks, including family support and peer mentorship, have been shown to enhance entrepreneurial capacity by building social capital and reducing perceived risks [3, 7]. Nevertheless, systemic variations in university-level entrepreneurial ecosystems highlight the need for a more uniform approach to fostering entrepreneurship across higher education institutions [4].

Despite the growing emphasis on entrepreneurship education, significant challenges remain. For instance, while entrepreneurial intentions serve as a precursor to entrepreneurial action, the transition from intention to venture creation is often fraught with barriers, emphasizing the need for longitudinal studies to understand the intention-action gap [2]. Additionally, disparities in entrepreneurial intentions and outcomes between men and women, as well as across cultural and economic contexts, underscore the need for further targeted research and policy interventions [5, 8].

This study, conducted within the framework of the GUESSS Spain 2023, seeks to contribute to this ongoing discourse by analyzing university-related predictors of entrepreneurial intentions. Specifically, the research examines how variables such as university climate, entrepreneurship education, and program learning influence students' entrepreneurial aspirations. By addressing these factors, this study aims to offer actionable insights for educators, policymakers, and university administrators striving to strengthen entrepreneurial ecosystems, with the ultimate goal of fostering innovation and sustainable economic development.

The objectives of the study are to identify the university-related factors that significantly predict entrepreneurial intentions; to examine the influence of university climate and entrepreneurship education on entrepreneurial aspirations and provide actionable recommendations for enhancing entrepreneurship education and fostering an entrepreneurial ecosystem within higher education institutions.

The article is structured as follows: Sect. 2 provides a comprehensive review of the relevant literature, establishing the theoretical foundation for the study. Section 3 outlines the research methodology and data collection procedures, detailing the approach used to investigate the research questions. Section 4 discusses the results and offers a thorough analysis of the data. Finally, the article concludes with a discussion of the key findings, practical implications, and recommendations, while also acknowledging the study's limitations and suggesting directions for future research.

2 Literature Review

2.1 The Role of Higher Education Institutions

Higher education institutions play a pivotal role in fostering entrepreneurial intention by providing the knowledge, skills, and support necessary for entrepreneurial ventures. Universities are increasingly seen as key players in nurturing entrepreneurial talent and fostering innovation, making them integral to regional and national economic development. Research shows that universities contribute to entrepreneurial intentions through their curricula, pedagogical approaches, and supportive environments [9]. Structured entrepreneurship courses equip students with theoretical and practical knowledge about business creation, while experiential learning activities, such as startup simulations and business plan competitions, enable application in real-world contexts [10].

Several studies have identified the digital divide as a key barrier to entrepreneurial development, especially in the Global South. This divide not only limits access to technology but also reflects broader social and economic inequalities, directly affecting the capacity for innovation and entrepreneurship [11]. Moreover, addressing this issue in contexts with significant gender and socioeconomic disparities, such as Pakistan, requires a holistic approach that combines digital literacy initiatives, gender equality promotion, and targeted entrepreneurial education [12].

Universities also act as agents of cultural transformation, promoting an entrepreneurial mindset and cultivating a pro-entrepreneurship climate. This climate encourages innovation, risk-taking, and creativity [13]. A supportive university environment positively influences students' attitudes toward entrepreneurship and fosters their perceived behavioral control, key determinants of entrepreneurial intention [14].

Extracurricular activities, such as innovation challenges, entrepreneurial clubs, and hackathons, further enhance students' entrepreneurial self-efficacy [15]. These activities build critical competencies such as problem-solving, opportunity recognition, and leadership, contributing to shaping entrepreneurial aspirations. By fulfilling these roles, higher education institutions bridge the gap between education and business creation, supporting broader societal goals, including employment generation, innovation diffusion, and regional economic growth.

2.2 University Environment

The university environment significantly influences entrepreneurial intention by shaping students' attitudes, self-efficacy, and opportunity identification. Studies suggest that universities fostering a supportive environment enhance entrepreneurial intentions by mediating attitudes toward entrepreneurship and perceived behavioral control [16, 17]. Faculty and institutional leadership play crucial roles by integrating real-world examples, case studies, and mentorship opportunities into teaching, enhancing students' entrepreneurial readiness.

Policies such as startup grants, venture capital access, and flexible academic schedules for student entrepreneurs create a more empowering university environment. Beyond fostering individual entrepreneurial ambitions, universities contribute to societal economic development by equipping students with entrepreneurial skills and encouraging risk-taking. Findings underscore the need to invest in university entrepreneurship programs aligned with national economic priorities [4].

Interactive learning spaces and activities like business plan competitions and incubators allow students to practice entrepreneurial skills in realistic settings. These environments enhance technical skills and build confidence, turning entrepreneurial ideas into action. Integrating policy support, mentorship programs, and resource allocation can significantly enhance students' entrepreneurial engagement.

2.3 Entrepreneurship Education

Entrepreneurship education is critical in shaping students' entrepreneurial intentions by equipping them with the knowledge, skills, and mindset necessary to identify and act upon opportunities. By focusing on theoretical and practical applications, entrepreneurship education fosters confidence and builds core competencies for entrepreneurial activities [16]. It motivates students by visualizing the feasibility and desirability of entrepreneurial careers, enhancing ambition and self-efficacy [18].

Experiential learning, such as real-world simulations, case studies, and projects, promotes creativity and problem-solving skills while also fostering resilience and adaptability, essential traits for overcoming the challenges of starting a business [19, 20]. Additionally, structured mentorship programs, combined with tailored support, further boost students' confidence in seizing entrepreneurial opportunities [21].

Program learning plays a pivotal role in enhancing entrepreneurial intentions through structured, interdisciplinary, and experiential approaches. Immersing students in real-world entrepreneurial scenarios and challenges related with Sustainable Development Goals improved critical thinking and technical skills demonstrating an interest in solving these problems [22, 23]. Moreover, collaborative projects and teamwork also build interpersonal competencies, leveraging diverse perspectives that are vital for entrepreneurship [19].

Personalized mentorship and university-led initiatives connect students with entrepreneurial ecosystems, such as accelerators and incubators, providing essential resources. Incorporating digital tools and adaptive learning technologies customizes the educational experience, ensuring better engagement and comprehension. The integration of methodologies like simulations and e-learning creates a holistic approach to entrepreneurship education, addressing diverse learning needs and strengthening entrepreneurial intentions [21].

3 Methodology

3.1 *Data*

This study utilized data from the GUESSS Spain project, one of the largest entrepreneurship research initiatives in the world. This article specifically draws on data collected from Spanish students and corresponds to the tenth data collection wave, in which 57 countries participated. A total of 44,815 students participated in the Spanish survey and a total of 35,815 answered the questions related to the study. Table 1 summarizes the profile of the respondents.

Table 1 Respondents' profile

		Frequency	Percentage
Gender	Male	12,337	34.4
	Female	23,152	64.6
	Other	326	0.9
Degree	Bachelor's degree	28,437	79.4
	Master's degree	5338	14.9
	PhD	1754	4.9
	Other	286	0.8
	Total	35,815	100

3.2 Measurement of Variables

In this study, we selected a series of variables to analyze the factors influencing entrepreneurial intention among university students: University Environment, Entrepreneurship Education and Program Learning. These variables were chosen based on their theoretical significance and prior empirical evidence in the field of entrepreneurship research.

The first set of variables focuses on the university environment, as conceptualized by Franke and Lüthje [8]. This dimension reflects students' perceptions of how the university environment supports entrepreneurial activities. It is measured using a Likert scale ranging from 1 to 7 across the following items:

- UE1: "The climate of my university inspires me to develop ideas for new businesses."
- UE2: "My university offers an environment conducive to becoming an entrepreneur."
- UE3: "My university motivates me to participate in entrepreneurial activities."
- UE4: "I can easily obtain advice and guidance on entrepreneurship at my university."

The set of entrepreneurship education variables was adapted from the GUESSS project. This is a binary variable (1: Agreement with the statement, 0: No agreement) capturing students' engagement with entrepreneurship courses, with the following categories:

- EE1: "I have never attended entrepreneurship courses until now." (Note: EE1 is reverse-coded so that 1 indicates entrepreneurship education exposure)
- EE2: "I have attended at least one voluntary course on entrepreneurship."
- EE3: "I have attended at least one compulsory course on entrepreneurship that was part of my study plan."
- EE4: "I am currently doing a specific course on entrepreneurship."

The set of Program learning variables, adapted from [2], measures the perceived effectiveness of entrepreneurship courses or services that students receive in terms of practical skills development and knowledge acquisition. This variable is operationalized using a Likert scale ranging from 1 to 7 and includes the following items:

- PL1: "The courses increased my knowledge of the attitudes, values, and motivations needed for entrepreneurship."
- PL2: "The courses increased my knowledge of the activities necessary to start a business."
- PL3: "The courses improved my practical business management skills required to create a company."
- PL4: "The courses improved my ability to create business networks."
- PL5: "The courses helped me to identify business opportunities aligned with my interests."

- PL6: "The courses helped me to identify business opportunities related to trends, discontinuities, technological impacts, and other changes in the environment."

A multiple linear regression analysis was conducted using SPSS to assess the influence of the above-mentioned variables on students' overall entrepreneurial intention. The dependent variable, entrepreneurial intention (EI), was computed as the average of the six intention-related items (Likert scale from 1 to 7), based on [1]. Aggregating these items into a total variable provides a more robust measure of entrepreneurial intention by capturing different aspects of students' intentions. The items included are:

- EI1: "I am willing to do anything to be an entrepreneur."
- EI2: "My professional goal is to become an entrepreneur."
- EI3: "I will do everything possible to create and manage my own business."
- EI4: "I am determined to create a company in the future."
- EI5: "I have thought very seriously about starting a business."
- EI6: "I have the firm intention of creating a business one day."

4 Data Analysis

Entrepreneurial education has been widely recognized as a key driver of entrepreneurial intentions among university students. The present study aims to analyze the relationship between entrepreneurial education experiences and students' willingness to engage in entrepreneurship.

To ensure the internal consistency of the EI measure, a Cronbach's Alpha test was conducted. The results showed $\alpha = 0.960$, indicating excellent reliability. This justifies the use of the aggregated mean score as a valid measure of entrepreneurial intention.

The F-test was highly significant ($F = 484.130$, $p < 0.001$), confirming that the independent variables collectively contribute to predicting entrepreneurial intention.

Table 2 presents the unstandardized and standardized regression coefficients along with their significance levels and VIF values.

The regression analysis reveals key insights into the factors influencing entrepreneurial intention (EI). Regarding the university environment, having a climate that inspires entrepreneurial ideas (UE1, $B = 0.130$, $p < 0.001$) significantly fosters EI, whereas merely providing a general entrepreneurial-friendly atmosphere (UE2, $B = 0.062$, $p < 0.001$) has a weaker impact. Surprisingly, university-driven entrepreneurial motivation (UE3, $B = -0.049$, $p < 0.001$) and accessibility to entrepreneurship guidance (UE4, $B = -0.052$, $p < 0.001$) negatively affect EI, suggesting that these initiatives may not be effectively implemented or perceived as valuable. Concerning entrepreneurship education, having never attended such courses (EE1, reverse-coded, $B = -0.222$, $p < 0.001$) show significantly higher entrepreneurial intention, reinforcing the importance of exposure to entrepreneurship

Table 2 Regression analysis

Model	B	Beta	t	Sig	VIF
Constant	2.109		44.606	<0.001	
UE1	0.130	0.127	14.318	<0.001	3.232
UE2	0.062	0.061	5.971	<0.001	4.274
UE3	−0.049	−0.051	−5.979	<0.001	2.962
UE4	−0.052	−0.053	−6.625	<0.001	2.674
EE1	−0.222	−0.058	−5.628	<0.001	4.323
EE2	0.160	0.037	4.162	<0.001	3.223
EE3	−0.258	−0.044	−6.762	<0.001	1.740
EE4	0.406	0.031	6.149	<0.001	1.058
PL1	0.106	0.112	12.258	<0.001	3.423
PL2	0.044	0.048	4.528	<0.001	4.593
PL3	0.117	0.124	11.701	<0.001	4.614
PL4	−0.057	−0.061	−7.315	<0.001	2.860
PL5	0.017	0.018	1.684	0.092	4.600
PL6	0.076	0.081	7.767	<0.001	4.466

Dependent variable: EI

training. Voluntary courses (EE2, $B = 0.160$, $p < 0.001$) and ongoing entrepreneurship education (EE4, $B = 0.406$, $p < 0.001$) positively contribute to EI, whereas mandatory entrepreneurship courses (EE3, $B = -0.258$, $p < 0.001$) have a negative impact, possibly due to their structure or perceived relevance. The effect of perceived learning (PL) from entrepreneurship courses highlights that increased knowledge of attitudes, values, and motivations (PL1, $B = 0.106$, $p < 0.001$), the knowledge of the activities to start a business (PL2, $B = 0.044$, $p > 0.001$) and practical business management skills (PL3, $B = 0.117$, $p < 0.001$) significantly enhance EI. Conversely, improved networking skills (PL4, $B = -0.057$, $p < 0.001$) negatively impact EI, suggesting that while networking is beneficial, it may not directly translate into entrepreneurial intention. Identifying business opportunities based on trends and technological changes (PL6, $B = 0.076$, $p < 0.001$) has a stronger effect on EI than recognizing opportunities aligned with personal interests (PL5, $B = 0.017$, $p = 0.092$), which is not statistically significant. Additionally, multicollinearity diagnostics indicate acceptable variance inflation factors (VIF < 5), suggesting that collinearity is not a major issue in the model. Overall, these findings suggest that a university's role in fostering EI is highly dependent on the effectiveness of its entrepreneurial climate and educational programs, emphasizing the importance of well-structured and engaging entrepreneurship education that aligns with students' aspirations and needs.

5 Discussion and Conclusions

This study provides valuable insights into the factors influencing entrepreneurial intentions among university students, emphasizing the critical roles of the university environment, entrepreneurship education, and experiential learning.

Regarding the university environment, a supportive university environment (UE1, UE2) is pivotal in nurturing students' entrepreneurial aspirations. Prior research suggests that an entrepreneurial-oriented university environment positively correlates with increased entrepreneurial intentions among students [24]. Conversely, environments lacking entrepreneurial support (UE3, UE4) may lead to heightened perceptions of barriers to entrepreneurship [24]. Therefore, universities aiming to foster entrepreneurship should cultivate an ecosystem that encourages innovation, provides access to resources, and offers mentorship opportunities.

With respect to entrepreneurship education, the evidence shows that its effectiveness depends on its structure, delivery, and relevance to students' needs. Voluntary participation in entrepreneurship courses (EE2) appears to have a more positive impact than mandatory programs (EE3), highlighting the importance of student engagement and motivation in entrepreneurial learning.

The study also underscores the significance of practical, hands-on learning experiences in strengthening entrepreneurial intention. Programs that focus on the development of entrepreneurial attitudes (PL1), business management skills (PL3), and opportunity recognition (PL5, PL6) contribute more effectively to fostering entrepreneurial ambition. However, merely providing resources or networking opportunities (PL4) does not necessarily translate into increased entrepreneurial motivation, indicating that the quality and perception of these initiatives play a crucial role. Additionally, integrating environmental awareness into entrepreneurship curricula has been found to increase students' intentions to engage in sustainable entrepreneurial ventures [5]. These findings suggest that entrepreneurship education should be dynamic, incorporating practical experiences and addressing contemporary issues to remain relevant and impactful.

With regard to the role of experiential learning, engaging students in hands-on projects, simulations, and real-world problem-solving not only enhances their entrepreneurial skills but also strengthens their confidence and motivation to pursue entrepreneurial careers [25]. For example, programs that involve students in developing solutions to societal challenges have been effective in fostering entrepreneurial mindsets [26]. Consequently, integrating experiential learning opportunities into the university curriculum is essential for translating entrepreneurial intentions into actions.

These findings reinforce the need for universities to develop dynamic and student-centered entrepreneurship programs that go beyond theoretical instruction. Institutions should focus on creating an ecosystem where students perceive entrepreneurial support as accessible, relevant, and actionable. Actively engaging students' participation and embedding experiential learning components in entrepreneurship education can substantially enhance its overall impact. By adopting these strategies, universities can play a crucial role in cultivating entrepreneurial talent and driving innovation-led economic development.

While this study offers important contributions, several limitations must be acknowledged. The data is based on subjective self-reported responses, which may introduce biases. While the data is considered representative of the university community, there is limited contextual generalizability; the findings are context-dependent and may not be applicable across different cultural or institutional settings. Additionally, the study is confined to a specific educational context, which may restrict generalizability to other institutions and cultural settings.

Future research should prioritize longitudinal studies to examine the sustained impact of entrepreneurship education on students' entrepreneurial engagement beyond their academic years. Comparative studies across multiple institutional contexts, including universities with varying levels of entrepreneurial support and infrastructure, would provide deeper insights into best practices for fostering entrepreneurial ambition. Additionally, further exploration of the role of specific pedagogical interventions, such as mentorship programs, experiential learning opportunities, and the integration of digital tools in entrepreneurship training, could enhance our understanding of effective educational strategies.

By addressing these gaps, future research can contribute to a more nuanced and globally relevant framework for entrepreneurship education. Universities, policymakers, and educators can leverage these insights to develop evidence-based strategies that empower students to transition from entrepreneurial intention to successful venture creation, ultimately fostering a culture of innovation and economic growth.

References

1. Liñán, F., Chen, Y.W.: Development and cross-cultural application of a specific instrument to measure entrepreneurial intentions. Entrep. Theory Pract. **33**(3), 593–617 (2009)
2. Souitaris, V., Zerbinati, S., Al-Laham, A.: Do entrepreneurship programmes raise entrepreneurial intention of science and engineering students? The effect of learning, inspiration, and resources. J. Bus. Ventur. **22**(4), 566–591 (2007). https://doi.org/10.1016/j.jbusvent.2006.05.002
3. Laspita, S., Sitaridis, I., Sarri, K.: The effect of sustainable development goals and subjective well-being on art nascent entrepreneurship: the moderating role of entrepreneurship education. Educ. Sci. **14**(5), 491 (2024). https://doi.org/10.3390/educsci14050491
4. Nguyen, T., Phan, T.: The role of university entrepreneurial ecosystems in fostering students' startup intentions. J. Small Bus. Enterp. Dev. **31**(2), 187–209 (2024)
5. Moreno-Gómez, J., Skrbková, D., Rydvalová, P.: The entrepreneurial intention of university students: a comparative perspective. Eur. Res. Manag. Bus. Econ. **28**(1), 43–59 (2021)
6. Skrbková, D., Rydvalová, P.: Entrepreneurial intentions: a comparative study of African and European university economics students. J. Afr. Bus. (2023). https://doi.org/10.1080/15228916.2023.2276541
7. Edelman, L.F., Manolova, T.S., Shirokova, G.: The impact of family support on young entrepreneurs' start-up activities. J. Bus. Ventur. **31**(4), 428–448 (2016)
8. Franke, N., Lüthje, C.: Entrepreneurial intentions of business students-a benchmarking study. Int. J. Innov. Technol. Manag. **1**(3), 269–288 (2004)

9. Fayolle, A., Gailly, B.: From craft to science: teaching models and learning processes in entrepreneurship education. J. Eur. Ind. Train. **32**(7), 569–593 (2008). https://doi.org/10.1108/03090590810899838
10. Rideout, E.C., Gray, D.O.: Does entrepreneurship education really work? A review and methodological critique. J. Bus. Ventur. **28**(1), 106–121 (2013). https://doi.org/10.1016/j.jbusvent.2012.02.004
11. Visvizi, A., Troisi, O., Grimaldi, M.. Digital divide and entrepreneurial orientation in the global south: quantifying and explaining the nexus. In: Visvizi, A., Troisi, O., Grimaldi, M. (eds.) Research and Innovation Forum 2022. RIIFORUM 2022. Springer Proceedings in Complexity. Springer, Cham (2023). https://doi.org/10.1007/978-3-031-19560-0_62
12. Barra, C., Grimaldi, M., Muazzam, A., Troisi, O., Visvizi, A.: Digital divide, gender gap, and entrepreneurial orientation: how to foster technology adoption among Pakistani higher education students? Socioecon. Plann. Sci. **93**, 101904 (2024)
13. Guerrero, M., Urbano, D.: The development of an entrepreneurial university. J. Technol. Transf. **37**(1), 43–74 (2012). https://doi.org/10.1007/s10961-010-9171-x
14. Liñán, F., Nabi, G., Krueger, N.: Regional variations in entrepreneurial cognitions: start-up intentions of university students in Spain. Entrep. Reg. Dev. **23**(3–4), 187–215 (2011). https://doi.org/10.1080/08985626.2011.557739
15. Sánchez, J.C.: The impact of an entrepreneurship education program on entrepreneurial competencies and intention. J. Small Bus. Manage. **51**(3), 447–465 (2013). https://doi.org/10.1111/jsbm.12025
16. Ilman, A.H., Ananda, N.A., Pohan, A.S.: Entrepreneurial education and university environment toward students entrepreneurial intention. Int. J. Innov. Sci. Res. Technol. **5**(1), 1271–1278 (2020)
17. Ebewo, P.E., Nesamvuni, A.E.: Entrepreneurial environment as an antecedent of university students' entrepreneurship intentions. J. Entrep. Innov. **1**(2) (2020)
18. Wannamakok, W., Liang, W.K.: Entrepreneurship education and entrepreneurial intention: perspectives on institutional theory. J. Entrep. Bus. Econ. **7**(2), 106–129 (2019)
19. de Krosse, L.E., van der Werf, J.J., van der Meer, H.J., van Geert, P.L.: Learning entrepreneurship by entrepreneurial learning?. In: 18th Annual High Technology Small Firms Conference, HTSF 2010. University of Twente (2010)
20. Nugroho, N., Suryanto, S., Irianto, J.: Analysis of the effect of entrepreneurship education, family environment. In: RSF Conference Series: Business, Management and Social Sciences (2023). https://doi.org/10.31098/bmss.v3i3.701
21. Ku, W., Jeon, B.: The effect of the university entrepreneurial-environment on entrepreneurship: mediating effects of discovering-entrepreneurial-opportunities. The Korean Career, Entrepreneurship & Business Association (2024). https://doi.org/10.48206/kceba.2024.8.5.71
22. Jaenudin, A., Mulyono, K., Widodo, J.: The role of environmental learning to improve students' entrepreneurial intention. In: Proceedings of the 3rd International Conference on Economics, Business and Economic Education Science, ICE-BEES 2020, 22–23 July 2020, Semarang, Indonesia (2021). https://doi.org/10.4108/EAI.22-7-2020.2307889
23. Ferreras-García, R., Sales-Zaguirre, J., Serradell-Lopez, E.: Impact on student learning of a change in course design. In: Springer Proceedings in Complexity—Rii Forum 2024 (2025)
24. Shava, H., Ganda, F., Dzimunya, L.: The impact of university environment and culture on entrepreneurial barriers and intentions: Investigating the mediation role of public infrastructure. ResearchGate Working Paper (2021)
25. Maritz, A., Jones, C., Shulman, A.: The impact of experiential learning on entrepreneurial skills development: a longitudinal study. Springer J. Entrep. Innov. **5**(2), 132–148 (2024)
26. Giones, F., Brem, A., Berger, E.S.: Exploring the effectiveness of entrepreneurship education programs: a systematic review. arXiv:2012.03746 (2020)

The Contribution of Hungarian Universities to the UN Sustainable Development Goals: A Comparative Analysis of State-Run, Church-Affiliated, and Model-Transition Institutions (2019–2023)

Emese Belenyesi and Peter Sasvari

Abstract Sustainability and the United Nations' Sustainable Development Goals (SDGs) are increasingly significant in academic research, particularly in funding opportunities and scientific visibility. SDG categorization provides universities with a pathway to enhance international recognition and improve rankings. This study aims to analyze how Hungarian universities contribute to achieving the SDGs through SDG-related publications. The research examines 18 Hungarian universities, including state-run, church-affiliated, and model-transition institutions. It also assesses which SDGs are prioritized within these institutions' research activities and identifies trends in SDG-related publications between 2019 and 2023. The study utilizes data from the Scopus and SciVal databases, highlighting SDG-related publications, their proportion, and citation impact relative to the total academic output. Particular attention is paid to the thematic distribution of SDGs, the research priorities of institutions, and the scientific and societal impact of the publications. Initial results indicate that SDG-related publications strengthen Hungary's international recognition while contributing to the global implementation of sustainable development goals. The findings provide valuable insights for shaping the future research strategies of Hungarian universities, with a special focus on sustainability.

Keywords Hungarian universities · Sustainable development goals · SDG-3 · Scopus · SciVal

E. Belenyesi (✉) · P. Sasvari
Ludovika University of Public Service, Budapest, Hungary
e-mail: belenyesi.emese@uni-nke.hu

P. Sasvari
e-mail: sasvari.peter@uni-nke.hu

P. Sasvari
University of Miskolc, Miskolc, Hungary

A. Visvizi et al. (eds.), *Research and Innovation Forum 2025 Volume 1*,
Springer Proceedings in Complexity, https://doi.org/10.1007/978-3-032-09066-9_17

1 Introduction

In 2015, the United Nations (UN) adopted the Sustainable Development Goals (SDGs), comprising 17 goals and 169 targets aimed at eradicating poverty, reducing inequalities, promoting economic growth, and protecting the environment by 2030 [1]. Higher education institutions (HEIs) play a vital role in advancing these goals through education, research, and community engagement.

With diverse institutional structures, Hungarian universities offer a distinctive case for studying SDG integration in higher education. This study analyzes 18 universities—categorized as state-run, church-affiliated, and model-transition institutions (see Annex, Table 1)—focusing on SDG-related research outputs, thematic priorities, and societal contributions within a European context.

Recent global developments, including the COVID-19 pandemic and the intensification of climate and social justice crises, have amplified the urgency to address sustainability challenges through higher education. [2] HEIs are under increasing pressure to not only produce high-quality research but also to demonstrate tangible impact aligned with the SDGs. In this context, understanding how universities strategically prioritize SDG themes in their research portfolios has become a critical area of inquiry.

Although various global reports and case studies emphasize the importance of university contributions to sustainability, the literature reveals several gaps. Most existing research tends to focus on leading institutions in Western Europe, North America, or East Asia, with limited comparative analyses from Central and Eastern European (CEE) contexts. Furthermore, while studies often assess institutional sustainability performance through rankings and indicators, few delve deeply into the thematic distribution of SDG-related publications across different types of universities, particularly the emerging "model-transition" category in Hungary. This gap calls for systematic research that not only quantifies publication trends but also maps the alignment between institutional profiles and SDG priorities.

By examining publication trends between 2019 and 2023, the study evaluates Hungarian universities' contributions to the SDGs, highlighting key thematic areas, engagement differences, and citation impact. It also explores systemic challenges and opportunities to improve alignment with the SDGs. The findings aim to inform policy and practice and promote stronger sustainability strategies across the sector.

2 Theoretical Background

Academic research and policy debates have increasingly focused on the role of HEIs in achieving the SDGs. Universities serve as knowledge hubs that drive societal transformation through education, research, and community engagement [3]. An SDG-oriented culture can be promoted by embedding sustainability into curricula, research, and governance structures, as highlighted in recent literature [4].

Globally, institutions are increasingly aligning their operations with the SDGs, often assessed through frameworks like the Times Higher Education Impact Rankings and QS Sustainability Ranking [5]. Leading universities prioritize themes such as SDG 4 (Quality Education), SDG 13 (Climate Action), and SDG 11 (Sustainable Cities and Communities) [6]. This has reinforced the concept of universities as "living laboratories" for sustainability, integrating these principles into institutional strategies [7].

In Central and Eastern Europe, including Hungary, the integration of SDGs varies significantly due to governance models, funding, and institutional priorities [8]. State-run, church-affiliated, and model-transition universities each adopt distinct approaches, necessitating a comparative analysis of their engagement.

3 Literature Review

HEIs play a pivotal role in achieving the SDGs by leveraging research, innovation, and policy influence to address global challenges. Many universities now embed sustainability into their academic missions and advance interdisciplinary research in areas such as climate change, poverty, and sustainable development [9].

An increasing number of institutions participate in global rankings, such as the Times Higher Education Impact Rankings and QS Sustainability Rankings, to demonstrate their contributions to the SDGs [7]. However, integration levels vary globally. Institutions in Western Europe, North America, and parts of Asia show stronger engagement than those in developing regions [7], highlighting the need for equitable funding and capacity-building in low-income countries.

Three key areas of university contribution include:

(1) Research and Innovation—Leading universities such as Cambridge and Stanford have pioneered work on carbon neutrality and sustainable cities [4].
(2) Education and Curriculum—Institutions like Harvard and the University of Sydney integrate sustainability across disciplines to ensure broad engagement with SDG themes [10].
(3) Community Engagement—Universities often partner with governments, NGOs, and industry. For example, the University of Cape Town contributes to sustainable water management in African communities [8].

SDG-related research shows thematic concentration, particularly on SDG13 (Climate Action), SDG4 (Quality Education), and SDG9 (Industry, Innovation and Infrastructure). Other goals, like SDG14 (Life Below Water) and SDG16 (Peace and Justice), remain underrepresented [9].

Despite progress, universities face challenges in fully integrating SDGs due to financial limitations, bureaucratic barriers, and inconsistent reporting frameworks. Moreover, the competitive nature of sustainability rankings can skew focus toward selected SDGs rather than comprehensive strategies [7].

To strengthen sustainability efforts, the Hungarian Universities Sustainability Platform (MEFP) was launched in September 2022, with 14 institutions—including state, church, and model-transition universities—joining the initiative. Its goals are to share best practices and coordinate collaborative sustainability actions [11].

Data show that Hungarian universities particularly focus on SDG3 (Good Health), SDG4, SDG11 (Sustainable Cities), and SDG13. These priorities are reflected in research on public health, renewable energy, urban development, and inclusive education [12].

However, several systemic challenges hinder deeper SDG integration. These include limited resources, fragmented coordination, and inconsistent reporting frameworks, which affect both research scope and international visibility [10].

While SDG-related research improves global visibility and rankings, its overall impact in Hungary remains modest compared to Western Europe, revealing disparities in institutional support and societal engagement [13]. Western universities benefit from strong funding and established sustainability cultures, while Hungary, though showing progress, faces structural barriers [5].

In conclusion, universities play a vital role in advancing the SDGs. Hungarian institutions contribute meaningfully through research and engagement, but further alignment with global best practices, enhanced funding, and stronger collaborations are needed to strengthen their impact on sustainability.

4 Research Design

This study applies a *mixed-methods approach* combining *quantitative bibliometric analysis* with a *comparative institutional framework*. The central aim is to identify how Hungarian universities contribute to the SDGs through research outputs and to assess institutional variations in terms of volume, thematic focus, and impact. University of Veterinary Medicine Budapest (ÁOE) was excluded due to lack of classification in the database. Some private Hungarian universities were omitted from the analysis due to unavailable data.

The research is concerned with the following questions:

RQ1: How has the number of SDG-related publications changed between 2019 and 2023? Which universities have shown the most significant increase?

RQ2: Which SDGs generate the highest volume and citation rates? How are these outputs distributed across institutions?

RQ3: What factors influence citation impact? How can less visible universities improve their SDG-related performance?

Data were collected from *Scopus* and *SciVal* databases using keywords matching the 16 SDGs (SDG1 to SDG16) as defined by Elsevier's SDG mapping methodology. Publications were filtered to include only peer-reviewed articles and reviews affiliated with Hungarian universities during the period from 2019 to 2023.

Each publication was assigned to one or more SDG categories based on metadata tagging and keyword-based matching. For every university, three key indicators were extracted:

(1) Total number of publications
(2) Number of publications linked to SDG1–SDG16
(3) Citation impact (average citations per publication) for SDG-tagged outputs.

Statistical methods, including standard deviation, trend analysis, and proportional comparisons, were used to assess differences across institutional types. A *dendrogram (hierarchical cluster analysis)* was conducted to classify institutions based on their SDG research profiles and publication volumes.

The results were interpreted in relation to institutional characteristics (e.g., governance model, research orientation, size), national policy developments (e.g., sustainability platforms), and global SDG priorities. Citation impact was used as a proxy for visibility and research influence. To strengthen qualitative interpretation, institutional missions, previous SDG-related initiatives, and participation in sustainability platforms were also reviewed.

This enhanced methodology ensures transparency and replicability while allowing for nuanced insights into SDG integration across Hungary's complex university landscape.

5 Findings

The fundamental statistical analysis yielded several noteworthy insights into the publication activity of universities. On average, a university published 3,865 articles, approximately 1,279 of which were related to the SDGs 1–16 goals. This implies that publications associated with SDGs 1–16 topics accounted for an average of 33.3% of the total output.

The analysis revealed significant variations among universities. The standard deviation for the total number of publications was 3,575, while for publications related to SDGs 1–16, this value was 1,203. These findings indicate considerable diversity in publication performance across institutions.

The minimum and maximum values further illustrate these disparities, with the university with the fewest publications producing 396 articles and the institution with the highest output publishing 11,227 articles. The number of SDGs 1–16-related publications ranged from 84 to 4,268, demonstrating similarly substantial differences. In terms of proportions, SDGs 1–16 publications accounted for between 18.8% and 49.7% of the total publications, suggesting that while some universities place significant emphasis on these topics, others assign them less priority.

The analysis provided substantial information about the publication activity of universities and its correlation with the SDGs 1–16 goals, as illustrated in Fig. 1.

As demonstrated in Fig. 1, it is evident that Eötvös Loránd University (ELTE) has the highest number of publications. However, it is also obvious that several

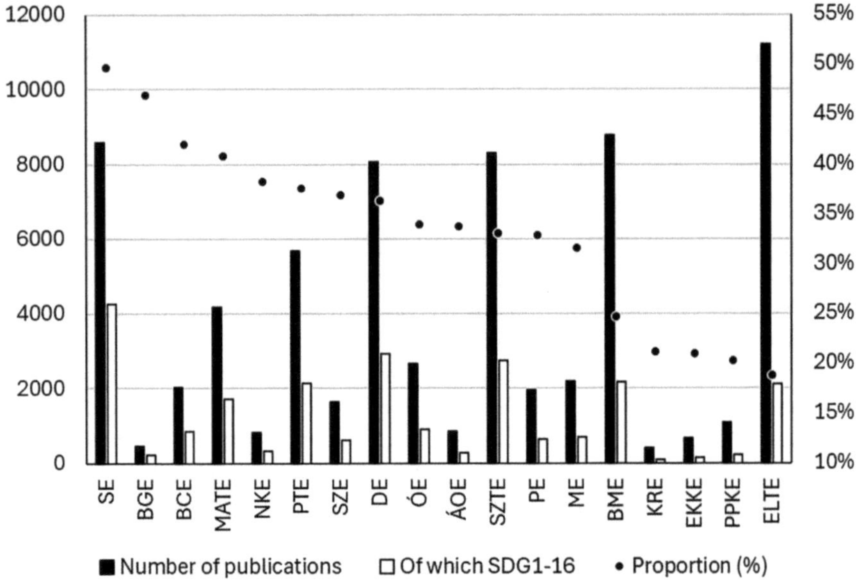

Fig. 1 The total number of publications across universities, the number of publications related to SDGs 1–16 by each university, and the proportion of publications related to SDGs 1–16 between 2019–2023. *Source* Authors ["Data set with Authors, data available upon request"]

smaller universities have produced significantly fewer publications, highlighting considerable disparities in research activity levels among institutions.

The focus is on the number of publications related to SDGs 1–16 goals by each university (see Annex, Table 2). Semmelweis University (SE) leads in absolute terms. However, a closer examination of the proportions, as demonstrated in the figure, reveals that not all universities dedicate equal attention to SDGs 1–16 topics.

The proportion of publications related to SDGs 1–16 is presented as a percentage of the total number of publications (see Annex, Table 3). This analysis indicates that certain universities with a comparatively limited overall publication output, such as Budapest Business School (BGE), allocate a higher proportion of their efforts to these objectives. Conversely, larger institutions, such as ELTE, demonstrate a comparatively lower proportional emphasis on publications related to SDGs 1–16. These observations indicate varied research priorities within academic institutions, despite comparable overall publication performance.

The trend analysis reveals marked variations in the growth patterns of total publications across universities from 2019 to 2023. Budapest University of Technology and Economics (BME) has demonstrated consistent leadership in total research output, indicative of its robust research capacity. Conversely, smaller institutions such as the ÁOE and BGE have exhibited moderate growth, signifying their steady advancement in research activities.

Concerning the number of publications related to SDGs, significant variations are observed among universities. Some institutions, such as the Corvinus University of Budapest (BCE), show a consistent and steady increase over time, while others, like BGE, initially display a gradual pace but later accelerate their focus on SDG-related research.

In terms of the proportion of publications related to the SDGs in comparison to the total output of research, there is a mixture of stability and fluctuation among universities. It is noteworthy that BGE consistently maintains a high proportion of publications related to the SDGs, thereby highlighting its significant commitment to sustainability.

The majority of universities exhibit an unmistakable upward trajectory in both total and SDG-related publications, albeit with fluctuations in the rate of growth across different institutions. Universities such as BGE and BCE place a strong emphasis on SDG-related research, as evidenced by their significant and growing contributions in this area. Nevertheless, there is still scope for enhancement. For instance, institutions with a high total output, such as the BME, could further bolster their commitment to sustainability by augmenting the proportion of SDG-related research within their overall output, as illustrated in Fig. 2.

Notably, the most significant increase of 20% was observed at Károli Gáspár University of the Reformed Church (KRE), whereas Ludovika University of Public Service (NKE) experienced a 4% decline.

The aggregation of data reveals that the most popular research area is SDG3, with a total of 11,656 publications. In second place is SDG9, with 2,273 publications. Furthermore, SDG7 (Affordable and Clean Energy), SDG8 (Decent Work and Economic Growth), and SDG13 have also garnered significant interest, with each exceeding 1,000 publications. Conversely, SDG1 (No Poverty) and SDG5 (Gender Equality) are among the least researched areas, with fewer than 400 publications.

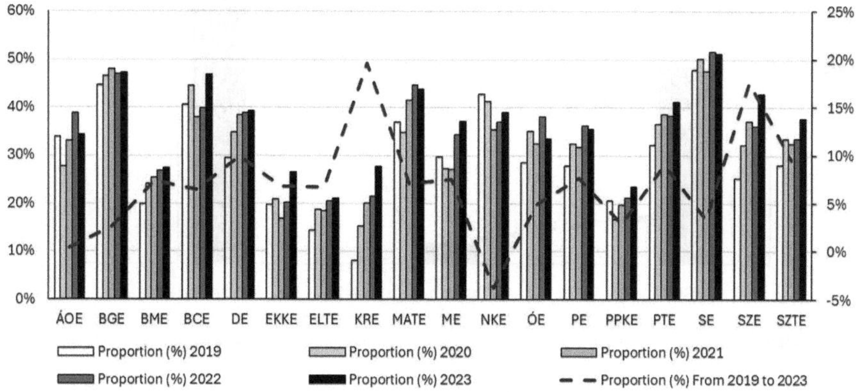

Fig. 2 Trends in the proportion of publications related to SDGs 1–16 across universities between 2019–2023. *Source* Authors ["Data set with Authors, data available upon request"]

At the university level, the field of SDG3 research is dominated by SE, with a total of 4,000 publications, reflecting the institution's healthcare-focused profile. The field of the University of Debrecen (DE) is also prominent in this research field, with 1,756 publications, and has achieved significant results in the areas of SDG13 and SDG15 (Life on Land). Research on SDG7 and SDG9 is particularly strong at the BME, aligning with the institution's strength in engineering and technical sciences.

ELTE boasts a diverse research portfolio, with particular strengths in SDG3, SDG12 (Responsible Consumption and Production), and SDG16. The Hungarian University of Agriculture and Life Sciences (MATE), with its focus on agricultural and life sciences, demonstrates significant publication activity in SDG2 (Zero Hunger) and SDG15, consistent with the institution's agricultural profile.

It is interesting to note that smaller universities, such as Eszterházy Károly Catholic University (EKKE) and KRE, have contributed fewer publications to SDG-related research. However, among them, EKKE has shown some activity in SDG13. In contrast, NKE has achieved notable results in SDG16, which aligns with its social and governance-oriented focus.

Between 2019 and 2023, there was a substantial increase in both the total number of publications and the proportion of those addressing SDGs 1–16 across various university types. The data indicates significant trends in publication activity and a growing emphasis on sustainability-related research, as illustrated in Fig. 3.

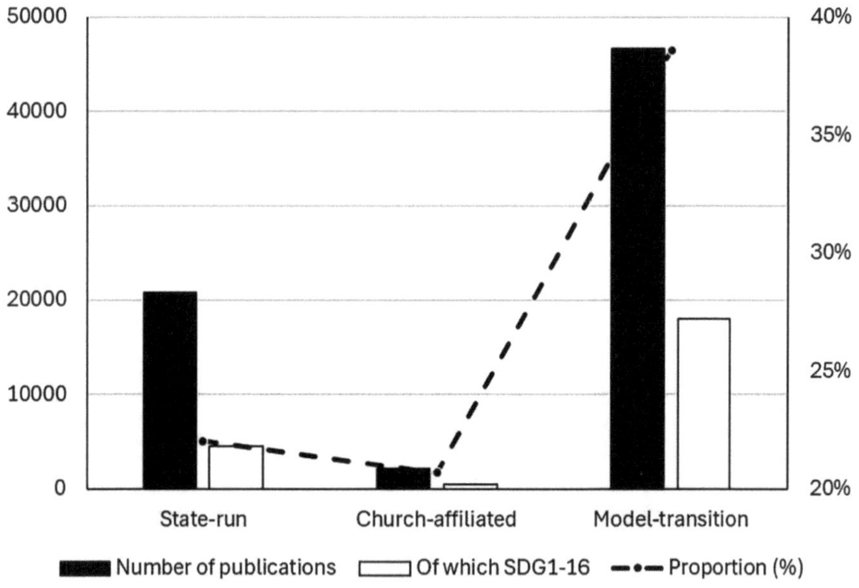

Fig. 3 The total number of publications, the number of publications related to SDGs 1–16, and the proportion of SDG-related publications between 2019–2023 based on the type of university. *Source* Authors ["Data set with Authors, data available upon request"]

The model-transition universities demonstrated the most significant progress, with their publications increasing from 7,335 in 2019 to 10,811 in 2023, an increase of approximately 47%. State-run universities also showed an increase in publications, rising from 3,820 in 2019 to 4,491 in 2023, which represents an approximate 17% growth. A similar trend was observed in church-affiliated universities, which exhibited an increase in output from 312 publications in 2019 to 537 in 2023, signifying a 72% growth. The aggregate number of publications across all university types exhibited an increase from 11,467 in 2019 to 15,839 in 2023, signifying an overall growth of 38%.

The proportion of publications related to SDGs 1–16 also exhibited an upward trend, with state-run universities experiencing an increase from 17.6% in 2019 to 24.3% in 2023, and church-affiliated universities demonstrating a comparable rise, with their proportion rising from 18.9% in 2019 to 25.5% in 2023. Model-transition universities exhibited a higher initial proportion of SDGs 1–16 publications, at 33.9% in 2019, which increased further to 41.5% in 2023. The proportion of SDG 1–16-related publications increased from 28% in 2019 to 36% in 2023, indicating a clear shift toward sustainability-focused research, as illustrated in Fig. 4.

This analysis reveals several key trends. Firstly, the overall growth in both total publications and SDGs 1–16-related research underscores the increasing productivity and relevance of universities in addressing global challenges. Secondly, differences among university types reveal that model-transition universities lead in terms of both publication output and focus on sustainability. State-run universities displayed steady growth, while church-affiliated universities showed smaller yet noteworthy improvements. Finally, the increasing proportion of SDGs 1–16-related publications across all university types highlights academia's growing commitment to contributing to the SDGs.

In conclusion, the data reveals a consistent and significant increase in sustainability-related research alongside overall publication activity. It is evident that

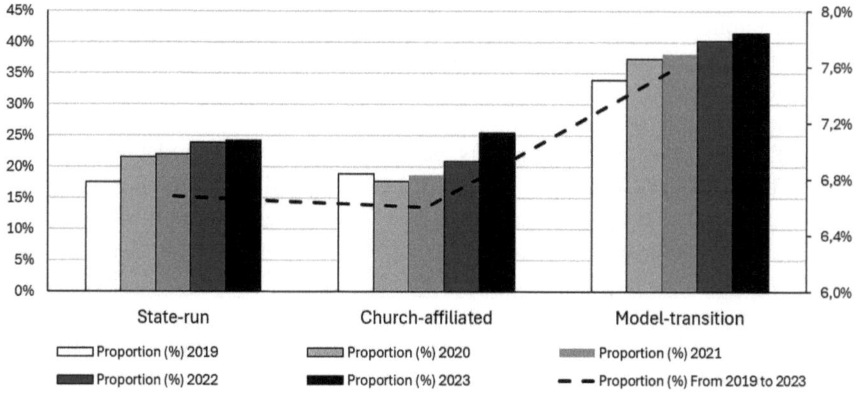

Fig. 4 Trends in the proportion of publications related to SDGs 1–16 between 2019–2023 based on the type of university. *Source* Authors ["Data set with Authors, data available upon request"]

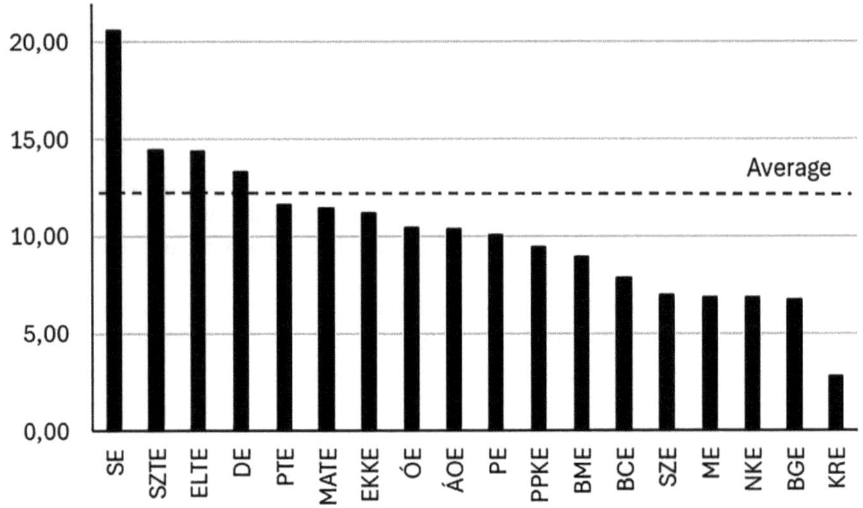

Fig. 5 Trends in the number of citations to publications related to SDGs 1–16 across universities between 2019–2023. *Source* Authors ["Data set with Authors, data available upon request"]

universities are playing a vital role in advancing the global sustainability agenda, with each type contributing to this progress in unique ways.

In analyzing the citation performance of Hungarian universities concerning the SDGs for publications from 2019 to 2023, the data reveal substantial variations both in average citations per publication and across specific SDG domains. On average, Hungarian universities achieved a citation rate of 12.7 citations per publication for SDG-related research. However, significant disparities are evident among institutions and across individual SDG targets, as illustrated in Fig. 5.

SE achieved an average of 20.56 citations per publication, the highest among all universities analyzed. Notably, the university excelled in SDG3, achieving 29.32 citations per publication, highlighting the global importance of health-related research. Additionally, SE showcased exceptional performance in sustainable urban development related to SDG11, achieving an extraordinary 73.18 citations per publication, the highest value across the entire dataset.

The DE also achieved strong results, particularly in SDG2 and SDG3, with averages of 16.83 and 18.67 citations per publication, respectively. These figures underscore the importance of agricultural and life sciences research in the institution's profile. ELTE displayed balanced performance, with notably high citation values across multiple SDG targets, including SDG3 at 16.63, SDG13 at 18.41, and SDG15 at 11.54.

In contrast, the University of Pannonia (PE) achieved remarkable results in SDG5, with a citation-to-publication ratio of 50, the highest across all SDG categories. This reflects the institution's dedicated research efforts in this area, which have garnered significant international recognition.

Among smaller institutions, Óbuda University (ÓE) demonstrated strong performance in SDG12 and SDG15, achieving citation ratios of 22.15 and 31.59, respectively. Conversely, institutions such as KRE and EKKE exhibited lower average citation ratios, potentially reflecting reduced research activity or limited international presence.

The analysis indicates that SDG3 consistently demonstrates exceptional citation metrics across universities, emphasizing the global significance of health-related research. Similar trends are observed in SDG6 (Clean Water and Sanitation) and SDG13, which exhibit high citation values, highlighting the growing importance of environmental research. In contrast, SDG4 and SDG5 exhibit comparatively lower citation levels across many institutions, reflecting differences in research priorities and resource allocation (see Annex, Table 4).

To summarize, the data presented in the table reveals clear variations in research focus and university contributions to different SDGs. Notably, larger institutions, such as SE, the DE, and ELTE, have achieved exceptional international citation performance in specific areas. In contrast, other universities have focused on more targeted, yet narrower-spectrum research. Therefore, enhancing research activity in SDGs with lower citation rates could contribute to broader progress toward sustainability goals and improve the international visibility of Hungarian universities.

The dendrogram analysis, which represents a hierarchical cluster analysis of university publications, and the proportion of articles related to SDGs 1–16, identified three main clusters, as illustrated in Fig. 6.

The *first cluster* includes institutions with lower publication outputs and a relatively weaker emphasis on sustainability. Such institutions include the ÁOE, NKE and Pázmány Péter Catholic University (PPKE). These institutions evince reduced research activity and are responsible for more circumscribed contributions to the SDG targets.

The *second cluster* includes universities with medium publication performance, whose research moderately aligns with sustainability goals. Institutions such as the MATE and the University of Pécs (PTE) fall into this category. These universities possess diversified research portfolios but do not demonstrate outstanding dominance in any specific SDG area.

The *third cluster* includes institutions that show both high research performance and strong alignment with the SDG targets. For instance, SE exhibits exceptional performance in SDG3, a reflection of its healthcare-focused profile. BME and ELTE are closely associated within this cluster, as both exhibit high publication output and extend their research across multiple SDG targets. The University of Szeged (SZTE), the PTE, and the DE are also part of this group, demonstrating significant research activity related to the SDGs.

The distances in the dendrogram indicate the degree of similarity between institutions within the clusters. Smaller distances, such as between ELTE and BME, suggest a closer research focus and similar publication strategies. Conversely, a larger distance, such as between the ÁOE and SE, reflects differing research priorities and performance.

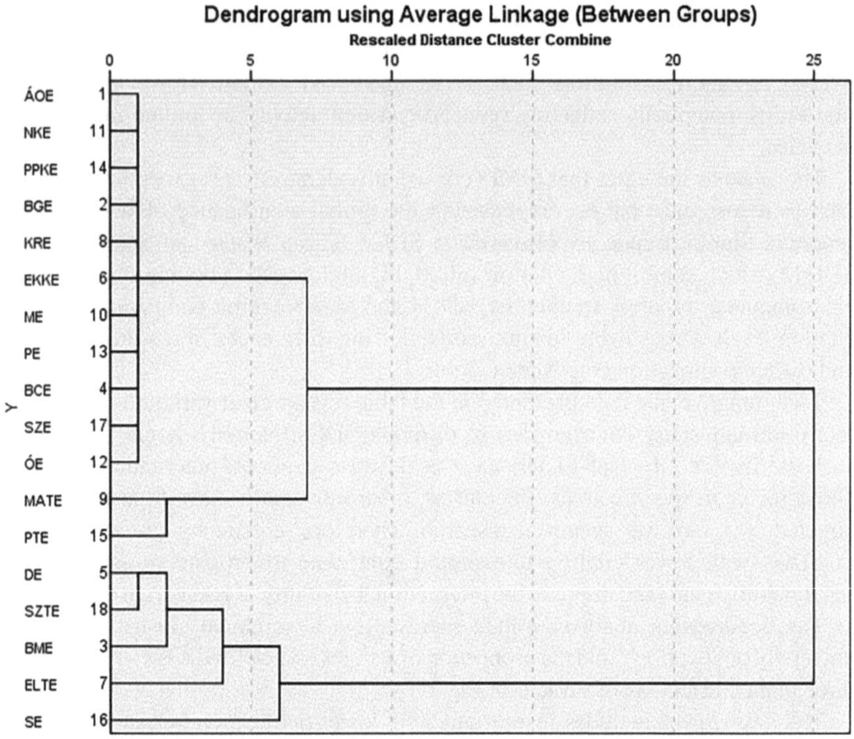

Fig. 6 The dendrogram of publications by universities and the proportion of articles related to SDGs 1–16. *Source* Authors ["Data set with Authors, data available upon request"]

The dendrogram comprehensively represents the disparities in research foci and contributions to sustainability objectives among universities. The clusters demonstrate that variations between institutions are predominantly influenced by the extent of research activity and the concentration of SDG-related research domains. These findings are essential for evaluating the research strategies of universities and understanding their contributions to achieving sustainability goals.

A systematic return to the research questions is made, with each question being revisited and the findings aligned with the original objectives. This process is used to evaluate the progress and contributions of Hungarian universities to the SDGs.

RQ1a. To what extent has there been an increase in the number of publications related to the SDGs in Hungarian universities between 2019 and 2023?

The study identified a significant increase in SDG-related publications across Hungarian universities. The total output rose from 3,218 publications in 2019 to 5,715 in 2023, reflecting a growth of 77%. The most significant growth was observed in church-affiliated universities (132%), followed by model-transition universities (80%) and state-run universities (62%). These findings suggest a discernible trend of Hungarian universities progressively aligning their research output with the SDGs.

RQ1b. Which universities have demonstrated the most significant increase in the number of publications related to the SDGs?

SE, BME, and SZTE showed the most substantial growth in SDG-related research output. Notably, SE excelled in SDG3, contributing significantly to the number of SDG-related publications. In contrast, smaller institutions such as KRE and EKKE demonstrated a more gradual increase in SDG-related research output.

RQ2a: Which publication stands out among all SDG publications, both in terms of quantity and citations?

The most prominent publications were those under SDG3, which comprised 11,656 papers and achieved the highest citation impact. SE and the DE contributed the most to this domain. The focus of SDG3 on healthcare is in alignment with global academic priorities, resulting in broader visibility and greater societal impact. Furthermore, the publications concerning SDG4 and SDG13 were distinguished by their interdisciplinary character and high citation rates.

RQ2b: How is the number of publications on this SDG distributed across universities?

SE has been identified as a leading institution in the field of SDG3 research, with a total of 4,088 publications. This is followed by the DE (1,756 publications). These figures reflect the institutions' strategic focus on healthcare and biomedical research. Other universities, such as the SZTE and the PTE, have also made significant contributions to this field, though their publications are comparatively fewer in number.

RQ3a. What factors influence whether a university's SDG-related publications achieve a high citation rate?

The study identified several factors that contribute to high citation impact, including strong international collaborations, multidisciplinary research approaches, and alignment with global priorities. Institutions such as SE, SZTE, and DE) demonstrated excellence due to their robust research networks and focus on highly cited fields such as healthcare and climate action.

RQ3b. What strategies could be used to increase the visibility and impact of universities with fewer citations in the SDG field?

Smaller academic institutions, such as KRE and BGE, could benefit from adopting interdisciplinary approaches, expanding international collaborations, and focusing on underrepresented SDGs such as SDG1 and SDG5. Furthermore, the establishment of partnerships with more prominent universities has the potential to amplify the research impact of these institutions.

RQ3c. How can the profile or research focus of universities contribute to the citation rate of SDG publications on specific topics?

It has been demonstrated that universities with specialized research profiles, such as healthcare at SE (SDG3) or technical sciences at BME (SDG7 and SDG9), have exhibited higher citation performance. Conversely, institutions with broader or less-targeted research focus have demonstrated a lower impact. This indicates that aligning institutional strengths with specific SDGs improves both output and citation performance.

6 Discussion and Conclusions

In conclusion, a combination of thematic alignment, interdisciplinary approaches, global collaboration, and institutional support are critical to achieving high citation rates in SDG-related research. Institutions should tailor their research strategies to align with global priorities and leverage collaborative opportunities to enhance their citation performance.

Hungarian universities can be grouped into three distinct clusters based on their research focus and output. The first cluster encompasses high-performing universities, such as SE) and BME, which exhibit extensive research portfolios, robust international collaborations, and a pronounced focus on sustainability. The second cluster comprises moderate performers, including institutions like the PTE and the MATE, which demonstrate diversified but less concentrated efforts. Finally, the third cluster includes low-output institutions such as KRE and PPKE, which contribute minimally to the SDG agenda.

In order to enhance their contributions to the SDGs, it is recommended that Hungarian universities give due consideration to the following recommendations. Firstly, it is essential that institutions align their research priorities more closely with national and global SDG targets. This requires strong state-run support and proactive leadership to embed sustainability within academic missions and curricula. Secondly, it is imperative to invest in research infrastructure and interdisciplinary programs to strengthen the underrepresented SDGs, such as SDG1 and SDG5. Targeted funding could also enable smaller universities to improve their research contributions.

Collaboration represents a further pivotal domain for enhancement. The establishment of partnerships between universities, industries, and communities has the potential to augment the impact of research endeavors. Hungarian universities are encouraged to adopt the successful practices of Western European institutions, such as multi-stakeholder collaborations and robust sustainability networks, to enhance their societal relevance. It is particularly recommended that smaller universities focus on niche areas of sustainability and align their strengths with underexplored SDGs to increase visibility. Enhanced international collaborations and improved reporting mechanisms have the capacity to further boost their research impact.

Policymakers undoubtedly play a pivotal role in this process. They should provide targeted incentives for SDG-related research, such as the allocation of competitive grants, the establishment of international exchange programs, and the recognition of impactful sustainability projects through awards. These measures can encourage universities to broaden their focus on addressing pressing global challenges.

Building on the empirical findings regarding the Hungarian universities' contributions to the Sustainable Development Goals (SDGs), it is worth considering how these efforts align with emerging innovation frameworks, particularly the Industry 5.0 (I5.0) paradigm and smart city concepts [14]. Their study emphasizes a transition from a technology-centric Industry 4.0 model toward a more human-centric, sustainable, and resilient I5.0 approach, which prioritizes not only technological advancement but also societal well-being and environmental responsibility.

From this perspective, Hungarian universities—especially model-transition institutions—are well-positioned to contribute to the knowledge infrastructure necessary for implementing I5.0 ideals. The increased research focus on SDG3, SDG7, and SDG11 directly supports the pillars of smart urban development and the digital–ecological convergence envisioned in the I5.0 framework. However, recent research also highlights critical enablers such as blockchain-based applications, digital trust, and multi-stakeholder collaboration—areas that are relatively underexplored within the analyzed university outputs [14].

This reveals an opportunity—and a gap. While Hungarian universities are expanding their SDG research portfolios, especially in public health and technical disciplines, they may need to strategically *broaden their interdisciplinary engagement* toward fields like digital governance, blockchain ethics, and socio-technical systems. Doing so would align their sustainability strategies not only with the 2030 Agenda but also with forward-looking paradigms such as smart cities and Industry 5.0, which demand *a co-evolution of innovation, inclusivity, and institutional transformation.*

In this light, future sustainability strategies in higher education should go beyond traditional SDG alignment and embrace *system-level thinking*, fostering *resilience, human-centric innovation*, and *technological foresight*. Collaborative research hubs, urban testbeds, and stronger integration with industry and local government ecosystems could serve as critical accelerators in this transition.

Comparative analyses of the EU27 countries also confirm the significant growth in SDG-related research between 2019 and 2023, alongside marked differences among countries [15]. While Germany, France, and Italy lead in overall publication output, smaller countries such as Cyprus and Latvia stand out in terms of the relative share of SDG publications. Hungary—similar to most Central and Eastern European countries—shows a moderate but growing commitment to sustainability research, particularly in areas related to SDG3, SDG7, and SDG12. These findings underline the importance of targeted policy support and enhanced international collaboration to improve the effectiveness of sustainability-focused research within the European Higher Education Area.

By implementing these recommendations, Hungarian universities can significantly enhance their contributions to the global SDG agenda, ensuring a greater impact on sustainability both nationally and internationally.

Acknowledgements Project no. TKP2021-NKTA-51 has been implemented with the support provided by the Ministry of Culture and Innovation of Hungary from the National Research, Development and Innovation Fund, financed under the TKP2021-NKTA funding scheme.

Annex

See Tables 1, 2, 3, 4.

Table 1 The names and abbreviations of the universities included in the research. *Source* Own editing

University's name abbreviation	Name of university	Institution type
ÁOE	University of Veterinary Medicine Budapest	Model-transition
BGE	Budapest Business School	Model-transition
BME	University of Technology and Economics	State-run
BCE	Corvinus University of Budapest	Model-transition
DE	University of Debrecen	Model-transition
EKKE	Eszterházy Károly Catholic University	Church-affiliated
ELTE	Eötvös Loránd University	State-run
KRE	Károli Gáspár University of the Reformed Church	Church-affiliated
MATE	Hungarian University of Agriculture and Life Sciences	Model-transition
ME	University of Miskolc	Model-transition
NKE	Ludovika University of Public Service	State-run
ÓE	Óbuda University	Model-transition
PE	University of Pannonia	Model-transition
PPKE	Pázmány Péter Catholic University	Church-affiliated
PTE	University of Pécs	Model-transition
SE	Semmelweis University	Model-transition
SZE	University of Győr	Model-transition
SZTE	University of Szeged	Model-transition

Table 2 The total number of publications, the number of publications related to SDGs 1–16, and the proportion of publications related to SDGs 1–16 between 2019–2023

Name	Number of publications					Of which SDGs 1-16					Proportion %					
	2019	2020	2021	2022	2023	2019	2020	2021	2022	2023	2019	2020	2021	2022	2023	2019–2023
ÁOE	121	152	158	196	222	41	42	52	76	76	34%	28%	33%	39%	34%	0%
BGE	56	58	100	98	146	25	27	48	46	69	45%	47%	48%	47%	47%	3%
BME	1782	1689	1703	1845	1744	355	406	431	494	476	20%	24%	25%	27%	27%	7%
BCE	312	322	396	488	509	126	143	150	194	238	40%	44%	38%	40%	47%	6%
DE	1412	1559	1650	1720	1731	414	540	631	665	678	29%	35%	38%	39%	39%	10%
EKKE	81	125	172	119	166	16	26	29	24	44	20%	21%	17%	20%	27%	7%
ELTE	1949	2149	2157	2436	2536	279	401	398	498	532	14%	19%	18%	20%	21%	7%
KRE	37	59	75	84	141	3	9	15	18	39	8%	15%	20%	21%	28%	*20%*
MATE	571	749	928	951	991	210	260	385	424	434	37%	35%	41%	45%	44%	7%
ME	338	382	433	465	583	100	104	117	159	216	30%	27%	27%	34%	37%	7%
NKE	89	119	173	239	211	38	49	61	88	82	43%	41%	35%	37%	39%	**-4%**
ÓE	316	414	490	631	815	90	145	159	240	272	28%	35%	32%	38%	33%	5%
PE	321	396	413	407	407	89	128	131	147	144	28%	32%	32%	36%	35%	8%
PPKE	194	228	218	222	230	40	38	43	47	54	21%	17%	20%	21%	23%	3%
PTE	892	1059	1204	1294	1220	286	386	463	494	501	32%	36%	38%	38%	41%	9%
SE	1375	1591	1859	1874	1885	658	796	884	966	964	48%	50%	48%	52%	51%	3%
SZE	195	215	260	353	625	49	69	96	127	267	25%	32%	37%	36%	43%	18%
SZTE	1426	1622	1786	1790	1677	399	542	579	598	629	28%	33%	32%	33%	38%	10%
Total/Average	11,467	12,888	14,175	15,212	15,839	3218	4111	4672	5305	5715	28%	32%	33%	35%	36%	8%

Table 3 The proportion of publications related to SDGs 1–16 across universities between 2019–2023

Name	SDG 1	SDG 2	SDG 3	SDG 4	SDG 5	SDG 6	SDG 7	SDG 8	SDG 9	SDG 10	SDG 11	SDG 12	SDG 13	SDG 14	SDG 15	SDG 16
ÁOE	0	25	166	4	0	9	23	5	1	1	20	12	11	7	42	2
BGE	7	7	29	38	3	10	7	82	37	24	20	57	17	1	6	14
BME	18	27	356	82	15	111	608	135	575	38	533	209	186	26	27	30
BCE	67	51	127	76	46	16	58	224	152	178	60	105	77	3	30	106
DE	38	248	1756	117	44	134	172	216	165	72	141	159	183	38	254	70
EKKE	4	21	18	21	4	9	16	14	5	4	11	21	26	2	44	5
ELTE	28	83	804	173	83	122	178	113	129	109	178	110	153	26	193	178
KRE	2	1	32	19	5	0	0	13	1	12	2	1	0	0	0	15
MATE	38	372	223	75	16	218	218	295	206	64	155	266	210	110	321	30
ME	17	21	119	25	0	84	193	65	192	30	53	76	59	13	14	31
NKE	7	14	48	17	9	43	37	55	42	34	41	24	24	1	19	63
ÓE	3	41	266	75	4	48	197	89	193	15	115	62	52	9	22	17
PE	5	21	131	36	3	85	160	54	136	16	67	73	65	24	36	9
PPKE	3	1	101	6	8	5	18	18	14	24	13	4	2	0	1	37
PTE	23	53	1460	76	26	61	77	152	95	66	133	92	44	15	111	62
SE	16	27	4088	57	24	12	18	16	25	52	11	18	8	1	8	52
SZE	16	61	92	47	3	34	73	114	197	36	123	111	63	4	22	16
SZTE	32	85	1840	99	46	134	177	97	108	81	127	87	135	27	139	58
Total	324	1159	11,656	1043	339	1135	2230	1757	2273	856	1803	1487	1315	307	1289	795

Table 4 The number of citations to publications related to SDGs 1–16 across universities between 2019–2023

Name	SDG 1	SDG 2	SDG 3	SDG 4	SDG 5	SDG 6	SDG 7	SDG 8	SDG 9	SDG 10	SDG 11	SDG 12	SDG 13	SDG 14	SDG 15	SDG 16	Average
ÁOE	ND	5.88	9.66	3.00	10.67	10.56	9.65	13.00	5.00	7.00	49.25	17.00	10.91	17.71	11.88	15.50	10.39
BGE	6.57	4.43	13.76	4.18	4.33	7.30	**4.86**	9.76	12.08	5.21	9.80	6.96	1.24	6.00	13.83	11.36	6.76
BME	4.67	9.04	11.79	6.74	3.80	14.99	13.94	14.76	13.13	6.21	11.30	13.89	16.20	23.92	9.81	18.93	8.94
BCE	7.48	16.76	15.69	6.59	4.93	13.63	16.31	9.83	13.93	12.30	12.75	14.02	15.29		12.73	6.50	7.88
DE	9.74	16.83	18.67	6.87	3.91	12.59	11.57	17.09	13.95	8.79	8.43	16.66	16.16	10.32	17.72	4.99	13.30
EKKE	3.00	5.19	4.22	4.05	**0.00**	**3.67**	8.56	8.79			6.36		10.15	4.50	10.66	2.80	11.20
ELTE	11.04	8.93	16.63	5.84		17.00	11.34	16.44	9.71	8.14	8.76	10.85	9.30	18.41	14.12	11.54	14.38
KRE	11.50	**0.00**	**2.88**	**1.58**	0.80	NA	NA	2.31	6.00	2.92		6.00	NA	NA	NA	3.20	
MATE	13.63	11.04	11.01	8.21	9.50	10.43	13.60	15.53	18.58	9.20	11.59	17.17	11.52	11.57	10.08	7.17	11.46
ME	12.88	3.62	13.22	11.08	NA	14.60	9.53	7.00	7.91	5.07	10.04	10.33	15.24	8.38	12.71	2.77	6.86
NKE	4.00	14.50	16.85	6.59	7.44	8.86	15.70	7.00	16.40	6.97	10.78	18.00	17.58	7.00	10.53	2.92	6.84
ÓE	11.67	*17.71*	13.03	7.79	3.75	*45.50*	16.80	13.56	10.88	3.93	29.10	22.15	27.23	10.22	31.59	5.65	10.43
PE	21.00	16.57	13.68	8.97	*50.00*	11.68	16.44	15.24	12.70	12.38	10.67	11.07	15.17	15.67	18.00	5.44	10.06
PPKE	**1.00**	1.00	9.84	10.00	9.50	6.80	8.50	**1.83**	14.07	5.25	8.85	20.25		NA			9.42
PTE	7.83	17.64	16.60	5.76	12.88	17.72	9.01	6.01	8.08	6.74	10.26	6.40	18.14	38.87	14.24	8.42	11.59
SE	*28.69*	13.07	*29.32*	*46.96*	7.79	14.00	19.56	*33.31*	13.36	20.10	73.18	46.17	8.63	7.00	9.38	197.08	20.56
SZE	6.25	11.93	7.10	3.53	NA	9.74	5.45	8.39	12.30	12.61	7.16	10.63	5.10	21.00	10.36	10.00	6.99
SZTE	7.78	13.07	16.87	8.57	7.59	12.26	*20.51*	11.05	21.66	6.23	15.11	15.68	21.61	10.89	15.87	5.28	14.43
Average	10.00	12.85	20.71	8.87	9.58	13.63	14.03	11.93	12.94	9.34	12.80	13.95	15.68	14.19	13.56	19.44	12.70
Min	1.00	0.00	2.88	1.58	0.00	3.67	4.86	1.83	3.00	0.50	1.50	4.05	0.50	4.00	4.00	1.24	2.76
Max	28.69	17.71	29.32	46.96	50.00	45.50	20.51	33.31	21.66	20.10	73.18	46.17	27.23	38.87	31.59	197.08	20.56

References

1. United Nations (2015). Resolution adopted by the General Assembly on 25 September 2015. Transforming our world: the 2030 Agenda for Sustainable Development.
2. Loia, F., Capolupo, N., Gravili, G., & Adinolfi, P.: Bridging sustainable higher education and Metaverse during the post-pandemic era: Highlights from a multiple case study. In: Visvizi, A., Troisi, O., Corvello, V. (eds.) Research and Innovation Forum 2023. Springer Proceedings in Complexity. Springer (2024). https://doi.org/10.1007/978-3-031-44721-1_27
3. Cembranel, P., Gewehr, L., Dal Moro, L., Fuchs, P.G., Birch, R.S., Andrade Guerra, J.B.S.O.d.A.: The pivotal role of higher education institutions in cultivating a sustainable development goals-centric culture. Int. J. Sustain. High. Educ. **25**(7), 1385–1411 (2024). https://doi.org/10.1108/IJSHE-01-2024-0057
4. Khare, A., Stewart, B.: Guest editorial: Making an impact –UN sustainable development goals and university performance. Int. J. Sustain. High. Educ. **25**(5), 901–902 (2024). https://doi.org/10.1108/IJSHE-07-2024-607
5. Szemlér, T.: The Performance of EU CEE Universities Against the UN Sustainable Development Goals. In: Nedelko, Z., Vide R.K. (eds.) 7th FEB International Scientific Conference: Strengthening Resilience by Sustainable Economy and Business—Towards the SDGs, pp. 147–154) Univerzitetna založba Univerze v Mariboru (2023). https://doi.org/10.18690/um.epf.3.2023
6. Hernádi, H., Bettina és Bereczk, Á., Sasvári, P.L.: A visegrádi országok felsőoktatási intézményeinek kutatási tevékenysége a fenntartható városok és közösségek témakörében (SDG-11). Multidiszciplináris Tudományok: A Miskolci Egyetem Közleménye, **13**(1), 398–410 (2023). https://doi.org/10.35925/j.multi.2023.1.30
7. Purcell, W.M., Henriksen, H.A., Spengler, J.D.: Universities as the engine of transformational sustainability toward delivering the sustainable development goals: "Living labs" for sustainability. Int. J. Sustain. High. Educ. (2019). https://doi.org/10.1108/IJSHE-02-2019-0103
8. Dumitraşcu, L.M.: Embedding the sustainability development goals in universities. Audit financiar (2023). https://doi.org/10.20869/AUDITF/2023/171/020
9. Filho, W.L., Eustachio, J.H.P.P., Ávila, L.V., Dinis, M.A.P., Hernandez-Diaz, P.M., Batista, K., Borsari, B., Abubakar, I.R.: Enhancing the contribution of higher education institutions to sustainable development research: A focus on post-2015 SDGs. Sustain. Dev. 1–13 (2024). https://doi.org/10.1002/sd.3184
10. Hong, X., Calderon, A., Coates, H.: Universities and SDGs: evidence of engagement and contributions, and pathways for development. Policy Reviews in Higher Education **7**(1), 56–77 (2022). https://doi.org/10.1080/23322969.2022.2121311
11. Univ Pécs online magazine: Homepage, https://univpecs.com/egyetemi_elet/megalakult_magyar_egyetemek_fenntarthatosagi_platformja (2022). Accessed 1 Jan 2025
12. Náhlik, A., Bertalan, L., Polgár, A.: Mélyinterjús kutatás a fenntarthatósági attitűdökről egy hazai egyetemen. E-Conom **13**(2), 21–40 (2024). https://doi.org/10.17836/EC.2024.2.002
13. Smolennikov, D., Makarenko I., Bacho, R., Makarovych V., Oleksich Z., Gorodysky M., Polishchuk I.: Do higher education institutions contribute to countries' SDG progress: Evidence from university rankings. Knowl. Perform. Manag. **8**(1), 133–148 (2024). https://doi.org/10.21511/kpm.08(1).2024.10
14. Visvizi, A., Malik, R., Guazzo, G.M., Çekani, V.: The Industry 5.0 (I50) paradigm, blockchain-based applications and the smart city. Eur. J. Innov. Manag. **28**(1), 5–26 (2025). https://doi.org/10.1108/EJIM-09-2023-0826
15. Belényesi, E., Sasvári, P.: Research performance on the UN Sustainable Development Goals in the EU27 (2019–2023). Adm. Sci. **15**(9), 361 (2025). https://doi.org/10.3390/admsci15090361

Redefining Architecture Pedagogy for GenAI-Enhanced Classrooms: Impacts on Student Engagement and Learning Outcomes

Mohammad Salman

Abstract The integration of Generative Artificial Intelligence (GenAI) in architecture education presents transformative opportunities for design processes, visualization, and creative problem-solving. GenAI, with its ability to generate design alternatives, automate repetitive tasks, and enhance decision-making, offers a paradigm shift in architectural pedagogy. To fully harness these benefits, architecture education must evolve, embedding GenAI tools strategically to enhance student engagement and optimize learning outcomes. This study investigates the potential of GenAI to redefine teaching methodologies in architecture by analyzing its effects on academic experiences and student achievements. Using a mixed-methods approach that includes survey questionnaire and recently published literature on GenAI, identifies key pedagogical shifts and proposes best practices for integrating GenAI into architecture education. The findings will provide insights into how GenAI can be used as a pedagogical tool to cultivate creativity, improve design efficiency, and enhance critical thinking skills among architecture students.

Keywords GenAI · Generative design · Architecture pedagogy · Learning outcome · Student engagement · Classroom management · SDG-4 (quality education)

1 Introduction

Architecture education has long been rooted in a pedagogical framework that emphasizes a combination of traditional hand-drafting techniques, physical model-making, and digital design tools. This approach, while effective in fostering a fundamental understanding of architectural principles, spatial reasoning, and construction methodologies, is increasingly being challenged by rapid technological advancements [1]. The emergence of digital design software such as Computer-Aided Design (CAD)

M. Salman (✉)
Faculty of Architecture, Planning & Design, Integral University, Lucknow, India
e-mail: mdsalman@iul.ac.in

A. Visvizi et al. (eds.), *Research and Innovation Forum 2025 Volume 1*,
Springer Proceedings in Complexity, https://doi.org/10.1007/978-3-032-09066-9_18

and Building Information Modeling (BIM) has already revolutionized the way architectural concepts are developed, refined, and presented. However, the advent of GenAI represents a new frontier in architectural education, offering possibilities that extend beyond the capabilities of conventional digital tools [2]. GenAI refers to artificial intelligence systems capable of autonomously generating design alternatives based on specific parameters, learning from data inputs, and iterating solutions through advanced machine-learning algorithms. Unlike traditional design software, which requires direct user intervention to manipulate forms and structures, GenAI-driven tools can analyze complex design constraints, generate multiple iterations, and optimize design outcomes in real-time. This transformative capability has the potential to redefine architecture pedagogy, enabling students to explore more innovative and computationally complex design solutions while focusing on critical thinking and creative decision-making [3]. Given these advancements, architecture education must adapt to effectively integrate GenAI into its curriculum, ensuring that students are well-equipped with the necessary skills to navigate an AI-augmented design landscape. The incorporation of GenAI tools can significantly impact student engagement by fostering creativity, enhancing collaborative learning, and streamlining the design process. Research suggests that students who engage with AI-driven design tools are more likely to develop problem-solving skills, experiment with unconventional design solutions, and refine their conceptual thinking abilities more efficiently than those relying solely on traditional methods [4]. However, despite its promising potential, the integration of GenAI into architecture education is not without challenges. Concerns regarding over-reliance on AI, ethical considerations surrounding authorship, and the potential dilution of fundamental architectural skills remain key areas of debate. It is crucial to strike a balance between AI-assisted and manual design approaches, ensuring that GenAI serves as an enhancement rather than a replacement for traditional pedagogical methods [5]. Furthermore, architecture educators must develop new instructional strategies that guide students in the responsible and ethical use of AI-generated designs, fostering an understanding of AI's limitations and implications in the creative process. This study explores the role of GenAI in redefining architecture pedagogy, with a particular focus on its impact on student engagement and learning outcomes [6]. This study builds on the ongoing discourse within the RII Forum community regarding digital transformations in architecture education. Notably, earlier editions of the forum have examined pedagogical innovation in the context of architecture and design education, addressing compatibility issues, experiential limitations in virtual studios, and the changing role of user experience in digital classrooms. The present study aims to deepen the understanding of how such frameworks can be adapted to meet the challenges of GenAI-enhanced learning environments while remaining pedagogically robust. By answering these questions, this research aims to provide insights into integrating GenAI into architecture education, ensuring that future architects are equipped with the skills needed to thrive in an AI-enhanced professional landscape.

2 Generative AI (GenAI): The Evolutionary Leap in AI

The development of artificial intelligence (AI) has unfolded in a chronological and layered manner, evolving from foundational AI concepts in the 1950s to the cutting-edge domain of Generative AI (GenAI) in the 2020s. Generative AI represents the most recent and transformative phase within the AI continuum. It is built upon the principles of deep learning, a subfield of machine learning that emerged prominently in the 2010s and aims to mimic the neural structure and functioning of the human brain. While earlier AI systems primarily focused on analytical or decision-making tasks using historical data, GenAI moves a step further by enabling machines to generate original, creative outputs. GenAI leverages advanced deep learning architectures, particularly foundation models such as large language models (e.g., GPT) and diffusion models, to produce content that did not previously exist. This includes generating text, images, audio, video, and even code. Unlike traditional AI, which is task-specific, GenAI models are trained on massive datasets and exhibit remarkable versatility across multiple domains [7]. The distinguishing characteristic of GenAI lies in its capacity for creativity and synthesis, enabling novel applications in architecture, design, education, healthcare, and beyond. In essence, GenAI marks a pivotal shift from intelligent automation to creative augmentation. It empowers users not merely to process or analyze data but to co-create with machines, thus redefining the relationship between human cognition and computational intelligence. As a product of decades of research and innovation—beginning from symbolic AI in the 1950s, through rule-based and statistical learning in the 1980s, to neural networks in the 2010s—Generative AI stands at the forefront of a new era of intelligent systems that transcend predefined outputs and actively contribute to the evolution of knowledge, art, and design [8].

2.1 GenAI in Education

GenAI is reshaping education by advancing beyond traditional AI tools such as adaptive learning platforms and intelligent tutoring systems to enable dynamic content creation, personalised instruction, and real-time student interaction [9, 10]. Unlike earlier AI systems that primarily categorised and responded to data, GenAI actively generates original educational content, such as study materials, immersive simulations, and virtual tutors, transforming educators into facilitators of AI-enhanced learning environments [11]. GenAI promotes inclusivity by tailoring instructional content to individual student profiles, adjusting for skill levels, cognitive preferences, and linguistic diversity [10]. Its ability to deliver instant feedback through AI-driven formative assessments supports personalised learning loops, helping students refine learning strategies in real-time [12]. Furthermore, the integration of GenAI with immersive technologies like Virtual and Augmented Reality (VR/AR) enables experiential learning through interactive environments and collaborative simulations [13].

Table 1 Recent advancements in GenAI

Advancements	Description
AI-generated learning materials	GenAI enables real-time content creation, including lecture notes, quizzes, and interactive case studies. This reduces instructor workload and enhances accessibility for diverse learners [9]
Personalized learning pathways	AI can analyze student performance and learning styles, tailoring instruction to individual needs. This enhances engagement and improves learning outcomes [10]
AI-powered design and simulation for architecture education	AI-powered visualization tools like Midjourney, DALL·E, and AI-driven CAD software allow students to explore complex design concepts in real time, bridging the gap between theoretical knowledge and practical application [12]
Conversational AI and virtual tutors	AI chatbots such as ChatGPT and Claude assist students with academic writing, problem-solving, and brainstorming, fostering critical thinking and reducing dependency on static learning materials [11]
Assessment automation and feedback	AI can assess assignments, detect plagiarism, and provide real-time feedback, making evaluations more efficient and objective [15]

However, the adoption of GenAI also introduces pedagogical and ethical challenges. Concerns regarding academic integrity, intellectual authorship, data privacy, and algorithmic transparency must be critically addressed to ensure that AI augments—rather than replaces—the human-centric ethos of education [14, 15]. A balanced approach to implementation, supported by institutional guidelines and inclusive design, is necessary to ensure equitable and responsible GenAI deployment across learning contexts (Table 1).

2.2 AI and Design Thinking: Role of AI in Creative Disciplines

Design thinking is an iterative, user-centred approach to problem-solving that emphasises creativity, prototyping, and user feedback. Traditionally, this process relied heavily on human intuition and experience. However, the integration of Artificial Intelligence (AI) in design thinking has introduced new ways to enhance creativity, streamline ideation, and improve decision-making in creative disciplines. AI is no longer just a supporting tool; it is now an active collaborator in design processes, product development, architecture, and the arts [16]. AI enables designers to push creative boundaries by generating unique design variations and offering real-time insights. AI-powered tools such as DALL·E, Midjourney, and Stable Diffusion allow for rapid visualisation of ideas, enhancing conceptualisation and prototyping. In addition, machine learning (ML) algorithms analyze vast datasets to suggest optimised design solutions that align with user preferences and industry trends [13]. Moreover, AI-powered generative design tools such as Autodesk Dreamcatcher and Rhino Grasshopper assist architects and product designers in creating data-driven, structurally optimized solutions. These tools automate aspects of the iterative design process, allowing designers to focus on refining creative and functional aspects rather than technical constraints [17]. In creative education, AI enhances design learning experiences by enabling interactive design simulations, AI-assisted brainstorming, and real-time design validation. AI-driven platforms assist students by suggesting alternative approaches, refining artistic concepts, and offering instant feedback on design choices [18]. Furthermore, AI tutors and co-creators help students develop a design-thinking mindset by promoting experimentation and critical reflection. In disciplines such as architecture, graphic design, and industrial design, AI facilitates collaborative creative environments where students and AI systems work together to generate and refine ideas [19] (Table 2).

Table 2 GenAI tools for creative disciplines

AI-advancements	Description
AI-generated learning materials	GenAI enables real-time content creation, including lecture notes, quizzes, and interactive case studies. This reduces instructor workload and enhances accessibility for diverse learners [9]
Personalized learning pathways	AI can analyze student performance and learning styles, tailoring instruction to individual needs. This enhances engagement and improves learning outcomes [10]
AI-powered design and simulation for architecture education	AI-powered visualization tools like Midjourney, DALL·E, and AI-driven CAD software allow students to explore complex design concepts in real time, bridging the gap between theoretical knowledge and practical application [12]
Conversational AI and virtual tutors	AI chatbots such as ChatGPT and Claude assist students with academic writing, problem-solving, and brainstorming, fostering critical thinking and reducing dependency on static learning materials [11]
Assessment automation and feedback	AI can assess assignments, detect plagiarism, and provide real-time feedback, making evaluations more efficient and objective [15]

2.3 Current Methods in Architectural Pedagogy: Traditional Versus AI-Enhanced Approaches

Traditional Approach: Architectural education has traditionally been centred around studio-based learning, manual drafting, and critique-intensive design processes, where students develop conceptual thinking, creativity, and spatial awareness through hands-on engagement with physical materials and peer feedback [4, 20]. This model emphasises iterative learning, where design solutions are refined through a cycle of critiques and revisions, guided by both faculty and peers. Manual drawing and physical model-making encourage deep cognitive processing, fostering intuitive design development and problem-solving grounded in material and spatial understanding [4]. However, these approaches are often time-intensive, resource-heavy, and less adaptable to rapid iteration or interdisciplinary integration. Furthermore, traditional methods provide limited opportunity for data-driven performance evaluation or immersive spatial simulation, which are increasingly relevant in a technologically evolving design context. As educational technologies evolve, there is growing interest in complementing but not replacing traditional pedagogy with computational and AI-assisted tools [20] (Table 3).

AI-Enhanced Approaches: AI has introduced a technology-driven transformation in architectural education, offering dynamic, data-driven, and computationally enhanced learning methods that redefine traditional studio-based approaches [4]. Unlike conventional methods that rely on manual drafting and human critique, AI-powered tools enable automated generative design, real-time performance analysis, and immersive visualisation techniques. One of the most significant advancements is AI-driven generative design, where machine learning algorithms such as GANs (Generative Adversarial Networks), Autodesk Dreamcatcher, and Grasshopper AI assist students in rapidly exploring multiple design alternatives [17]. This enables them to optimise structural integrity, material efficiency, and sustainability considerations factors increasingly central to responsible design education [13]. Additionally, AI-powered Building Information Modelling (BIM) tools such as Revit AI and Smart BIM provide data-driven insights into energy efficiency, material usage, and structural feasibility. These systems help students make informed decisions earlier in the design process and ensure that projects align with sustainability parameters [12]. Virtual and Augmented Reality (VR/AR) simulations, enhanced by AI, offer immersive experiences that allow students to explore architectural spaces at full scale, thereby improving their understanding of spatial relationships and user experience [13]. AI is also revolutionising critique and feedback systems in architectural education. While traditional critique depends on subjective human evaluation, AI tools now offer real-time, objective design assessments using image recognition and predefined principles [20]. Moreover, AI-based learning platforms deliver adaptive coursework, personalised feedback, and interactive design tutorials, enabling more equitable and self-paced learning experiences [18]. As summarised in Table 4, these innovations support interdisciplinary and collaborative studio learning, integrating

Table 3 Traditional approaches in architecture education

Traditional approaches	Description
Manual drafting and sketching	Encourages students to develop spatial visualization skills through hand-drawn sketches and conceptual design exercises. This method fosters creativity but is time-intensive and limits rapid iterations [4]
Critique-based learning	Involves professor and peer reviews, where students refine designs based on constructive feedback. This method strengthens design reasoning and problem-solving, but the feedback process can be subjective and time-consuming [20]
Physical model making	Hands-on model building enhances spatial understanding, materiality, and construction techniques. However, it requires significant time, effort, and resources, making it less adaptable to fast-paced design iterations [20]
Lecture-based theoretical learning	Focus on architectural history, materials, structural mechanics, and urban planning through classroom-based instruction. While essential for foundational knowledge, it lacks interactive and experiential engagement [4]
Software-based design (AutoCAD, Revit, Rhino, SketchUp)	Used for 2D and 3D modelling, allowing for detailed technical drawings and visualisation. However, design decisions still rely heavily on the student's intuition and expertise, and software alone does not provide real-time generative solutions [4]

data from urban analytics, material science, and sustainability models to help students develop design solutions that are contextually aware and future-ready [12, 18].

Traditional Versus AI-Enhanced Approaches: Architectural education is undergoing a significant transformation as AI-enhanced methodologies challenge the dominance of traditional studio-based learning. Traditional approaches such as manual drafting, critique-based feedback, and physical modelling have long fostered creativity, spatial reasoning, and conceptual development [4]. However, these methods can be time-consuming, resource-intensive, and limited in their capacity for rapid iteration and data-driven refinement [20]. They often lack adaptability and do not easily integrate real-time performance analysis or sustainability metrics. In contrast, AI-enhanced pedagogy introduces algorithmic design tools, generative parametric models, and immersive simulations that significantly accelerate both design iteration and visualisation [17]. Technologies such as AI-powered VR/AR environments, real-time performance feedback systems, and personalised learning platforms support students with adaptive coursework, iterative feedback, and contextual design decision-making [18]. These tools enhance efficiency and accuracy while

Table 4 AI enhanced approaches in architecture education

AI-enhanced approaches	Description
AI-generated parametric design	AI-driven generative algorithms (e.g., Grasshopper AI, Dreamcatcher) assist students in automating complex design iterations, allowing them to explore multiple design alternatives instantly [17]
Virtual and augmented reality (VR/AR) in design studios	AI-enhanced mixed-reality systems provide students with immersive design experiences, enabling them to walk through 3D models and simulate real-world architectural spaces [13]
AI-driven performance analysis	AI tools assess structural integrity, material sustainability, and environmental impact, helping students integrate eco-friendly and energy-efficient design principles into their projects [12]
AI-powered personalised learning	AI-based educational platforms provide adaptive coursework, design tutorials, and personalized feedback, tailoring learning to each student's skill level and progress [18]
AI-based automated design critiques	Machine-learning models analyse student projects and offer instant feedback on aesthetics, functionality, and structural integrity, supplementing traditional human-led critiques [20]

allowing students to consider sustainability, material optimisation, and structural integrity earlier in the design process [12]. As outlined in Table 5, while traditional techniques remain fundamental for cultivating intuitive design judgment and critical creativity, AI-driven methods offer scalability, precision, and interdisciplinary insight. Striking a balance between human creativity and machine efficiency is essential to ensure that architectural originality and holistic problem-solving are not compromised in increasing digitised learning environments [4].

3 Research Methodology

This study adopts a mixed-methods research design, combining a systematic literature review with empirical data collection through a structured questionnaire survey, in alignment with best practices in educational research that emphasise methodological triangulation for enhancing validity [21]. The literature review followed a structured protocol using databases such as Scopus, Web of Science, and Google Scholar, with keywords including "Generative AI," "architecture education," "AI in pedagogy," "student engagement," and "learning outcomes." Articles were selected based on relevance, academic rigour, and recency (2020–2025), resulting in a final

Table 5 Traditional versus AI-enhanced approaches

Criteria	Traditional pedagogy	AI-enhanced pedagogy
Design process	Manual drafting, hand-drawn sketches [4]	AI-generated parametric models, algorithmic design [17]
Iteration speed	Time-consuming revisions [4]	Rapid iteration with AI-driven suggestions [20]
Critique and feedback	Professor/peer-based feedback [4]	AI-assisted design evaluation with real-time feedback [12]
Visualization and presentation	Physical models & 2D drawings [4]	VR/AR immersive experiences and AI-generated renders [13]
Sustainability integration	Limited predictive analysis [20]	AI-driven performance and sustainability analysis [12]
Learning personalization	One-size-fits-all instruction [4]	Adaptive, AI-driven personalized learning experiences [18]

sample of 47 peer-reviewed publications focused on AI-enhanced design education. The empirical component involved a structured survey titled "AI in Architectural Pedagogy," targeting architecture students (undergraduate, postgraduate, doctoral), educators, and early-career design professionals. The study employed purposive sampling to select participants with prior exposure to AI tools, a common approach in domain-specific educational inquiries [22]. The online survey was administered via Google Forms and included 32 questions organised into six thematic sections: AI tool exposure, student engagement, impact on creativity and learning, ethical concerns, curriculum alignment, and pedagogical suggestions. The survey incorporated both closed-ended Likert-scale items and open-ended questions to facilitate quantitative and qualitative analysis. A total of 389 valid responses were obtained, comprising 52.4% undergraduate students, 30.3% postgraduates, 11% doctoral candidates, and 6.3% educators and professionals, primarily from India. Quantitative data were analysed using descriptive statistics to identify prevalent patterns, while qualitative responses were subjected to thematic analysis to extract key themes and concerns [23]. This integrated approach enabled a comprehensive understanding of the pedagogical impacts of Generative AI in architectural education, grounded in both scholarly literature and experiential insights.

4 Findings

The findings of this study provide a comprehensive understanding of the impact of GenAI on architectural pedagogy, particularly in terms of student engagement, creativity, and learning outcomes. Data collected from the survey indicate that GenAI significantly enhances student motivation, curiosity, and participation by enabling rapid visualisation and iterative design exploration. Approximately 84% of students using AI-assisted tools reported higher enthusiasm compared to 62% in traditional settings, citing reduced creative blocks and greater freedom to experiment. This observation aligns with previous research indicating that AI-supported environments can boost learner confidence and design autonomy [4]. AI-supported brainstorming also fostered collaborative learning, with team-based projects showing increased peer interaction, active critique, and improved confidence during design reviews. Faculty respondents similarly observed heightened student involvement in discussions and critiques within AI-integrated studios. However, concerns emerged regarding over-dependence on AI, as some students struggled to generate original ideas without algorithmic support, echoing concerns raised regarding authorship dilution and aesthetic homogenization [8].

With respect to engagement, the results indicate a strong positive orientation towards AI-assisted design practices. Approximately 64% of respondents rated AI-assisted design iterations as either 'Engaging' or 'Highly Engaging' (ratings 4 and 5 on a 5-point Likert scale), while only 15% perceived such iterations as minimally engaging (ratings 1 or 2). Furthermore, 59% of the participants reported that AI-generated suggestions significantly inspired creativity and encouraged exploration of novel design ideas. Peer collaboration and discourse were also reported to be positively influenced by AI integration; nearly 73% either agreed or strongly agreed that AI tools enhanced collaboration in design studios, a finding that aligns with the literature on AI-supported co-creation environments [18]. Regarding learning outcomes, AI's contribution appears substantial. When asked whether AI-assisted learning enhanced their ability to iterate and refine design outputs, a combined 78% responded affirmatively, with 38.8% reporting significant improvement. Despite this generally positive orientation, ethical and cognitive concerns emerged. Nearly half of the respondents (48.8%) agreed or strongly agreed that AI-generated outputs could reduce originality and independent thinking. This perception underscores a critical tension within the discourse on AI in design education—while AI enhances productivity and exploration, it may also risk homogenizing creativity and undermining authorship. Nevertheless, 64.3% of respondents reported feeling comfortable or very comfortable relying on AI for making design decisions, suggesting a pragmatic acceptance of AI tools despite ethical reservations. Crucially, a significant majority (70%) supported the integration of AI ethics and responsible AI usage modules within the architecture curriculum, reinforcing the need for structured ethical frameworks in digital pedagogy [5].

Impact of GenAI on Design Skills and Creativity: The study revealed that GenAI significantly enhanced students' capacity to produce complex, innovative, and diverse architectural designs. The experimental group, using AI tools, outperformed the control group in creativity and conceptual originality, scoring 22% higher in design assessments [19]. Faculty observed that AI-enabled students explored unconventional geometries and advanced visualization techniques, benefiting from the speed and efficiency of rapid prototyping through iterative design development. GenAI facilitated a deeper understanding of form-finding and enabled site-specific solutions by generating contextual variations, fostering a more data-driven and responsive design approach [13]. However, the study also identified concerns over potential design homogenization, as some students overly relied on AI-generated outputs, leading to repetitive patterns and diminished conceptual depth. These findings highlight the need for a balanced pedagogical framework that integrates GenAI as an augmentative tool while reinforcing critical thinking and originality in architectural education.

Effect of GenAI on Learning Outcomes: The study demonstrated a strong correlation between GenAI integration and enhanced learning outcomes in architectural education. Students using AI tools showed a 25% improvement in proficiency with advanced digital modeling, rendering, and generative design techniques, including parametric form exploration and environmental simulations [20]. AI-assisted learners also exhibited stronger problem-solving and adaptive thinking, effectively modifying design parameters in response to real-time feedback, thereby preparing them for complex, data-driven challenges in professional contexts. However, concerns around authorship and originality emerged, prompting calls for AI literacy training to address ethical implications, dataset biases, and responsible use of AI-generated content in architectural practice [5].

Considerations for AI Integration in Architectural Pedagogy: While GenAI offers clear benefits, the study also identifies challenges requiring careful pedagogical attention. A key concern is balancing AI use with traditional design skills, as some students became overly reliant on AI, struggling with manual sketching and conceptual thinking without digital assistance [3]. Educators emphasized that foundational practices like hand drafting, physical modeling, and intuitive problem-solving must remain core to the curriculum. The risk of AI dependency was further compounded by a lack of critical engagement with AI outputs, underscoring the need for structured assignments that integrate both AI and non-AI design methods. Additionally, students often required guidance in crafting effective AI prompts, as generic inputs led to repetitive or conceptually weak results. Institutional disparities in access to high-performance computing and commercial AI tools also emerged, suggesting a pressing need for open-source solutions and collaborative frameworks to ensure equitable AI integration across architecture schools [12].

5 Discussion

The integration of GenAI in architectural education is fundamentally transforming traditional pedagogical practices by enabling design innovation, enhancing iterative workflows, and fostering collaborative studio environments [4]. This study corroborates findings from earlier works, showing that GenAI empowers students to produce complex and diverse design outcomes, but also underscores the necessity for a pedagogical shift that preserves core architectural principles such as critical thinking, spatial reasoning, and conceptual integrity [3]. The capacity of GenAI to facilitate rapid prototyping and real-time design exploration supports more adaptive and exploratory learning, as also observed [16], who emphasize that AI-enhanced environments foster experimentation and creative risk-taking. However, these opportunities are contingent upon educators being adequately prepared to integrate GenAI meaningfully into curricula, echoing the pedagogical imperatives raised in AI-supported design studios [20]. Concerns raised in this study regarding students' growing dependence on AI-generated content reflect the tensions identified [8], who caution that AI tools, if uncritically applied, may diminish conceptual depth and erode authorship. The risk of aesthetic homogenization where students replicate algorithmically generated outputs without critical interpretation aligns with critiques [13], who warns against the uncritical adoption of AI-generated patterns that undermine design originality. Consequently, fostering AI literacy becomes essential—not only in terms of technical proficiency, but in cultivating students' ability to question, evaluate, and refine AI-generated propositions [2]. Moreover, ethical considerations including authorship ambiguity, intellectual property concerns, and algorithmic bias must be addressed through clear institutional policies and inclusive datasets that reflect regional diversity and cultural contexts. The findings of this study reaffirm the need for blended pedagogical frameworks that combine analogy design methods (hand sketching, physical modelling) with computational design and GenAI platforms to retain educational depth while advancing digital fluency [17]. As a result, this study supports emerging best practices for GenAI integration: blended learning models, project-based implementation, integration of AI ethics into coursework, iterative AI-assisted design development, and collaborative studios enhanced by AI-supported feedback systems [7, 9]. These approaches align with broader educational goals of fostering reflective, ethically grounded, and critically engaged future architects equipped to navigate the evolving digital design landscape.

6 Conclusion

This study offers preliminary evidence that GenAI may enhance architecture education, showing how AI-enhanced environments can significantly boost student engagement, design creativity, and learning outcomes. Students using AI produced more innovative and complex solutions, benefiting from rapid iterations and exploratory workflows, which also increased their motivation. However, the findings stress the need to balance AI with traditional methods, as over-reliance may diminish critical thinking and foundational skills like sketching and physical modeling. The study also raises ethical concerns, including authorship, intellectual property, and algorithmic bias, highlighting the importance of AI literacy to ensure responsible use. It recommends future longitudinal research to assess the long-term impact of AI in education and practice. In conclusion, while GenAI has the potential to influence architectural learning through interactivity and innovation, its integration must be guided by structured frameworks that uphold core competencies, ethical awareness, and critical design thinking to prepare students for a technologically evolving profession. A key limitation of this study is its reliance on self-reported survey data without triangulation through interviews or classroom observations, which may limit the depth and generalizability of the findings. Nonetheless, the study highlights the practical need for faculty training and curriculum redesign to support the responsible use of GenAI in design studios. Future research should consider longitudinal studies and incorporate observational methods to better understand the long-term impacts of GenAI on architectural education.

References

1. Xu, C., Huang, Y.: Technological innovation in architectural design education: empirical analysis and future directions of midjourney intelligent drawing software. Buildings **14**(10), 3288 (2024). https://doi.org/10.3390/buildings14103288
2. Fan, H., Goyal, B., Ghafoor, K.Z.: Computer-aided architectural design optimization based on BIM technology. Informatica **46**(3) (2022). https://doi.org/10.31449/inf.v46i3.3935
3. Li, C., Zhang, T., Du, X., Zhang, Y., Xie, H.: Generative AI models for different steps in architectural design: a literature review. Front. Archit. Res. **14**(3), 759–783 (2025). https://doi.org/10.1016/j.foar.2024.10.001
4. Kee, T., Kuys, B., King, R.: GenAI to enhance architecture education to develop digital literacy and holistic competency. J. Artif. Intell. Archit. **3**(1), 24–41 (2024). https://doi.org/10.24002/jarina.v3i1.8347
5. Gupta, U., Khurana, A., Kaur, H.: Exploring the intersection of AI and ethics in architecture: implication for design, design thinking and built environment. Int. Res. J. Adv. Eng. Manag. **3**(03), 641–646 (2025). https://doi.org/10.47392/IRJAEM.2025.0105
6. Kehinde-Awoyele, A.A., Adeowu, W.A., Oladejo, B.: Enhancing classroom learning: the impact of AI-based instructional strategies on student engagement and outcomes. Int. J. Res. Innov. Soc. Sci. **VIII**(IIIS), 5732–5742 (2024). https://doi.org/10.47772/IJRISS.2024.803429S
7. Dasgupta, D., Venugopal, D., Gupta, K.D.: A Review of Generative AI from Historical Perspectives (2023). https://doi.org/10.36227/techrxiv.22097942
8. Onatayo, D., Onososen, A., Oyediran, A.O., Oyediran, H., Arowoiya, V., Onatayo, E.: Generative AI applications in architecture, engineering, and construction: trends, implications for practice, education & imperatives for upskilling—a review. Architecture **4**(4), 877–902 (2024). https://doi.org/10.3390/architecture4040046
9. Zawacki-Richter, O., Marín, V.I., Bond, M., Gouverneur, F.: Systematic review of research on artificial intelligence applications in higher education—where are the educators? Int. J. Educ. Technol. High. Educ. **16**(1), 39 (2019). https://doi.org/10.1186/s41239-019-0171-0
10. Zaharuddin, C.Y., Yao, G.: Enhancing student engagement with AI-driven personalized learning systems. Int. Trans. Educ. Technol. **3**(1), 1–8 (2024). https://doi.org/10.33050/itee.v3i1.662
11. Chauhan, D., Singh, C., Rawat, R., Chouhan, M.: Conversational AI applications in ed-tech industry. In: Conversational Artificial Intelligence, pp. 411–433. Wiley (2024). https://doi.org/10.1002/9781394200801.ch25
12. Vashishth, T.K., Sharma, V., Sharma, K.K., Kumar, B., Panwar, R., Chaudhary, S.: AI-Driven Learning Analytics for Personalized Feedback and Assessment in Higher Education, pp. 206–230 (2024). https://doi.org/10.4018/979-8-3693-0639-0.ch009
13. Agboola, O.P.: The role of artificial intelligence in enhancing design innovation and sustainability. Smart Des. Policies **1**(1), 6–14 (2024). https://doi.org/10.38027/smart-v1n1-2
14. Chris, E., John, M., Mercy, G.: Generative AI for Educational Content Creation (2024)

15. Buzick, H.M., Casabianca, J.M., Gholson, M.L.: Personalizing large-scale assessment in practice. Educ. Meas. Issues Pract. **42**(2), 5–11 (2023). https://doi.org/10.1111/emip.12551
16. Chang, Y.-S., Tsai, M.-C.: Effects of design thinking on artificial intelligence learning and creativity. Educ. Stud. **50**(5), 763–780 (2024). https://doi.org/10.1080/03055698.2021.1999213
17. Channi, H.K., Kaur, A., Kaur, S.: AI-driven generative design redefines the engineering process. In: GenAI in Finance, pp. 327–359. Wiley (2025). https://doi.org/10.1002/9781394271078.ch17
18. Noor Afshan, S.S.: Exploring the impact of AI on architectural creativity and efficiency. Int. J. Multidiscip. Res. **6**(2), 1–18 (2024). https://doi.org/10.36948/ijfmr.2024.v06i02.15753
19. Gmeiner, F., Yang, H., Yao, L., Holstein, K., Martelaro, N.: Exploring challenges and opportunities to support designers in learning to co-create with AI-based manufacturing design tools. In: Proceedings of the 2023 CHI Conference on Human Factors in Computing Systems, pp. 1–20 (2023). https://doi.org/10.1145/3544548.3580999
20. Jin, S., et al.: Enhancing architectural education through artificial intelligence: a case study of an AI-assisted architectural programming and design course. Buildings **14**(6), 1613 (2024). https://doi.org/10.3390/buildings14061613
21. Creswell, J.W.: Research Design: Qualitative, Quantitative, and Mixed Methods Approaches, 4th edn. SAGE Publications, Inc., All, United States of America Library (2014)
22. Etikan, I.: Comparison of convenience sampling and purposive sampling. Am. J. Theor. Appl. Stat. **5**(1), 1 (2016). https://doi.org/10.11648/j.ajtas.20160501.11
23. Braun, V., Clarke, V.: Using thematic analysis in psychology. Qual. Res. Psychol. **3**(2), 77–101 (2006). https://doi.org/10.1191/1478088706qp063oa

Reaching the Parts Others Cannot: Embedding Further Education in the English Innovation Ecosystem for Inclusive Growth and Place-Based Innovation

Coral Grainger, Kate Webb, and Mandy Parkinson

Abstract This paper explores the potential role of Further Education (FE) colleges in the United Kingdom's innovation ecosystem, focusing on their potential to drive inclusive growth and place-based innovation. FE refers to the sector of education that serves learners beyond the compulsory schooling age of sixteen and encompasses academic, vocational and recreational courses, delivered by FE colleges that are often hyper-local and deeply embedded in their communities. Despite their significant reach and contributions to local economies, FE institutions have been underutilised in national innovation strategies. The study examines the early implementation of the Greater Manchester Further Education Innovation Programme (GMFEIP), a £2.5 million initiative funded by InnovateUK's Further Education Innovation Fund. The programme aims to strengthen the role of FE colleges as key enablers of innovation through the establishment of ten Innovation Centres across Greater Manchester and tests a novel conceptual framework: the radial innovation network. The early findings suggest that FE institutions can serve as effective innovation intermediaries, leveraging their strong industry partnerships, vocational focus, and community links to support business innovation and regional economic development. The paper contributes to the broader debate on inclusive innovation and regional economic policy, highlighting the potential of FE-led initiatives to foster sustainable economic growth.

Keywords Knowledge transfer · Radial · Diffusion · Networks · Inclusive growth · Tertiary education · Further education · SDG9 industry · Innovation · Infrastructure · SDG-8: decent work and economic growth

---

C. Grainger (✉)
Manchester Metropolitan University, Manchester, UK
e-mail: coralgrainger@gmail.com

K. Webb
The KWP Ltd, Compton, UK

M. Parkinson
University of Salford, Salford, UK

A. Visvizi et al. (eds.), *Research and Innovation Forum 2025 Volume 1*,
Springer Proceedings in Complexity, https://doi.org/10.1007/978-3-032-09066-9_19

1 Introduction and Literature Review

Further Education (FE) in England represents a substantial yet often under-recognised part of the education system, comprising hyper-local institutions that serve over three million learners annually across academic, vocational, and professional disciplines [1–3]. Despite its wide reach, FE is often marginalised in research, policy and media discourse, partly due to its complex remit, vocational orientation, and constant adaptation to shifting policy and market demands [3–5].

This paper addresses this gap by exploring how Further Education (FE) colleges can function as innovation intermediaries in regional economies. Drawing on the Greater Manchester Further Education Innovation Programme (GMFEIP), delivered by Greater Manchester Colleges (GM Colleges), it presents a novel conceptual framework, the radial innovation network, and evaluates its early implementation and implications for inclusive, place-based innovation.

FE has been overlooked in national innovation strategies despite local ties, applied curricula, and industry partnerships. The United Kingdom (UK) Innovation Strategy [6], which pertains to England, acknowledges the importance of diverse stakeholders, including FE, but focusses on the role of FE as a training provider, rather than as an intermediary for the diffusion of innovation in local economies. Furthermore, greater economic benefits could be gained by prioritising innovation across traditional sectors rather than creating new innovations in just a few areas [7]. Using universities and hospitals as anchor institutions to aid regional economies through their knowledge and expertise, has been recommended [7]. However, this paper posits that FE colleges, being accessible, diverse and hyper-local, can implement these ideas at a local level.

The marginalisation of FE in English innovation policy has been bolstered by the Triple Helix concept, in which industry, government and academia deliver wealth generation, public control, and novelty production positions Higher Education (HE) as the principal locus for research and development (R&D) and knowledge transfer activities [8–10]. The Triple Helix [10] provides a potential framework for growing economies and state intervention in times of economic crisis such as the Covid-19 pandemic [11]. However, it is often misunderstood and narrowly interpreted [12]. It is this narrow interpretation, of universities as the central nodes of innovation creation, which has marginalised other institutions and undermined inclusive growth.

An alternative model for consideration is the Innovation Ecosystem Stakeholder model [13, 14], based on the MIT innovation landscape, which identifies five key actors, and emphasises universities as coordinating hubs linking industry, government, entrepreneurs, and investors to support interdisciplinary innovation and commercialisation. Although this model is compelling for its expanded approach, it remains situated within the paradigm of the innovation ecosystem, which is not without complication.

The conceptual clarity of 'innovation ecosystem' is sometimes questioned [15–17] and this ambiguity can undermine the usefulness of ecosystem models across varied contexts [15, 16, 18]. Described as networks of diverse groups, actions, objects, institutions, and connections that influence innovation [10, 15], innovation ecosystems require both precise and adaptable interpretations. This paper's inclusion of complementary and substitution frameworks acknowledges the flow of knowledge across society, potentially allowing more actors like further education institutions to participate.

Adapting innovation ecosystems to different contexts is crucial [7, 9, 18]. The structural differences between US and English education systems complicate direct comparisons. In the US, colleges are part of university systems focused on undergraduate education, while in England, 'colleges' provide further education and vocational training [19]. This paper refers to British standalone FE institutions as 'colleges'.

Current models often overlook the needs of less connected communities and SMEs (small and medium size enterprises) in low productivity regions [20, 21]. By integrating FE colleges into national and regional innovation ecosystems, this paper suggests disparities in innovation-led growth might be reduced. FE, with local connections and community-focused services, can act as intermediaries and catalysts for economic development, fostering inclusive growth and addressing regional economic imbalances.

2 Research Gap

The dominant innovation infrastructure remains centred on higher education-led models, often clustered in innovation districts or knowledge quarters [22]. This risks overlooking place-based innovation in less connected communities [7, 20, 23] and excluding some SMEs, particularly in less productive regions [7, 8, 10].

Further education (FE) colleges are underexplored within regional innovation ecosystems, with limited recognition of their potential as direct enablers of SME innovation [24, 25]. Research on how devolution and place-based policies could empower FE colleges, distinct from higher education (HE), to drive regional innovation is scarce. Likewise, evidence on how FE–business collaboration enhances local innovation capacity is limited. More investigation is needed into the effectiveness of this approach in contributing to sustainable economic growth in underperforming regions.

3 Approach

To present and test a novel conceptual framework, the radial innovation network (Fig. 2), this study examined the early implementation of the Greater Manchester Further Education Innovation Programme (GMFEIP), a £2.5 million initiative funded by InnovateUK's Further Education Innovation Fund (FEIF). Greater Manchester, a metropolitan area in northern England with nearly three million residents and a diverse economy, faces significant deprivation, with some localities among the most deprived nationally [26, 27]. It also holds devolved powers across several policy domains [28, 29]. New government policy [28–30] strengthens the role of local and regional institutions in economic development, and Greater Manchester continues to lead in this area [30].

Launched in April 2024, the GMFEIP aimed to build FE colleges' capacity as enablers of innovation and productivity, particularly by supporting businesses with limited prior access to innovation services. Data from 2015–2021 showed that such access declined with distance from Greater Manchester's central innovation districts (Fig. 1). In response, a new conceptual framework was developed to address this gap: the radial innovation network (Fig. 2).

The model established ten Innovation Centres, one in each borough, to connect colleges and businesses. Roles were created, including Business Innovation Advisors (BIAs), Coordinators (BICs), Innovators in Residence (IiRs), Innovation Literacy Co-ordinators (ILCs) who train apprentices in innovation literacy, so they are developed as Apprentice Innovation Ambassadors (AIAs), to promote innovation diffusion and adoption.

The conceptual framework builds on the innovation stakeholder ecosystem model [13, 14] adding FE as an additional and complementary stakeholder in a place-based innovation ecosystem [24].

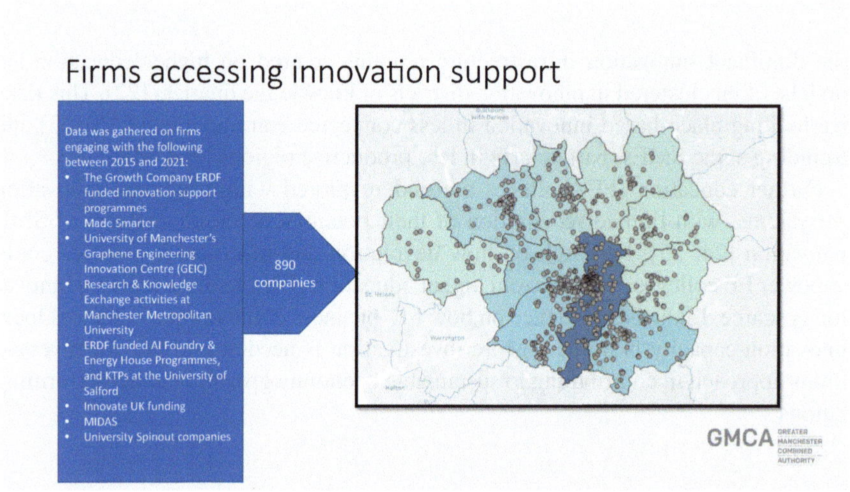

Fig. 1 Firms accessing innovation support. *Source* GMCA & The Data City, (2022) unpublished presentation

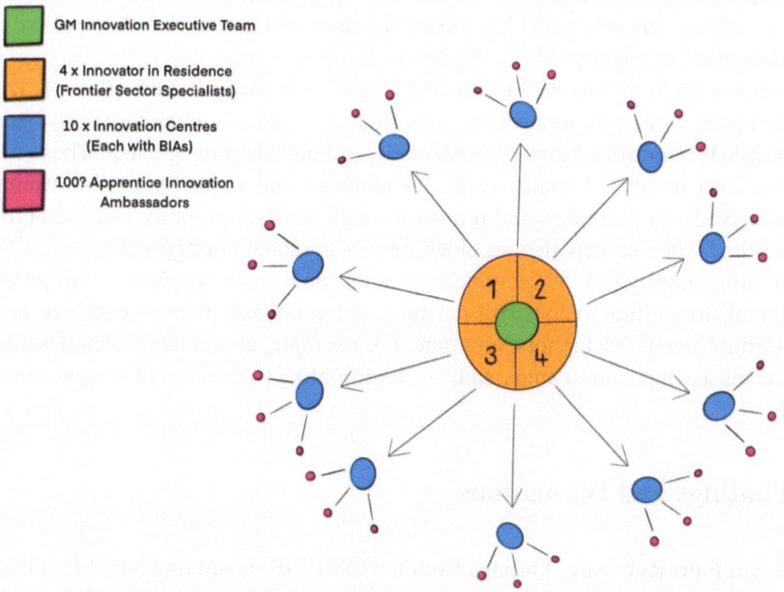

Fig. 2 Conceptual Framework- Radial Innovation Network. *Source* Grainger et al. (2025) unpublished conceptual framework

4 Methodology

This paper presents an early evaluation of the programme through a qualitative approach [31] to examine emerging impacts, challenges, and areas for further inquiry. Summary case studies contextualise the findings and highlight key themes.

The research was carried out in two phases. In the first, twenty-four semi-structured interviews were conducted (Sept–Oct 2024) with stakeholders and programme staff. Purposive sampling captured diverse perspectives across FE colleges, SMEs, apprentices, and partners, recruited via programme networks. This phase identified early insights for further enquiries resulting in the two research questions (RQ) below.

RQ1 How can FE institutions act as innovation intermediaries and place-based economic catalysts?

RQ2 How does a radial innovation network (Fig. 2) improve SME access to innovation?

The second phase (Jan–Feb 2024) involved thirty-seven semi-structured interviews with staff, stakeholders, and SMEs, further exploring the research questions across roles and engagement levels. Sampling remained purposive, with a focus on institutional diversity. Employers and apprentices, recruited through Business

Innovation Advisors and Innovation Literacy Coordinators, offered valuable early insights despite self-selection bias. Future research will aim to include more critical or disengaged perspectives.

Thematic analysis followed Braun and Clarke's six-phase framework [32]. Transcripts were reviewed for immersion, with inductive manual coding revealing themes such as SME innovation barriers, confidence, and network engagement. These were grouped into broader domains (e.g., 'Confidence and Capacity for Innovation', 'Cultural Shifts in Attitude') and refined through team discussions. Data saturation was reached when no new themes emerged across stakeholder groups.

Narrative inquiry [33, 34] complemented this analysis by exploring how participants used storytelling to construct meaning. Attention was given to narrative structure, turning points, metaphors, and tone. For example, apprentices often described themselves as "innovation ambassadors," highlighting their sense of empowerment.

5 Findings and Discussions

This section presents early findings from the GMFEIP, examining how FE colleges are emerging as innovation intermediaries through the application of the radial innovation network (Fig. 2) in a live, place-based context. Given the programme's formative stage, findings reflect evolving roles and structures rather than final outcomes. The analysis offers early insight into a systems-level intervention in regional innovation, not a summative impact assessment.

Initial quantitative data indicates strong demand. By December 2024, 842 businesses (up from 199 in October 2024) had engaged across all ten Greater Manchester boroughs. Around 70% were new to structured innovation support, suggesting reach into underserved SMEs. Engagements included advisory support and digital transformation events. These figures show promise, though sustained innovation outcomes are yet to be evidenced.

Rooted in regional innovation theory [13, 14, 23] the radial innovation network underpins GMFEIP. The model fosters decentralised, embedded collaboration and represents a novel FE-led approach to regional innovation. Though still context-specific and untested at scale, it offers potential for more inclusive, place-sensitive support. Broader evaluation is needed to assess generalisability and long-term impact.

Six emerging themes illustrate the network's potential to redefine FE's role in economic development and indicate areas for further evidence and refinement:

- FE as innovation intermediaries
- Radial network design
- Innovators in Residence
- Empowerment via Innovation Literacy
- Reaching the SME "missing middle"
- Inclusive, place-based digital transformation

The model aligns with literature on interactive learning in innovation systems [35] and decentralised innovation ecosystems [13, 14, 23] and responds to calls from the Gatsby Foundation [24] and United Kingdom Research and Innovation (UKRI) [36] for FE-led, locally embedded support for SMEs excluded from traditional R&D routes. While still evolving, this study contributes to these debates by examining the framework in practice.

5.1 FE as Innovation Intermediaries

GMFEIP illustrates how colleges can act as innovation intermediaries [37] addressing gaps in traditional research-led innovation models. The IiRs bring knowledge from the established innovation ecosystem into FE colleges, from where knowledge is diffused to business through the BIAs and ILCs based in the Innovation Hubs. The BIAs offer businesses training, tailored support and referrals within the innovation ecosystem, while the ILCs deliver Innovation Literacy Training to apprentices within local businesses, enabling them to continue diffusion and adoption of innovation in their workplaces.

5.2 Greater Manchester FE Model: Radial Innovation Network

The GMFEIP is different from innovation models such as the Triple Helix [10–12] and builds upon the MIT Innovation Stakeholder model [13, 14]. It is intended to ensure that innovation reaches businesses and communities that might otherwise remain disconnected from HE led knowledge exchange. The idea of "radial personal networks" plays a significant role in social systems in which a personal network comprises individuals connected through regular communication patterns [38]. When networks are radial, they are more open to an individual's environment, facilitating a broader exchange of information. This openness enhances the diffusion of innovations, as individuals are exposed to diverse ideas and influences.

The network (Fig. 2) established multiple diffusion points across a network of FE colleges, each embedded within its local economy and acting as a conduit for business support, skills development, and technology adoption. Rather than relying on a single knowledge hub, this approach aims to spread innovation outward from multiple locations, increasing accessibility for businesses, particularly SMEs. By leveraging a decentralised structure, this model aligns with Greater Manchester's devolved governance and economic landscape [26, 29, 30] providing targeted innovation support, particularly to SMEs.

The model challenges the assumption that innovation diffusion follows a sequential, hierarchical process. The approach has an interconnected and multi-directional process [38]. Innovators in Residence, Innovation Centre Teams, and Apprentice Innovation Ambassadors act as facilitators, ensuring that knowledge transfer occurs both within and across networks. This dynamic structure might allow for more fluid, organic innovation adoption that responds to regional economic needs rather than being dictated by centrally positioned institutions.

The radial innovation network (Fig. 2) positions colleges as 'boundary objects' [39], linking training, innovation support, research, and local businesses to provide accessible entry points for innovation. As boundary objects, they connect diverse fields and foster collaboration [39]. Serving deprived communities and a wide range of learners [1, 2, 4, 5] colleges operate across multiple spaces in society. While their diversity and complexity often lead to neglect in education policy [3–5], this framework reframes that diversity as a strength. It highlights FE colleges as key innovation intermediaries [37], bridging policy, industry, communities, and local economies. The model offers a scalable, practical approach to SME-led innovation, particularly in regions where traditional research-driven methods have struggled to achieve broad impact [40].

5.3 Innovators in Residence

Innovators in Residence (IiRs) were appointed based on their sector-specific expertise in one of Greater Manchester's four 'Frontier Sectors' [26]: Sustainable Advanced Materials and Manufacturing (SAMM), Digital, Creative & Media (DCM), Health Innovation & Life Sciences (HILS), and Net-Zero / Clean Growth. Unlike traditional academic or business consultancy models, the IiRs function as peripatetic sectoral experts and innovation intermediaries [37] diffusing the latest advancements in knowledge and skills to FE staff. Their work is significant to GMFEIP's radial innovation network (Fig. 2), wherein they gather knowledge from established ecosystem players and transfer that to the outer areas through the FE teams and GMFEIP hubs.

The IiRs engage and convene sector business leaders such as the Advanced Machinery and Productivity Institute (AMPI) and North of England Robotics Innovation Centre (NERIC at University of Salford) Amazon, and Siemens to inform the development of a programme of Continuing Professional Development (CDP) for college staff. Staff are more able to design and deliver training programmes that are aligned with the latest and future needs of the labour market. The IiRs establish relationships between the colleges and these innovation assets remain as a legacy after GMFEIP funding has ended.

5.4 Training and Innovation Literacy

The following case studies illustrate how the model is driving innovation through a range of activity, often in unexpected ways. For example, the GMFEIP Innovation

Literacy Training, targeted at apprentices working in SMEs, has been particularly impactful, fostering greater innovation capacity within businesses. Early feedback suggests this approach has a ripple effect, strengthening the confidence of apprentices to contribute to workplace innovation. Rather than simply receiving support, businesses and individuals are empowered to innovate on their own terms. The programme supports businesses in customising innovation strategies to their needs, whether through AI training, digital transformation, or skills development [20, 23].

Case Study 1—Apprentice

Alex had long been fascinated by artificial intelligence (AI), and his interest deepened after the launch of ChatGPT in 2022. While completing his Apprenticeship in Mechanical Engineering, he began self-studying AI tools and machine learning. Near the end of his apprenticeship, his college assessor introduced him to Innovation Literacy Training.

Seeing its relevance, Alex invited his managers to join. This led to the formation of a Digitalisation and AI Group at his workplace, aimed at educating staff, dispelling misconceptions, and exploring efficiency improvements. The programme gave Alex the confidence to lead a pilot project using AI to automate meeting minutes and streamline documentation. Its success prompted senior management to ask him to deliver a workshop for executives.

The programme not only expanded Alex's technical skills but positioned him as an 'Innovation Ambassador'. His ability to explain AI's benefits and support its implementation empowered colleagues to innovate. Collaborating with apprentices from varied fields also broadened his problem-solving perspective.

Despite initial resistance and misconceptions about AI, Alex persisted. He became an advocate for embedding innovation literacy early in apprenticeship training and for supporting those interested in emerging technologies. Now progressing to a Degree Apprenticeship [41] *in Digital and Technology Solutions, he continues to explore AI's business applications. His journey, from enthusiast to workplace innovator, shows how structured innovation training can enable individuals to shape their careers and drive organisational change.*

5.5 Supporting the 'Missing Middle' of SMEs

Traditional knowledge-transfer models often prioritise large firms in R&D partnerships, or early-stage start-ups supported by incubators and grants [42–44]. Mid-stage SMEs, lacking capacity to engage with research-intensive institutions but needing tailored support to scale, frequently face barriers to innovation access [44]. Innovation hubs such as those developed through GMFEIP, offer a practical alternative by embedding support across educational levels. These hubs combine sector strengths to help SMEs adopt technologies and transformation strategies without relying solely on university partnerships.

GMFEIP engaged 842 businesses, expanding outward from each hub (Fig. 3). Many were part of the "missing middle" [45]: too large for microfinance, not start-ups, yet too small to attract major investment [45, 46]. Business owners consistently cite funding and support limitations as growth barriers [46–50]. FE colleges help bridge this gap, offering innovation training, financial guidance, and digital strategies without the complexity of R&D partnerships. While the programme spans Greater Manchester, Fig. 3 shows that Innovation Support access has been strongest in more remote areas reversing the centralised access patterns shown in Fig. 1.

Instead of just focusing on apprenticeships and training, FE colleges are increasingly seen as partners in business productivity and innovation. Case Study 2 illustrates this.

Case Study 2: A Business Case Study

Frank was always looking to grow his business but found the landscape of support and funding difficult to navigate. The programme helped him reassess his business model, learn from peers at different stages, and refine his strategy. A key benefit was structured guidance that led to securing funding for a prototype, something he had previously struggled to find.

The programme also connected Frank with local colleges, enabling access to business support services. Collaborating with media students on a social media showcase demonstrated the value of colleges as partners in business development.

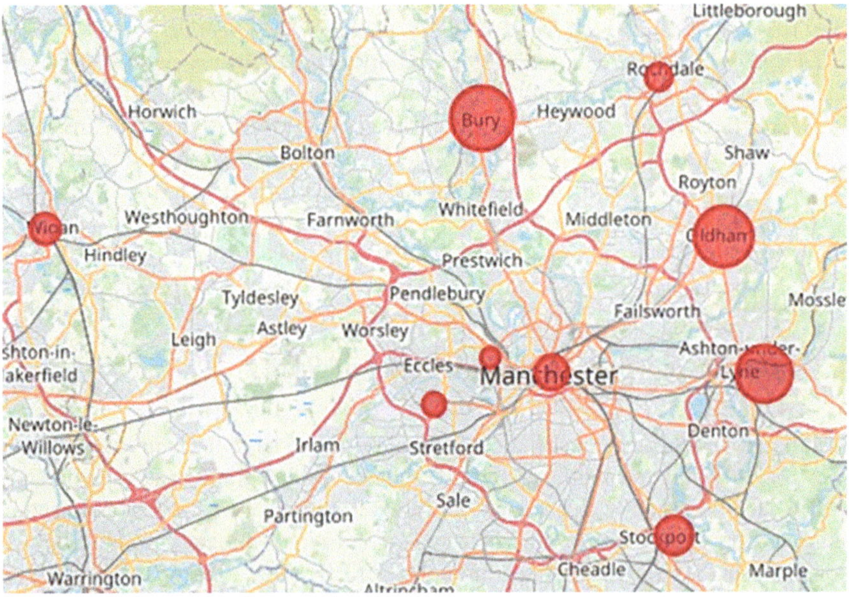

Fig. 3 Mapping of activity based on evaluation of the conceptual framework: Radial Innovation Network. *Source* Grainger and Webb (2025) unpublished data from GMFEIP Evaluation

Frank came to see innovation as a continuous process. Adopting a "test and learn" approach, his business began developing and refining products in real-world settings rather than aiming for perfection pre-launch. This agile mindset, reinforced by the programme, now underpins his growth strategy.

Challenges persist. Many SMEs remain unaware that colleges offer innovation support, often viewing them solely as vocational training providers. The term "innovation" itself can feel intimidating; simpler language like "good ideas" can make the concept more approachable.

Frank believes his experience shows that colleges can provide practical, collaborative support for entrepreneurs, bridging education and enterprises to help small businesses thrive.

5.6 Inclusive Place-Based Innovation

The GMFEIP presents a place-based innovation model that challenges the dominance of the London–Oxford–Cambridge golden triangle [51]. Northern regions like Greater Manchester, where productivity lags behind the national average [25, 26], often lack the interconnectedness needed for sustained growth, hindered by weaker university-business links, lower R&D investment, and limited SME support [45, 48, 50].

Building on the innovation stakeholder ecosystem model [13, 14] which connects HE, government, business, and investors, GMFEIP extends this framework by integrating FE colleges. Strengthening FE innovation capacity enables more tailored support for SMEs and underserved communities beyond traditional university-led systems [23].

By advancing innovation in foundational economy sectors [52], the programme promotes a more inclusive culture of innovation and accelerates its diffusion. FE colleges function as active hubs linking education and enterprise. Case Study 3 illustrates how this integrated approach helps distribute the benefits of innovation more equitably.

Case Study 3: A Foundational Economy Case Study

The hospice supports people with life-limiting illnesses and partnered with the GMFEIP to strengthen its operations and outreach. Through support in networking, events, and skills development, the hospice formed new partnerships with local businesses, expanding its reach.

The college provided event spaces, with media students promoting them and catering students delivering a high-quality experience. Workshops on artificial intelligence enabled the hospice to explore efficiencies, while staff accessed training to enhance skills and operational strategies.

This collaboration has reduced costs, improved visibility, and deepened community engagement. It has delivered practical benefits for both the hospice and the college, supporting long-term sustainability and opening new opportunities for partnership.

5.7 AI and Digital Transformation in SMEs

Many businesses engaged with GMFEIP reported that AI training demystified emerging technologies, allowing them to improve productivity without high investment costs (Case Study 4).

Case Study 4: An SME Case Study

Aktar, a business owner, joined the programme after attending a digital roadshow at his local college. Initially unaware of FE colleges' role in innovation, he quickly recognised their value as strategic partners for business growth and AI integration.

He arranged AI training for his senior team, introducing tools like ChatGPT and Synthesia. The sessions boosted productivity, eased concerns, and enabled automation of key processes. A major breakthrough was the creation of custom AI models, which significantly reduced manual workloads and embedded AI literacy across the business.

This engagement grew into an ongoing partnership with the college, leading to discussions on T-Level internships, student placements, and even donating training assets.

Aktar now views FE colleges as vital innovation hubs for SMEs, offering accessible, industry-relevant support. He advocates for future programmes to expand beyond AI into automation and intelligent workflows.

5.8 Strengthening FE's Standing in Regional Economic Policy

GMFEIP has highlighted the role of FE in Greater Manchester's innovation ecosystem. Innovation research stresses the importance of local, sector-specific knowledge transfer in promoting economic resilience [22]. As distributed innovation hubs, FE colleges support nearby businesses while enabling knowledge and skills to flow outward via the radial innovation network (Fig. 2).

This shift could reposition FE colleges from training providers to strategic partners in innovation ecosystems and local economies [23]. Early evaluation shows colleges are now engaged in economic forums they were previously excluded from. The Greater Manchester Combined Authority has described the programme as a potential "game changer" for integrating FE into regional innovation.

6 Conclusions

The English devolution policy [27] offers an opportunity to examine how FE colleges can play a significant role in place-based innovation ecosystems. Through GMFEIP, early findings suggest that embedding FE colleges into innovation networks can align with devolved economic strategies. This programme demonstrates the importance of locally responsive frameworks in driving innovation, building on the innovation stakeholder ecosystem model [13, 14].

While the initial results are promising, they are still exploratory. More comprehensive and long-term evaluations are necessary to understand whether these early outcomes translate into sustained innovation, economic resilience, and regional impact. This future research is crucial for assessing the consistent delivery of policy ambitions outlined in the devolution agenda.

GMFEIP provides a preliminary model of a place-based innovation policy that extends beyond university-centric interventions. By adopting a more inclusive and decentralised approach, the programme begins to address persistent regional inequalities that impact productivity and SME growth [46, 47]. However, the replicability and sustainability of this model remain to be determined. Policymakers should consider these findings as an early indication of potential, rather than a foundation for broad policy adoption without further evidence.

Recent shifts in the England's economic and governance policies [27, 53], emphasise decentralised innovation and workforce development. These trends recognise the strategic value of FE in providing skills training and SME support. The GMFEIP offers practical insights into how FE can also function as innovation hubs, fostering connections with businesses, supporting digital transformation, and contributing to local economic strategies.

FE colleges' geographical distribution, community trust, and diverse educational and industry roles [5, 23] position them well to support innovation in businesses outside traditional R&D ecosystems. Policymakers may want to explore how to support and fund FE-led innovation capacity as part of regional growth and industrial strategies. Ensuring that innovation-led economic growth reaches underperforming regions requires a comprehensive strategy, including research-intensive universities, metropolitan institutions, and smaller, locally embedded actors like FE colleges.

Organisations such as The Productivity Institute [51] and the Institute for Government [54] have emphasised the need for inclusive innovation approaches. The GMFEIP provides early insights into how FE-led models might contribute. Policymakers should consider these findings and explore ways to integrate FE colleges into innovation systems to foster inclusive and resilient economic growth.

6.1 Contribution to Knowledge and Recommendations for Future Research

This paper makes an early significant contribution to the body of knowledge by examining how FE colleges can extend the MIT innovation stakeholder ecosystem model [13, 14]. Instead of presenting a validated alternative, the paper introduces and tests a novel conceptual framework, the radial innovation network, to understand FE institutions as emerging innovation intermediaries within regional ecosystems.

The research enhances current thinking on inclusive innovation and regional economic policy by providing initial qualitative and engagement-based evidence from the Greater Manchester Further Education Innovation Programme (GMFEIP). The findings indicate that FE colleges, equipped with dedicated innovation roles and networks, are positioned to reach SMEs that fall into the "missing middle" [45], businesses that are too large for start-up support but too small or under-resourced to engage with university-led innovation. Although early engagement data shows strong uptake, further evaluation is needed to determine if these interactions lead to sustained innovation outcomes.

The paper explores how FE colleges might act as innovation intermediaries [37], providing access points to innovation literacy and applied support. By functioning in liminal spaces between education, training, research, and business, FE colleges serve as boundary objects [39], capable of brokering knowledge across sectors.

The early findings have potential policy relevance, especially given English devolution and industrial strategy reforms. The GMFEIP's decentralised and embedded structure aligns with emerging policy interest in place-based innovation. However, these insights should be viewed as indicative rather than definitive. Sustained policy impact will rely on further empirical research, broader implementation, and systematic measurement of long-term outcomes.

Challenging assumptions about where innovation occurs and who drives it, the study offers a tentative expansion of the theoretical and empirical understanding of FE's role within regional innovation systems. While the radial innovation network is promising, it remains in the early stages of development and testing. Ongoing evaluation and comparative study are crucial to determine if this approach can be scaled, sustained, and integrated into wider innovation policy in England and beyond.

The development of a collaboration, engagement, and impact index is a next step in evaluating the depth and sustainability of partnerships between FE colleges, businesses, and ecosystem stakeholders. This index is expected to provide a structured evidence base on how innovation relationships evolve over time. Additionally, a longer-term assessment framework is necessary to track business growth, workforce development, and broader regional economic outcomes.

Currently, the strongest evidence pertains to engagement and emerging cultural shifts within institutions and businesses. As the programme evolves, evaluation should focus on how these changes translate into tangible, sustained outcomes, and the policy mechanisms needed to support them.

Acknowledgements This study was partly-funded by InnovateUK (grant number 10105110).

Disclosure of Interests No Conflict of Interests.

References

1. Association of Colleges: College key facts (2024). Retrieved from https://www.aoc.co.uk/about/college-key-facts. Last accessed 11 May 2025
2. Wolf, A.: Review of vocational education – the wolf report (2011). UK Government. https://www.gov.uk/government/publications/review-of-vocational-education-the-wolf-report. Last accessed 2025/5/11 NOT USED
3. Hodgson, A., Spours, K.: Further education in England: At the crossroads between a national, competitive sector and a locally collaborative system? J. Educ. Work. **32**(3), 224–232 (2019). https://doi.org/10.1080/13639080.2019.1596231
4. Atkins, L., Flint, K.J.: Nothing changes: perceptions of vocational education in England. Int. J. Train. Res. **13**(1), 35–48 (2015). https://doi.org/10.1080/14480220.2015.1051344
5. Keep, E.: FE and skills in the UK: The neglected middle child of education policy? J. Educ. Policy **33**(5), 590–606 (2018). https://doi.org/10.4324/9781315142593-5
6. BEIS. (2022). The UK Innovation Strategy. Retrieved from https://assets.publishing.service.gov.uk/media/61110f2fd3bf7f04402446a8/uk-innovation-strategy.pdf. Last accessed 11 May 2025
7. Kitson, M.: Innovation policy and place: a critical assessment. Cambridge J. Regions, Economy Soc. 12 (2019). https://doi.org/10.1093/cjres/rsz007
8. de Vasconcelos Gomes, L.A., Facin, A.L.F., Salerno, M.S., Ikenami, R.K.: Unpacking the innovation ecosystem construct: Evolution, gaps and trends. Technol. Forecast. Soc. Chang. **136**, 30–48 (2018). https://doi.org/10.1016/j.techfore.2016.11.009
9. Etzkowitz, H., Leydesdorff, L.: The dynamics of innovation: From National Systems and “Mode 2” to a Triple Helix of university–industry–government relations. Res. Policy **29**(2), 109–123 (2000). https://doi.org/10.1016/S0048-7333(99)00055-4
10. Leydesdorff, L., Etzkowitz, H.: The triple helix as a model for innovation studies. Sci. Public Policy **25**(3), 195–203 (1998). https://doi.org/10.1093/spp/25.3.195
11. Cai, Y., Amaral, M.: The triple helix model and the future of innovation: a reflection on the Triple Helix research agenda. Triple Helix. **8**, 217–229 (2021). https://doi.org/10.1163/21971927-12340004
12. Cai, Y., Lattu, A.: Triple helix or quadruple helix: which model of innovation to choose for empirical studies? Minerva (2022). https://doi.org/10.1007/s11024-021-09453-6
13. Budden, P., & Murray, F.: MIT’s stakeholder framework for building & accelerating innovation ecosystems (2019). https://innovation.mit.edu/assets/Innovation-Stakeholder-Framework.pdf. Last accessed 11 May 2025
14. Budden, P., Murray, F.: Strategically engaging with innovation ecosystems. MIT Sloan Manage. Rev. 64(1), 38–43 (2022). https://www.proquest.com/openview/b5fd23c344a0c1ed54be01224ebc25c7/1?cbl=26142&pq-origsite=gscholar. Last accessed 11 May 2025
15. Granstrand, O., Holgersson, M.: Innovation ecosystems: a conceptual review and a new definition. Technovation **90**, 102098 (2020). https://doi.org/10.1016/j.technovation.2019.102098
16. Phillips, F., Park, S., Lee, E., Oh, D.-S.: Innovation ecosystems: a critical examination. Technovation 54 (2016). https://doi.org/10.1016/j.technovation.2016.02.004
17. Ritala, P., Almpanopoulou, A.: In defense of ‘eco’ in innovation ecosystem. Technovation **60–61**, 39–42 (2017). https://doi.org/10.1016/j.technovation.2017.01.004
18. Aggarwal, S., Sindakis, S.: Four pillars of quadruple helix innovation model: an approach for the future. In: Entrepreneurial rise in the Middle East and North Africa: the influence of quadruple helix on technological innovation (pp. 9–28). Emerald Publishing Limited. Web. https://doi.org/10.1108/978-1-80071-517-220221002

19. Cantor, L.M., Roberts, I.F.: Further education in England and Wales. Routledge (2021). https://doi.org/10.4324/9781003159445
20. Innovation Caucus. (2021). Rethinking the role of further education colleges in innovation ecosystems. https://innovationcaucus.co.uk/app/uploads/2022/08/Rethinking-the-Role-of-FECs-in-Innovation-Ecosystems.pdf. Last accessed 11 May 2025
21. Forth, T., Jones, R.: The Missing £4 billion. Making R&D work for the whole UK. https://media.nesta.org.uk/documents/The_Missing_4_Billion_Making_RD_work_for_the_whole_UK_v4.pdf Last accessed 2025/5/11
22. Katz, B., Wagner, J. The rise of innovation districts. Brookings.edu https://www.brookings.edu/articles/rise-of-innovation-districts/. Last accessed 11 May 2025
23. Cooke P. N., Heidenreich, M., Braczyk, H. J. (Eds.): Regional innovation systems: the role of governance in a globalized world (2004). Psychology Press. https://doi.org/10.4324/9780203330234
24. Baxter, E.: Further education colleges and innovation. Gatsby Charitable Foundation https://www.gatsby.org.uk. uk/uploads/education/reports/pdf/fe-collegesand-innovation.pdf. Last accessed 11 May 2025
25. Nelles, J., Walsh, K., Papazoglu, M., Vorley, T.: FECS, innovation, and skills: A literature review Productivity Insights Paper No. 012, The Productivity Institute. FECs, innovation, and skills: A literature review—The Productivity Institute. Last accessed 27 May 2025
26. Resolution Foundation & Economy 2030: A tale of two cities: productivity and economic performance in Manchester (2023)
27. Greater Manchester Combined Authority. (2022). Greater Manchester Independent Prosperity Review—Evidence Update: Reflections https://www.greatermanchester-ca.gov.uk/media/6715/gmipr-evidence-update-reflections-report-digital-version.pdf. Last accessed 2025/5/22
28. Greater Manchester Combined Authority (GMCA). (2018). Greater Manchester: The Emerging Impact of Devolution. Retrieved from greatermanchester-ca.gov.uk Last accessed 11 May 2025
29. HM Government. (2025). English devolution white paper: power and partnership—foundations for growth. https://www.gov.uk/government/publications/english-devolution-white-paper-power-and-partnership-foundations-for-growth/english-devolution-white-paper Last accessed 11 May 2025
30. Kenealy, D.: A tale of one city: the Devo Manc Deal and its implications for english devolution. The Political Quarterly **87**(4), 572–581 (2016)
31. Yin, R. K.: Case study research and applications: Design and methods (6th ed.) (2018). SAGE. https://doi.org/10.3138/cjpe.30.1.108
32. Braun, V., Clarke, V.: Using thematic analysis in psychology. Qual. Res. Psychol. **3**(2), 77–101 (2006). https://doi.org/10.1191/1478088706qp063oa
33. Bruner, J.: The narrative construction of reality. Crit. Inq. **18**(1), 1–21 (1991). https://doi.org/10.1086/448619
34. Riessman, C.K.: Narrative methods for the human sciences. SAGE (2008). https://doi.org/10.7202/015444ar
35. Lundvall, B.A.: National systems of innovation: Towards a theory of innovation and interactive learning. Francis Printer (1992). https://doi.org/10.7135/upo9781843318903
36. UK Research and Innovation (UKRI): Further Education colleges: catalysts for regional business innovation (2023). https://www.ukri.org/blog/education-colleges-catalysts-for-regional-business-innovation/. Last accessed 11 May 2025
37. Howells, J.: Intermediation and the role of intermediaries in innovation. Res. Policy **35**(5), 715–728 (2006). https://doi.org/10.1016/j.respol.2006.03.005
38. Rogers, E. M., Singhal, A., Quinlan, M. M.: Diffusion of innovations. In: An integrated approach to communication theory and research (pp. 432–448) (2014). Routledge. https://doi.org/10.4324/9780203710753-35
39. Fox, N.J.: Boundary objects, social meanings and the success of new technologies. Sociology **45**(1), 70–85 (2011). https://doi.org/10.1177/0038038510387196
40. Fitriatia, T. K., Purwanab, D., Buchdadic, A. D.: The role of innovation in improving small medium enterprise (SME) performance. Innovation 11(2), 232–250 (2020). https://ijicc.net/images/vol11iss2/11216_Fitriati_2020_E_R.pdf. Last accessed 11 May 2025

41. Nawaz, R., Editor, E.E., Holland, S.R., Cao, Q., Liu, L.S.: The impact of degree apprenticeships: analysis, insights and policy recommendations. Transforming Government: People, Process and Policy, Vol. 17 No. 3, pp. 372–386 (2023). https://doi.org/10.1108/TG-07-2022-0105
42. Brown, R.: Mission-oriented or mission adrift? A critical examination of mission-oriented innovation policies. Eur. Plan. Stud. **29**(4), 739–761 (2021). https://doi.org/10.1080/09654313.2020.1779189
43. Perkmann, M., et al.: Academic engagement and commercialisation: a review of university–industry relations. Res. Policy **42**(2), 423–442 (2013). https://doi.org/10.1016/j.respol.2012.09.007
44. Pattinson, S., Cunningham, J., Preece, D., Davies, M.A.: Trust building in science-based SMEs in the North East of England: an ecosystem perspective. J. Small Bus. Enterp. Dev. **29**(6), 902–919 (2022). https://doi.org/10.1108/JSBED-11-2019-0360
45. Brethenoux, J.: Financing missing middle SMEs in emerging markets. SME Finance Forum https://www.smefinanceforum.org/post/financing-missing-middle-smes-in-emerging-markets-insights-about-small-cap-mezzanine-finance. Last accessed 11 May 2025
46. ERC: The State of Small Business Britain: Enabling the triple transition. 2021, Enterprise Research Centre. https://www.enterpriseresearch.ac.uk/the-state-of-small-business-britain-2021-enabling-the-triple-transition/. Last accessed 11 May 2025
47. ERC, State of Small Business Britain Report, Enterprise Research Centre, Editor: Enterprise Research Centre (2018). https://www.enterpriseresearch.ac.uk/publications/the-state-of-small-business-britain-report-2024/. Last accessed 11 May 2025
48. Parkinson, M., Carter, J., Nawaz, R.: Leveraging artificial intelligence (AI) to build SMEs' resilience amid the global COVID-19 pandemic. In: The international research & innovation forum (pp. 547–556). Cham: Springer International Publishing. https://doi.org/10.1007/978-3-031-19560-0_46
49. Mulhaney, A., Parkinson, M., Abdulfatah, A.K.: A human-centric approach to digital transformation. Rii Forum, Ravello, Italy (2024)
50. Rodríguez-Pose, A.: The revenge of the places that don't matter (and what to do about it). Camb. J. Reg. Econ. Soc. **11**(1), 189–209 (2018). https://doi.org/10.1093/cjres/rsx024
51. Coyle, D., van Ark, B., Pendrill, J. (eds): The productivity agenda. Report No. 001 (2023). The Productivity Institute. https://www.productivity.ac.uk/wp-content/uploads/2023/11/TPI-Agenda-for-Productivity-2023-FINAL.pdf. Last accessed 11 May 2025
52. Froud, J.: Foundational economy: The infrastructure of everyday life. Manchester University Press (2022). https://doi.org/10.7765/9781526158727
53. HM Government: Invest 2035: The UK's Modern Industrial Strategy (2025). https://www.gov.uk/government/consultations/invest-2035-the-uks-modern-industrial-strategy/invest-2035-the-uks-modern-industrial-strategy Last accessed 2025/5/11
54. Institute for Government: 2025: A Make-or-Break Year for the Labour Government (2025). https://www.instituteforgovernment.org.uk/comment/2025-make-or-break-government Last accessed 2025/5/11

Cities, Urban Design and Architecture in Digital Era

Smart Transition in Historical Small Towns: First Insights from a Digital Material Flow Assessment in the Italian Case Study of Taurasi

Cristina Ciliberto, Grazia Calabrò, Giuseppe Caristi, and Giuseppe Ioppolo

Abstract This study investigates the integration of digital flows within Material Flow Analysis applied to Historical Small Towns (HSTs), using Taurasi, Italy as a case study. The research examines how digital technologies can promote responsible consumption while preserving the unique fabric of HSTs through Industry 5.0 principles and redefined Urban Metabolism (UM) frameworks. The methodology combines Strategic Environmental Assessment (SEA) with the DPSIR framework adapted for analyzing digital flows within traditional material systems. This methodology incorporates quantitative analysis of digital infrastructure alongside qualitative assessment through stakeholder engagement and expert consultation. Implementation of the MFA methodology in Taurasi reveals that digitization initiates a "smart metabolism" system that optimizes resource utilization while preserving heritage values. This integration yielded measurable outcomes, including a 15% reduction in energy consumption through smart monitoring systems, enhanced preservation of historical structures via sensor networks, and increased community participation in digital platforms. The study advances UM theory by demonstrating how digital and material flows can effectively coexist and interact within historical urban environments. The findings suggest that HSTs should develop strategic approaches to digital transformation that not only overcome geographical limitations but also support evidence-based decision-making. Such strategies should prioritize heritage-sensitive digital projects and community engagement to ensure technological advancement complements rather than compromises historical character. This research presents a novel application of MFA that systematically incorporates both physical and digital

C. Ciliberto (✉) · G. Calabrò · G. Caristi · G. Ioppolo
Department of Economics, University of Messina, Via Dei Verdi 75, Messina, Italy
e-mail: cristina.ciliberto@unime.it

G. Calabrò
e-mail: grazia.calabro@unime.it

G. Caristi
e-mail: giuseppe.caristi@unime.it

G. Ioppolo
e-mail: giuseppe.ioppolo@unime.it

A. Visvizi et al. (eds.), *Research and Innovation Forum 2025 Volume 1*,
Springer Proceedings in Complexity, https://doi.org/10.1007/978-3-032-09066-9_20

flows within urban systems. The proposed interoperable framework connects material metabolism with digital transformation concepts, offering a new paradigm for balancing technological innovation and heritage conservation in HSTs.

Keywords Material flow analysis · Digital flows · Historical small towns · Smart heritage · Industry 5.0 · Urban metabolism · SDG-9 · SDG-11 · SDG-12

1 Introduction

Historical Small Towns (HSTs) represent unique repositories of cultural heritage facing significant preservation challenges in the modern era. In Italy, these towns are particularly prevalent, with approximately 70% of municipalities having fewer than 5,000 inhabitants, covering roughly 54% of the country's territory. Among these, approximately 2500 small towns are classified as "Borghi", traditional settlements with distinctive historical and cultural characteristics [5].

These communities exist at a critical intersection where economic development, environmental conservation, and cultural identity preservation converge. As urbanization accelerates globally, HSTs must find sustainable pathways that balance resource protection with heritage conservation while maintaining their unique characteristics. This balancing act has become increasingly complex in the digital age, where technological solutions offer both opportunities and challenges for historical contexts.

Material Flow Analysis (MFA) has emerged as a valuable methodology for understanding resource utilization in urban environments, traditionally focusing on physical flows of energy, water, and materials. However, in the digital era, a significant gap exists in literature regarding how digital flows interact with and influence traditional material flows, particularly in sensitive historical environments. Digital transformation processes have become essential enablers for circular economy adoption in HSTs, offering potential for resource tracking, waste reduction, and process optimization. Nevertheless, implementing digital technologies in heritage contexts presents unique challenges related to architectural constraints, traditional economic systems, and cultural preservation requirements [3].

Industry 5.0 paradigm provides a promising framework for addressing these challenges by emphasizing human-centered, resilient, and sustainable approaches to technological integration [12]. This paradigm can be particularly valuable for small towns seeking equilibrium between cultural heritage preservation and development [4]. Recent studies have highlighted that conventional smart city initiatives often underestimate the social, cultural, and environmental implications of technological implementation [14], making a more refined approach necessary for HSTs.

This study focuses on Taurasi (Campania), a traditional Italian borgo, to investigate how digital flows can be integrated into Urban Metabolism (UM) frameworks within heritage preservation contexts. An integrated methodological approach that combines Strategic Environmental Assessment (SEA) with the DPSIR (Drivers-Pressures-State-Impact-Response) framework is employed to analyze digital flows within traditional material flow systems. By extending traditional MFA to incorporate digital dimensions, this research aims to explore how smart technologies can support responsible consumption and production while preserving historical urban fabric.

Considering this, this research addresses three primary questions: (1) How can digital flows be conceptualized and measured within historical urban contexts? (2) What specific challenges and opportunities arise when integrating digital technologies into the metabolism of HSTs? and (3) How can the Industry 5.0 paradigm guide the implementation of digital solutions that respect and enhance heritage values?

This research addresses a critical gap in current literature where the intersection of digital transformation and historical preservation remains understudied, particularly in the context of small historical towns. While extensive research exists on material flows in urban contexts and on digital transformation of modern cities, few studies have systematically investigated how digital and physical flows interact in heritage environments. The urgency of this investigation is underscored by the accelerating pace of digital adoption in historical areas, often implemented without adequate frameworks to assess impacts on historical fabric and community dynamics. By integrating digital flows into traditional MFA approaches, this study offers a timely response to the pressing need for methodologies that can guide sustainable technological implementation in heritage contexts.

The paper is structured as follows: Sect. 2 provides a theoretical background on UM, MFA, and digital transformation in heritage contexts. Section 3 details our methodological approach, while Sects. 4 and 5 present our findings and conclusions, respectively.

2 Theoretical Background

2.1 Urban Metabolism and Material Flow Analysis in Historical Contexts

UM represents the foundational system of any city, transcending sociotechnological, administrative, and economic domains. UM encompasses "the total sum of technical and socio-economic processes that occur in cities, resulting in growth, production of energy, and waste removal" [6]. This concept becomes particularly relevant for HSTs, where traditional urban fabrics must be integrated with modern resource management approaches while preserving cultural heritage.

Despite its importance, MFA, a key analytical framework within UM, has not been widely incorporated into urban planning processes for historical areas. MFA provides a systematic assessment of the flows and stocks of materials within a system defined in space and time [2]. When applied to HSTs, MFA must consider both tangible material flows (water, energy, waste) and increasingly important digital flows of information and services.

HSTs represent complex adaptive systems that can respond to changes in their fundamental elements while maintaining stability over extended periods [15]. These urban systems manage "selective and decentralized flows of matter, energy, and information among their parts", a characteristic that highlights the unique nature of historical settlements. Understanding how these flows interact and evolve is essential for developing sustainable management strategies that respect historical integrity while addressing contemporary challenges.

2.2 *Digital Transformation and Industry 5.0 in Heritage Contexts*

The integration of Industry 5.0 principles within historical contexts represents a significant transformation in urban development approaches. Unlike Industry 4.0, which primarily focused on technological advancement through automation and data exchange, Industry 5.0 emphasizes human-centricity, resilience, and sustainability [4]. This paradigm recognizes people as crucial elements in development processes rather than simply focusing on technological progress [10].

Industry 5.0 approach, illustrated in Fig. 1, is characterized by three key factors particularly relevant to historical contexts. First, it places humans at the center, viewing technology as an augmentation of human capabilities rather than a replacement. In heritage sites, digital tools should complement local traditional technologies, knowledge, and community participation rather than supplant them. Second, it emphasizes resilient solutions through developing robust technologies for protecting cultural heritage from various threats, including over-tourism, environmental degradation, and data loss. Third, it adopts a sustainability perspective that ensures long-term development while preserving visual identity through appropriate technological integration.

Digital transformation in heritage contexts, therefore, requires achieving a balance between modernization and preservation requirements. Implementing digital tools in historical settings demands a comprehensive strategy that addresses technological aspects while responding to cultural preservation needs [11].

Digital technology offers a glimpse of the future possibilities in historical spaces, provided technology is appropriately implemented and becomes vital to heritage, which is made more accessible, better preserved, and more deeply linked to contemporary life.

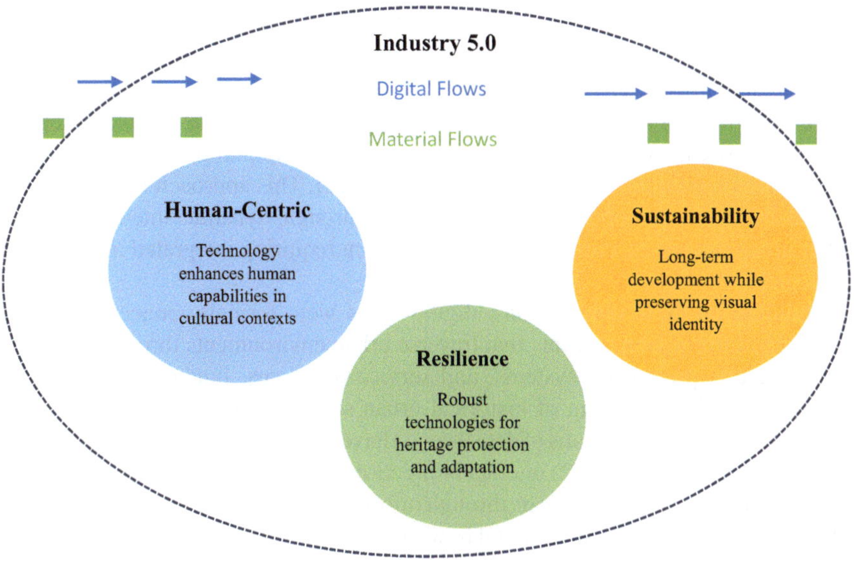

Fig. 1 Industry 5.0 Integration in Heritage Contexts. *Source* The Authors

2.3 Digital Flows in Historical Urban Contexts

In the digital age, information flows have become increasingly crucial, both as a new dimension of UM requiring in-depth analysis and as a driving force in urbanization processes. These digital flows are categorized into three main types, each characterized by specific nurturing forces and consequences affecting historical urban environments [7].

Infrastructure Flows constitute the backbone of digital connectivity in historical settings, including regional data networks connecting towns with broader digital ecosystems, local Internet of Things (IoT) systems designed for smart functioning under historical conditions, and smart grid initiatives modernizing energy management in heritage contexts.

Information Flows comprise digital data circulating through physical infrastructures, including environmental monitoring data detailing the effects of various variables on historical buildings and urban spaces, digital documentation of cultural heritage enabling information preservation for future generations, and resource management information optimizing resource allocation in historical urban contexts.

Service Flows represent functional applications built upon infrastructure and information layers, including e-government services enhancing administrative efficiency, digital preservation services supporting cultural heritage conservation, and smart city applications balancing authenticity with modern conveniences.

The integration of these digital flows with traditional material flows (water, energy, waste) creates new opportunities for enhancing resource management in historical towns while simultaneously presenting specific challenges for historical preservation.

Balancing technological intervention with historical conservation requires careful analysis of how digital flows can complement traditional conservation methods rather than threaten them.

A holistic approach to addressing digital flows within the broader scope of UM is proposed, enabling historical towns to undergo digital transformation while preserving their cultural and historical relevance [7]. This approach recognizes that digital flows do not operate in isolation but through dynamic interactions with other urban processes, creating a complex system requiring integrated management strategies.

Building on these observations, digital flows were formally conceptualized as structured information currents that traverse urban environments through dedicated infrastructure, processing systems, and service interfaces, fundamentally altering the operational metabolism of historical urban settings. Digital flows can thus be defined as measurable streams of data that: (1) have quantifiable resource implications (energy, space, materials); (2) interact bidirectionally with traditional material flows; and (3) transform urban functions through information-enhanced decision processes. This formalization establishes digital flows as a distinct yet interconnected dimension within the broader urban metabolism framework, providing a conceptual foundation applicable across diverse urban typologies.

3 Methods

3.1 Urban Metabolism and Material Flow Analysis in Historical Contexts

This study employs an integrated methodological approach that combines SEA with the DPSIR framework to analyze the socio-ecological changes catalyzed by digital transformation in historical urban areas. This mixed-methods design enables comprehensive quantitative and qualitative analysis of digital flows within the urban system of Taurasi, Italy.

In accordance with SEA principles, as prescribed in EU directives, this analytical framework addresses three primary dimensions. First, Environmental Integration focuses on evaluating the impact of digital infrastructure on historical fabric, assessing electromagnetic pollution levels, and analyzing energy consumption patterns associated with digital technologies. Second, Governance Alignment examines coordination between local, regional, and national digital initiatives, and evaluates compliance with heritage preservation regulations. Third, Stakeholder Engagement structures consultation processes with active community participation to ensure digital solutions address local needs while respecting cultural values.

The DPSIR framework was adapted to specifically address digital flows within historical urban contexts. This adaptation identifies Drivers as forces driving digital transformation, including smart city initiatives and heritage preservation imperatives.

Pressures represent the resulting demands on infrastructure, resources, and social systems. State describes the current condition of digital infrastructure and data flows. Impacts encompass effects on historical fabric, resource efficiency, and community dynamics. Responses include policy formulations, capacity building initiatives, and technical solutions.

3.2 Data Collection Methods

The research employed a sequential mixed-methods approach with quantitative data collection preceding qualitative investigation. Data analysis followed a convergent parallel design where quantitative measurements of infrastructure and digital flows were analyzed alongside qualitative stakeholder inputs, using triangulation to validate findings across multiple sources. Interpretation was guided by a participatory framework where preliminary findings were shared with key stakeholders to ensure contextual validity and alignment with local heritage preservation objectives.

The data collection strategy employs multiple complementary methods to ensure comprehensive coverage of both quantitative and qualitative dimensions.

Quantitative methods include digital infrastructure mapping through systematic documentation of network coverage, IoT deployment, and digital service access points within the historical center. Resource consumption metrics related to energy usage by digital infrastructure, data center operations, and associated cooling systems were measured. Usage analytics provide quantitative assessment of digital service utilization, data traffic volumes, and system performance indicators.

Qualitative methods comprise stakeholder interviews conducted with local government officials, heritage conservation experts, technology providers, and community representatives ($n = 18$). Expert consultations have been held with specialists in digital heritage, urban planning, and historical conservation ($n = 7$). Observational studies involve systematic observations of user interactions with digital services in public spaces.

Documentary analysis encompasses technical documentation analysis of digital system specifications, implementation reports, and maintenance records. Policy documents including local digital strategies has been reviewed, as well as heritage management plans, and regulatory frameworks. Historical records examination includes archival materials documenting the historical development of Taurasi.

3.3 Analytical Framework

The analytical framework assesses digital flows across four key dimensions [7]. The first dimension, Technical Infrastructure, evaluates network coverage and quality, IoT deployment mapping and functionality, and integration with existing urban systems. The second dimension, Data Systems, examines data collection mechanisms and their physical footprint, processing capabilities and energy requirements, and storage solutions and redundancy provisions. The third dimension, Service Delivery, analyzes e-government platform accessibility and usage, digital preservation tools and their effectiveness, and tourism information systems and community services. The fourth dimension, Knowledge Management, assesses digital literacy training programs, community engagement initiatives, and knowledge transfer mechanisms.

Industry 5.0 principles has been incorporated through metrics that address human-centricity, resilience, and sustainability [10]. Human-centricity is measured through user experience assessments, accessibility indicators, and inclusion metrics. Resilience is evaluated through system redundancy assessments, failure recovery capabilities, and adaptation capacity. Sustainability is gauged through energy efficiency ratios, material intensity of digital solutions, and long-term viability indicators.

This comprehensive methodological approach enables to analyze not only the technological aspects of digital flows in Taurasi but also their social, cultural, and environmental implications within the constraints of the historical urban context.

4 Results and Discussion

The analysis of the digital evolution process in Taurasi reveals a complex relationship between technological advancement and heritage preservation. Digital infrastructure deployment in historical contexts requires strategic planning to minimize physical intrusion while optimizing functional capacity [13]. The survey of Taurasi's digital infrastructure identified three principal support layers: Core Network Infrastructure, utilizing wireless technology with reduced ecological footprint and strategically positioned IoT sensors in heritage sites; Data Collection Systems, encompassing environmental monitoring networks and heritage preservation sensors; and Service Delivery Infrastructure, comprising digital tourist information systems and e-government platforms.

Successful integration of digital infrastructure in historical contexts depends on cautious approaches to physical and cultural preservation aspects [11]. The findings from Taurasi support this assertion, as infrastructure development prioritized non-invasive technologies adaptable to existing historical structures.

The examination of digital cycles in Taurasi reveals specific patterns of data generation, processing, and utilization within a three-step framework. Three major categories of digital flows were identified [7]: Environmental Monitoring Flows, providing real-time data on air quality and energy consumption; Cultural Heritage Flows, managing digital documentation of historical assets and visitor movements; and Transmission Flows, handling e-services usage and resource allocation data, as illustrated in Fig. 2.

The interconnection of these digital flows with traditional material flows has developed a "smart metabolism" system [5]. Within this system, embedded technologies enhance resource efficiency while maintaining heritage preservation objectives.

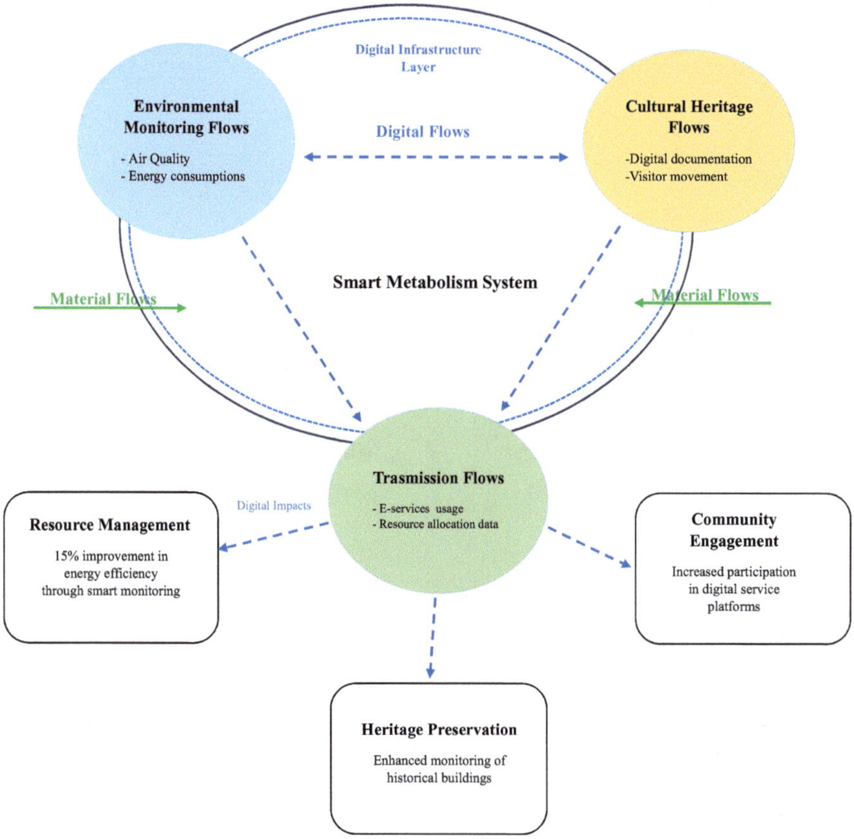

Fig. 2 Integration of digital flows in Taurasi's smart metabolism system. *Source* The Authors

Such a perspective, emphasizes the HST as dynamic ecosystem with its own metabolic processes, requiring collaborative efforts from various stakeholders. These stakeholders (policymakers, administrators and managers) should engage in the co creation and implementation of circular initiatives, by effectively planning and executing circular UM strategies (Fig. 3).

Moreover, the adoption of technology within Taurasi's traditional landscape presents significant challenges requiring innovative solutions. Physical Integration Challenges include space limitations for infrastructure installation and necessary preservation requirements restricting technology placement. Cultural Integration Issues involve the digitization of social services and balancing modernized lifestyles with historical practices. Technical Challenges encompass achieving reliable interoperability and implementing effective data protection measures. These findings align with who highlighted the complexity of digital transformation in heritage contexts [9].

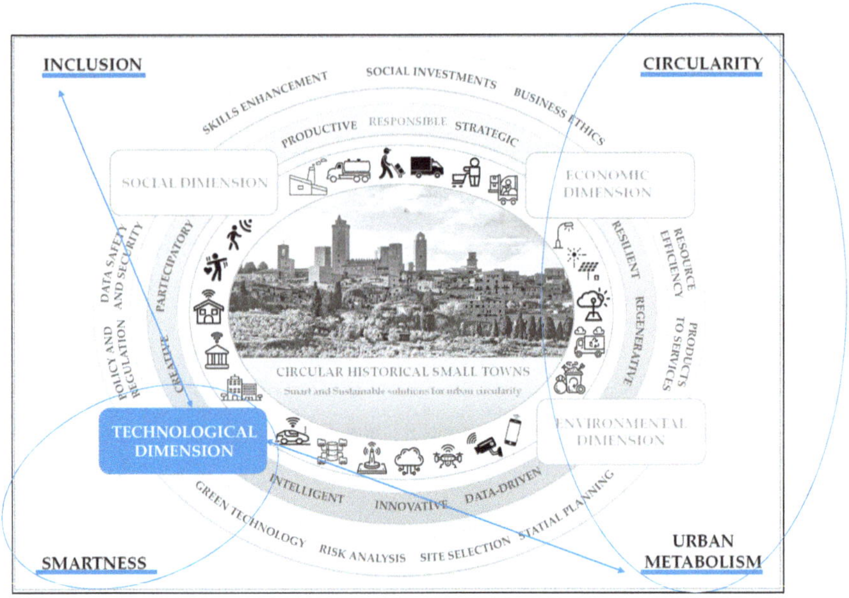

Fig. 3 Visual representation of circular HSTs. *Source* The Authors

The implementation of digital processes in Taurasi has produced measurable impacts across multiple domains. Resource Management data indicate a 15% improvement in energy efficiency attributed to enhanced monitoring systems. Heritage Preservation has benefited from more accurate building monitoring through sensor networks. Community Engagement has increased through participation in digital service platforms. These findings documented both immediate and long-term effects of digital transformation on heritage preservation and community development in historical towns [8].

This analysis contributes to the theoretical underpinnings of UM within a historical perspective by demonstrating the harmonious interaction between digitized flows and traditional material flows. The study provides empirical evidence of how digital flows function in real-world contexts [7], through a modeling framework applied in a specific historical urban location. Furthermore, the research advances the concept of "smart metabolism" [5], providing practical recommendations for urban planners and heritage site managers. The findings identify the importance of phased approaches to digital implementation that incorporate both technical and heritage considerations. Stakeholder engagement emerges as a critical factor throughout the digital transformation process, ensuring community acceptance and participation in technology-driven initiatives. Digital monitoring systems represent valuable additions to heritage preservation strategies, enabling more precise planning and efficient resource utilization.

Based on these findings, the establishment of specific digital transformation strategies for HSTs is recommended [8]. Such strategies should align with financial resources and cultural heritage considerations to implement heritage-sensitive digital initiatives that receive community support.

5 Conclusion and Limitations

The application of a framework to the specific historical urban setting of Taurasi provides empirical evidence of how digital flows function in real-world environments [7]. Furthermore, the study extends the "smart metabolism" [5] through case-specific examples of how digital landscapes can drive economic development while supporting long-term historical preservation.

The digital transformation of HSTs represents a challenging yet inevitable process in contemporary urban development. The Taurasi case demonstrates that successful transformation requires careful balancing of preservation and innovation, supported by appropriate technological solutions and comprehensive stakeholder involvement. The integration of digital flows into MFA offers a robust model for understanding and managing this transformation, creating pathways toward sustainable development that enhance rather than diminish historical character.

Future research should focus on how emerging technologies can simultaneously protect cultural heritage and reduce environmental impacts in historical areas. The convergence of digital innovation and heritage preservation creates promising opportunities for developing more resilient and sustainable historical urban environments [9].

Longitudinal studies tracking the long-term impacts of digital transformation on resource efficiency, heritage preservation, and community engagement would provide valuable insights into the sustainability of such interventions. Additionally, comparative analyses across multiple HSTs would help identify context-specific factors that influence the success of digital integration strategies.

Research investigating the potential of advanced technologies such as augmented reality, blockchain for heritage documentation, and artificial intelligence for predictive conservation would further expand understanding of how digital tools can support heritage preservation while enhancing visitor experiences and community engagement.

In HSTs digital transformation represents not merely a technological challenge but an opportunity to redefine UM through a sustainability perspective. The framework tested in Taurasi clarifies how the integration of material and digital operations can create a "smart metabolism" system that optimizes resource utilization while preserving historical and cultural heritage.

A three-stage implementation approach is recommended for future digital transformation initiatives in HSTs: (1) preliminary mapping of cultural and architectural characteristics specific to the site; (2) continuous engagement of local communities throughout all decision-making phases; and (3) establishment of integrated monitoring systems enabling longitudinal impact assessment.

By adopting this inclusive approach, HSTs can become centers of sustainable innovation where technology serves as a means to promote and enhance cultural heritage rather than replace it. This transformation requires recognition of the unique characteristics of each historical community and tailored strategies that respect both tangible and intangible heritage elements.

This research offers significant practical implications for various stakeholders. For urban managers and heritage site administrators, the integrated MFA framework provides an actionable tool for balancing technological advancement with preservation requirements, potentially reducing resource consumption costs by 10–15% while enhancing visitor experiences. Policy-makers can utilize these findings to develop evidence-based regulations that facilitate digital transformation while protecting cultural assets, creating opportunities for sustainable economic development through heritage tourism and innovative digital services. Local businesses in HSTs can benefit from the outlined digital integration strategies by adopting complementary technologies that enhance traditional offerings without compromising authenticity, opening new market opportunities in digital heritage experiences, smart tourism, and sustainable local products. The smart metabolism model demonstrated in Taurasi establishes a replicable approach that can be adapted across similar historical settlements, potentially transforming economic challenges of geographical isolation into market advantages through digital connectivity and heritage-sensitive innovation.

The integration of Industry 5.0 principles—emphasizing human-centricity, resilience, and sustainability—provides a valuable framework for guiding digital transformation in heritage contexts. By prioritizing these principles, HSTs can leverage digital technologies to address contemporary challenges while maintaining their distinctive historical identity and cultural significance.

The digital transformation of HSTs represents a complex yet essential process in ensuring their continued relevance and sustainability in the digital age. Through careful integration of digital flows within traditional material systems, these unique urban environments can preserve their cultural heritage while enhancing resource efficiency and community engagement. The case of Taurasi demonstrates that when implemented with sensitivity to historical context and community needs, digital technologies can support rather than threaten the distinctive character of HSTs, creating pathways toward sustainable development that honor the past while embracing the future.

Acknowledgements This research has received funding from PNRR—Missione 4, Componente 2, Investimento 1.1—Bando Prin 2022—Decreto Direttoriale n. 104 del 02-02-2022- Progetto "H-SMA-CE: a decision support system for circular economy transition" CUP J53D23009390006—codice identificativo 2022JZLL7J.

References

1. Agudelo-Vera, C.M., Leduc, W.R., Mels, A.R., Rijnaarts, H.H.: Harvesting urban resources towards more resilient cities. Resour. Conserv. Recycl. **64**, 3–12 (2012). https://doi.org/10.1016/j.resconrec.2011.09.014
2. Brunner, P.H., Rechberger, H.: Handbook of material flow analysis: for environmental, resource, and waste engineers, (2nd ed.). CRC Press (2016). https://doi.org/10.1201/9781315313450
3. Báldi, A., Pellaton, R., Bihaly, Á.D., Szigeti, V., Lellei-Kovács, E., Máté, A., Sárospataki, M., Soltész, Z., Somay, L., Kovács-Hostyánszki, A.: Improving ecosystem services in farmlands: beginning of a long-term ecological study with restored flower-rich grasslands. Ecosyst. Health Sustain. **8** (2022). https://doi.org/10.1080/20964129.2022.2090449
4. Industry 5.0 – Towards a sustainable, human-centric and resilient European industry. Publications Office of the European Union (2021). https://data.europa.eu/doi/10.2777/308407
5. Ioppolo, G., Cucurachi, S., Salomone, R., Shi, L., Yigitcanlar, T.: Integrating strategic environmental assessment and material flow accounting: A novel approach for moving towards sustainable urban futures. Int. J. Life Cycle Assess. **24**, 1269–1284 (2019). https://doi.org/10.1007/s11367-018-1494-0
6. Kennedy, C., Cuddihy, J., Engel-Yan, J.: The changing metabolism of cities. J. Ind. Ecol. **11**(2), 43–59 (2007). https://doi.org/10.1162/jiec.2007.1107
7. Li, X., Fong, P.S., Dai, S., Li, Y.: Towards sustainable smart cities: an empirical comparative assessment and development pattern optimization in China. J. Clean. Prod. **234**, 751–769 (2019). https://doi.org/10.1016/j.jclepro.2019.06.269
8. Longo, D., Zaccaria, V., Bonfiglio, A.: Cultural heritage and interoperable open platforms: strategies for knowledge, accessibility, enhancement and networking. WIT Trans. Ecol. Environ. **241**, 209–221 (2020). https://doi.org/10.2495/SDP200171
9. Masciotta, M.G., Morais, M.J., Ramos, L.F., Oliveira, D.V., Sánchez-Aparicio, L.J., González-Aguilera, D.: A digital-based integrated methodology for the preventive conservation of cultural heritage: the experience of HeritageCare project. Int. J. Architect. Heritage **15**(6), 844–863 (2021). https://doi.org/10.1080/15583058.2019.1668985
10. Murphy, C., Carew, P.J., Stapleton, L.: A human-centred systems manifesto for smart digital immersion in Industry 5.0: a case study of cultural heritage. AI & Soc. **38**, 1571–1586 (2023). https://doi.org/10.1007/s00146-022-01397-z
11. Nota, G., Petraglia, G.: Heritage buildings management: the role of situational awareness and cyber-physical systems. J. Ambient. Intell. Humaniz. Comput. **15**(1), 121–134 (2024). https://doi.org/10.1007/s12652-023-04678-9
12. Visvizi, A., Malik, R., Guazzo, G.M., Çekani, V.: The Industry 5.0 (I50) paradigm, blockchain-based applications and the smart city. Eur. J. Innov. Manag. **28**(1), 5–26 (2025). https://doi.org/10.1108/EJIM-09-2023-0826
13. Vuoto, A., Simonetti, M., De Fino, M., Fatiguso, F.: On the use of the digital twin concept for the structural integrity protection of architectural heritage. Infrastructure **8**(5), 71 (2023). https://doi.org/10.3390/infrastructures8050071
14. Yigitcanlar, T., Cugurullo, F.: The sustainability of artificial intelligence: an urbanistic viewpoint from the lens of smart and sustainable cities. Sustainability **12**(11), 4548 (2020). https://doi.org/10.3390/su12114548
15. Zellner M, Campbell SD (2015) Planning for deep-rooted problems: what can we learn from aligning complex systems and wicked problems? Plann. Theor. Pract. **16**, 457–478. https://doi.org/10.1080/14649357.2015.1084360

How to Facilitate Walkability in Hot-Airid Cities? The Case of Jeddah

Rawan Almesbahi and Asmaa Ibrahim

Abstract This study addresses the critical issue of walkability as a key element in promoting sustainable and healthy urban environments. It responds to the growing need for urban planning models to promote walkability in hot-arid cities such as Jeddah. The study sets out two research objectives: to identify the physical and social dimensions of walkability and to evaluate the effectiveness of existing pedestrian-related efforts in order to better align the existing strategies to promote walkability. The case study will tackle one of the major arterials in Jeddah, which is Al Tahlia Street. Using quantitative methods of the questionnaire, the study evaluates initiatives along Al-Tahlia Street, identifies their impact, and highlights gaps in design and planning. The results are analyzed using SPSS statistical analysis. The study concludes with recommendations aligned with Vision 2030 and the Quality-of-Life Program for enhancing walkability in Saudi cities.

Keywords Sustainable urban environments · Pedestrian-friendly projects · Urban walkability · Public health · Tahlia street · Sustainable cities and communities

1 Introduction

Walkability plays a vital role in creating sustainable and healthy cities. Globally, many urban areas have shifted toward pedestrian-friendly planning to reduce car dependency, improve public health, and enhance the quality of life [18]. Studies consistently show that how we design our cities directly affects our health. When cities are easy to walk in, people are more active and healthier.

R. Almesbahi
Effat College of Architecture and Design, Effat University, Jeddah, Saudi Arabia

A. Ibrahim (✉)
Urban Design Lab, College of Architecture and Design, Effat University, Jeddah, Saudi Arabia
e-mail: asibrahim@effatuniversity.edu.sa

A. Visvizi et al. (eds.), *Research and Innovation Forum 2025 Volume 1*,
Springer Proceedings in Complexity, https://doi.org/10.1007/978-3-032-09066-9_21

Researchers are actively exploring what makes a city walkable, from specific features that help older people get around [9] to how we define walkable neighborhoods for a sustainable future [1, 4]. It is clear that city planners and designers play a significant role in creating healthier, more vibrant places to live [20].

Jeddah, a major city in Saudi Arabia, reflects these challenges. Although some streets attract high foot traffic, they often lack safe, accessible, and comfortable pedestrian infrastructure. Al-Tahlia Street, a vibrant commercial zone in Jeddah, illustrates this tension. Despite being a popular destination for shopping and socializing, the street offers limited walkability due to issues like narrow sidewalks, poor accessibility, and minimal shading or seating.

This study explores how Al-Tahlia Street can be transformed into a more walkable urban space. Using a case study approach supported by questionnaires, the research identifies key physical and non-physical factors that affect walkability and assesses recent improvement initiatives. The goal is to provide data-driven recommendations that align with Vision 2030 and the Quality-of-Life Program, supporting efforts to create more inclusive and pedestrian-friendly cities in Saudi Arabia.

Accordingly, this research addresses two questions: What are the primary factors influencing walkability on Al-Tahlia Street? How effective are current initiatives in promoting walkable environments?

These questions are shaped by the broader urgency to transition to promote more pedestrian-oriented environments, in line with national sustainability goals. Given the nature of the data and the analysis conducted, the questions are approached with a descriptive intent at the beginning, whereas the second forms the empirical core of the research.

2 Literature Review

Walkability is a multifaceted concept central to sustainable urban design, encompassing both the physical characteristics of the built environment and the social, perceptual, and experiential aspects of walking. Walkability is defined not only in terms of infrastructure but also in relation to safety, accessibility, and street-level aesthetics. This comprehensive perspective positions walkability not just as a transport issue, but as a reflection of quality of life and urban inclusiveness [12].

Key trends in walkability research highlight the significance of mixed land uses, connectivity, and the integration of smart technologies in supporting pedestrian-friendly urban spaces [16]. The evolution of smart city frameworks suggests that walkability should be considered alongside broader digital and social dimensions of urban development, moving beyond purely technological interpretations to more inclusive, human-centered approaches [19]. In this context, the morphology of neighborhoods plays a critical role: Spatial multi-criteria analysis can be used to assess the impact of urban form on walkability, identifying structural barriers or enablers across different contexts [6]. Moreover, actively involving citizens in shaping walkable environments has become increasingly important. The citizen feedback in land use policy, especially within smart city initiatives, where digital tools can help democratize urban planning, is very valuable [7]. Together, these studies suggest that improving walkability requires an integrated, participatory, and context-aware approach that bridges urban morphology, smart technologies, and community engagement.

Accordingly, walkability is increasingly seen as a key indicator of sustainable urban design, public health, and social equity. Numerous studies have sought to define and measure walkability, exploring both physical and non-physical dimensions that contribute to the pedestrian experience in urban spaces.

The importance of urban design elements—like street width, lighting, and surface quality—in shaping walkable spaces is significant [3]. Researchers developed a framework for evaluating "urban design qualities" such as imageability, enclosure, and transparency, all of which influence pedestrian movement and perception [2]. These qualities are especially important in contexts where walking competes with dominant car-oriented infrastructure, such as in many parts of Saudi Arabia.

A review of global walkability trends identifies shifts in planning priorities from mere connectivity toward designing for accessibility, safety, and environmental comfort [16]. In Jeddah, these concerns are highly relevant due to extreme heat, fragmented pedestrian paths, and limited public shade infrastructure. The physical aspects of walkability—such as sidewalk width, presence of greenery, and lighting—are therefore crucial to enhancing comfort and encouraging walking in such hot-humid climates [17].

Beyond physical elements, several scholars have stressed the importance of non-physical or intangible factors, including cultural relevance, social safety, and inclusivity. Walkability also contributes to urban equity and cohesion by connecting communities and enabling access to public amenities [18]. These broader goals align with Saudi Vision 2030, which prioritizes livable, accessible, and culturally rich urban spaces.

Recent studies in comparable Middle Eastern or North African contexts provide additional insights. For example, several researchers employed spatial multi-criteria analysis to show how neighborhood morphology impacts walkability [6], while other studies explored the role of public participation in shaping inclusive land-use policies within smart city initiatives [7]. Both studies support the notion that localized data and citizen feedback are crucial for designing effective pedestrian environments.

To understand the spatial capacity of sidewalks and their effect on pedestrian movement, researchers have analyzed how different sidewalk widths correspond to pedestrian flow rates. This is particularly relevant in busy urban streets such as Al-Tahlia Street, where the volume of foot traffic can vary by location and time. Table 1 presents common sidewalk configurations and their associated capacity:

Accessibility plays a key role in walkability by ensuring that streets are usable for individuals of all abilities. Incorporating universal design features enhances safety, social inclusion, and comfort. Table 2 summarizes major accessibility elements and the value they add to pedestrian infrastructure:

Walkability is shaped by a range of physical and non-physical factors that affect how people interact with the urban environment. These factors influence social

Table 1 Pedestrian capacity based on sidewalk widths

Configuration	Sidewalk width (m)	Pedestrian capacity/hour	Description
Standard one-lane	2.0	20–25	Suitable for low traffic; allows single-file walking
Wider one-lane	3.0	25–30	More comfortable; allows passing
Standard two-lane	4.0	40–50	Good for moderate foot traffic and social interaction
Wider two-lane	5.0	50–60	Suitable for larger groups, improves flow
Pedestrian plaza	6.0+	60+	Ideal for high traffic areas, encourages events

Source Authors

Table 2 Key accessibility features and benefits

Feature	Description	Benefits
Tactile paving	Textured ground surface for the visually impaired	Enhances safety and orientation
Curb ramps	Sloped transitions between sidewalks and streets	Facilitate access for wheelchairs, strollers
Seating areas	Benches and shaded rest zones	Encourages longer visits and comfort
Wide pathways	Space for wheelchairs and assistive devices	Reduces congestion, supports mobility
Accessible signage	Includes braille, pictograms, and readable fonts	Improves navigation and clarity

Source Authors

behavior, cultural expression, and economic activity. Table 3 categorizes these impacts across different dimensions.

3 Materials and Methods

This study adopts a quantitative approach to provide a well-rounded understanding of walkability along Al-Tahlia Street in Jeddah. This approach enables analysis of measurable aspects of pedestrian infrastructure while also capturing lived experiences and local perceptions.

Table 3 Impact of physical and non-physical walkability factors

Factor	Social impact	Cultural impact	Economic impact
Lighting	Enhancing safety, reducing crime	Creates a sense of pride and activity	Lowers crime-related costs, attracts businesses
Sidewalk width	Encourages interaction and comfort	Supports communal space use	Boosts footfall and local commerce
Seating design	Promotes social gatherings	Reflects local identity and traditions	Draws users to linger, supports economic activity
Green spaces	Improves mental well-being, encourages use	Adds aesthetic and environmental value	Raises property values, supports tourism
Accessibility features	Inclusion for all community members	Embodies equality and public service	Reduces the cost of legal/policy non-compliance

Source Authors

Al-Tahlia Street was selected as a case study because of its central role as a commercial and social hub in Jeddah. As one of the most vibrant pedestrian corridors in the city, it exemplifies the tensions between high foot traffic and insufficient walkability infrastructure. Case study methodology is ideal in such situations where contextual depth and spatial specificity are essential for understanding the real-world dynamics of urban spaces.

Structured Questionnaire was used, consisting of a 13-item questionnaire, which was designed to assess key dimensions of walkability, such as sidewalk width, green spaces, seating comfort, accessibility, and cultural identity. All questions were closed-ended to allow for quantifiable analysis. The questionnaire was conducted in person with pedestrians near Al-Tahlia Park, a highly frequented area, using random sampling. Respondents represented a range of ages, genders, and visitor types. The final sample included 62 valid responses.

3.1 Data Analysis

Questionnaire data were analyzed using Principal Component Analysis (PCA) via SPSS software. To analyze the questionnaire data effectively using SPSS, responses must be coded into a numerical format that the software can interpret. Questions with non-numeric answers—typically categorical or nominal variables—require statistical transformation. This involves converting each category within a nominal variable into separate dummy variables. Each category is treated as a distinct variable and is assigned a numerical code (e.g., 1, 2, 3, 4, etc.) to facilitate analysis [14]. Once the dataset is appropriately coded, Principal Component Analysis (PCA) is used as a method of factor analysis to reduce data dimensionality while preserving as much variability as possible. PCA works by transforming the original variables into a new set of uncorrelated variables called principal components, which capture the essential structure of the data. Before applying PCA, all variables must be standardized—adjusted to have a mean of zero and a standard deviation of one—to ensure comparability across different scales.

The Total Variance Explained table is a crucial output of PCA, displaying the eigenvalues associated with each principal component. Eigenvalues reflect the amount of variance each component explains, and they are listed in descending order of importance. The sum of all eigenvalues represents the total variance in the dataset, and components with higher eigenvalues account for greater proportions of the dataset's variability.

Another key output, the Component Matrix, shows the loadings—essentially the correlation coefficients—between the original variables and the extracted principal components. Each column corresponds to a component, while each row represents a variable. These loadings reveal how strongly each variable influences a particular component: higher loadings denote a stronger relationship. Analyzing these outputs helps researchers identify underlying patterns in the data and determine which components are most significant for interpretation. These results are discussed in detail in the following sections.

4 Findings

PCA revealed that Component 1 accounted for 37.039% of the total variance, indicating that a single underlying dimension significantly shaped pedestrian perception, as shown in Table 4.

This result indicates that the first component (Component 1) captures the most important cluster of pedestrian-related perceptions on Al-Tahlia Street. Table 5 includes the variables shown in the successive table that illustrates the significant

Table 4 Total variance explained using the SPSS program

Component	Initial eigenvalues			Extraction sums of squared loadings		
	Total	% of Variance	Cumulative %	Total	% of Variance	Cumulative %
1	1.852	37.039	37.039	1.852	37.039	37.039
2	0.905	18.091	55.130	–	–	–
3	0.841	16.812	71.942	–	–	–
4	0.762	15.249	87.191	–	–	–
5	0.640	12.809	100.000	–	–	–

Extraction method: principal component analysis
Source Based on SPSS analysis

Table 5 Component matrix using SPSS program

Category	Variable	Component loading
Physical	The width of the sidewalks for walking and social interactions?	0.686
Non-physical	The visual appeal of the green spaces on Al Tahlia Street	0.630
	The impact of the green spaces on Al Tahlia Street in promoting community well-being and social gatherings?	0.632
	The representation of the cultural identity through seating designs on Al Tahlia Street	0.560
	The impact of the accessibility features in promoting inclusivity and equality on Al Tahlia Street	0.522

Extraction method: principal component analysis
Source Authors based on SPSS analysis

variables (with values of at least + or − 0.5 to be counted as significant). This means they are responsible for at least 50% of the change in either direction.

These results show that sidewalk width and green spaces positively contribute to the perceived walkability, while seating design and accessibility are negatively perceived and may need improvement.

Table 6 illustrates the correlations deduced between the different variables. The significant values that are more than 0.5 are identified below.

5 Discussion

The findings of the PCA point to several critical factors shaping how pedestrians experience Al-Tahlia Street. Among the most positively perceived were sidewalk width and the presence of green spaces. Participants showed a strong preference for wide sidewalks, associating them with greater comfort and social interaction, which echoes the findings of Kim [8]. Despite this, observations and feedback suggest that the current sidewalk conditions often fail to meet these expectations, resulting in crowding and limited ease of movement.

Green spaces were also perceived positively, particularly in terms of their visual appeal and their potential to enhance community well-being. This supports the conclusions of Martadinata and Permana [13], who emphasize the multifunctional role of greenery in dense urban contexts. However, the impact of green spaces on actual pedestrian comfort may be limited by climatic challenges such as the lack of adequate shading.

In contrast, seating design was among the elements rated negatively. Respondents expressed that current seating lacks cultural relevance and thermal comfort, often

Table 6 Correlation matrix

Variables	Respondent's gender	Age Group	Feeling safe based on lighting conditions	Satisfaction with the width of sidewalks for walking/ social interaction	Seating areas encourage social interaction	Green spaces promote community well-being and gatherings	Seating design reflects Jeddah's cultural identity	Visual appeal of green spaces	Accessibility features promote inclusivity and equality for all pedestrians	Weather impact on participation in outdoor/ cultural activities	Lighting reduces crime and supports local businesses	Wide sidewalks attract foot traffic and benefit businesses	Shade structures offer comfort and support for economic activity
Respondent's gender	1.00	−0.23	−0.05	0.17	0.02	0.33	−0.11	−0.10	0.07	0.21	−0.15	−0.22	0.02
Age Group	−0.23	1.00	0.30	−0.07	0.07	0.09	−0.11	0.09	0.18	0.08	0.01	0.22	0.14
Feeling safe based on lighting conditions	−0.05	0.30	1.00	−0.04	0.19	0.13	0.13	0.19	0.03	−0.06	0.36	0.24	0.15
Satisfaction with the width of sidewalks for walking/ social interaction	0.17	−0.07	−0.04	1.00	−0.19	0.29	−0.25	0.30	−0.14	−0.02	0.04	0.03	0.32
Seating areas encourage social interaction	0.02	0.07	0.19	−0.19	1.00	−0.13	0.05	−0.23	0.29	0.13	0.05	0.10	−0.12

(continued)

Table 6 (continued)

Variables	Respondent's gender	Age Group	Feeling safe based on lighting conditions	Satisfaction with the width of sidewalks for walking/ social interaction	Seating areas encourage social interaction	Green spaces promote community well-being and gatherings	Seating design reflects Jeddah's cultural identity	Visual appeal of green spaces	Accessibility features promote inclusivity and equality for all pedestrians	Weather impact on participation in outdoor/ cultural activities	Lighting reduces crime and supports local businesses	Wide sidewalks attract foot traffic and benefit businesses	Shade structures offer comfort and support for economic activity
Green spaces promote community well–being and gatherings	0.33	0.09	0.13	0.29	−0.13	1.00	−0.16	0.23	−0.20	0.18	0.12	−0.03	0.21
Seating design reflects Jeddah's cultural identity	−0.11	−0.11	0.13	−0.25	0.05	−0.16	1.00	−0.16	0.20	0.20	−0.01	0.07	−0.16
Visual appeal of green spaces	−0.10	0.09	0.19	0.30	−0.23	0.23	−0.16	1.00	−0.18	−0.16	0.07	0.13	0.09
Accessibility features promote inclusivity and equality for all pedestrians	0.07	0.18	0.03	−0.14	0.29	−0.20	0.20	−0.18	1.00	**0.50**[a]	−0.09	0.00	−0.06

(continued)

Table 6 (continued)

Variables	Respondent's gender	Age Group	Feeling safe based on lighting conditions	Satisfaction with the width of sidewalks for walking/social interaction	Seating areas encourage social interaction	Green spaces promote community well-being and gatherings	Seating design reflects Jeddah's cultural identity	Visual appeal of green spaces	Accessibility features promote inclusivity and equality for all pedestrians	Weather impact on participation in outdoor/cultural activities	Lighting reduces crime and supports local businesses	Wide sidewalks attract foot traffic and benefit businesses	Shade structures offer comfort and support for economic activity
Weather impact on participation in outdoor/cultural activities	0.21	0.08	−0.06	−0.02	0.13	0.18	0.20	−0.16	**0.50**[a]	1.00	−0.29	−0.16	−0.03
Lighting reduces crime and supports local businesses	−0.15	0.01	0.36	0.04	0.05	0.12	−0.01	0.07	−0.09	−0.29	1.00	0.04	0.05
Wide sidewalks attract foot traffic and benefit businesses	−0.22	0.22	0.24	0.03	0.10	−0.03	0.07	0.13	0.00	−0.16	0.04	1.00	−0.03
Shade structures offer comfort and support economic activity	0.02	–	0.15	0.32	−0.12	0.21	−0.16	0.09	−0.06	−0.03	0.05	−0.03	1.00

Source Authors based on SPSS analysis
[a] Indicates significant values

being exposed and unsuited for extended use. These results align with Gandawijaya et al. [5], who argue that culturally embedded and climatically adapted public furniture plays an essential role in encouraging pedestrian activity.

Accessibility, too, showed a negative loading. Although features like ramps and tactile indicators exist, many users reported that they are either poorly maintained, inconsistently implemented, or not intuitively placed. Kraus and Rosenbloom [10] emphasize that accessibility must go beyond minimum standards, requiring thoughtful integration into the flow of pedestrian movement.

Together, these findings suggest that while certain design elements are appreciated, there remains a disconnect between the presence of infrastructure and its effectiveness. A more holistic, user-centered approach that incorporates local culture, climate considerations, and everyday functionality is needed to create truly walkable urban environments.

Based on the correlation matrix, there is a strong relationship between the accessibility features that promote inclusivity and equality for all pedestrians and the impact of weather on participation in outdoor or cultural activities. This suggests an important intersection between urban design and environmental comfort. This connection implies that accessible environments, such as shaded pathways, ramps, and seating, may help mitigate the adverse effects of weather, making public spaces more usable in various conditions. Inaccessible spaces, on the other hand, may intensify the challenges posed by heat, rain, or other weather conditions, particularly for vulnerable groups such as the elderly, people with disabilities, or families with children. This correlation highlights the importance of considering universal design principles alongside climate responsiveness to ensure equitable use of public urban areas.

The implications of these findings are significant, highlighting the need for urban policies that go beyond physical infrastructure upgrades to include cultural and climatic adaptations. A strong relationship between accessibility features and weather-related participation in outdoor activities suggests that inclusive, climate-responsive design plays a critical role in enhancing walkability and public life. Urban planners can use these insights to prioritize investments in features that most directly impact pedestrian satisfaction and usability, such as shaded walkways, accessible routes, and weather-protective structures. By addressing the gap between the presence of infrastructure and the lived experience of users, municipalities can create public spaces that are not only physically navigable but also comfortable and inviting in varying weather conditions. From a planning and policy perspective, integrating accessibility and environmental resilience into walkability strategies promotes broader goals, supporting consistent civic engagement, social inclusion, public health, and economic vitality. Ultimately, cities that adopt this holistic approach to urban design are better positioned to foster livable, equitable, and culturally vibrant communities.

6 Conclusion

This study investigated walkability along Al-Tahlia Street in Jeddah using a mixed-methods approach that combined questionnaires, interviews, and statistical analysis through Principal Component Analysis (PCA). The findings reveal that the pedestrian experience is shaped by a dynamic interplay between physical design features and socio-cultural responsiveness. While elements like sidewalk width and green spaces were seen as positive contributors, infrastructure such as public seating and accessibility features received more critical evaluations from users. These insights emphasize that improving walkability requires more than the mere provision of pedestrian infrastructure; it calls for a shift toward thoughtful, inclusive design that reflects local cultural, environmental, and social realities. Enhancing Al-Tahlia Street's pedestrian experience is not just about convenience—it is a step toward healthier, more sustainable, and socially connected urban environments. A key takeaway from the analysis is the importance of sidewalk width. Narrow or congested sidewalks hinder pedestrian flow, especially during peak hours. Strategic expansion and redesign can ease movement and encourage street-level social interaction. Green spaces also emerged as valuable urban assets—not only for their aesthetic and environmental contributions but also for their role in supporting mental well-being. However, their effectiveness in Jeddah's climate depends on shade, culturally appropriate seating, and heat-tolerant vegetation. Public seating, when designed with local materials and social behaviors in mind, can activate street life. Participants highlighted a disconnect between existing seating options and actual user needs, underscoring the importance of culturally responsive and climate-adapted street furniture. Similarly, the study revealed gaps in accessibility infrastructure. Elements intended to be inclusive often fail in execution. Creating a truly accessible street requires continuous paths, intuitive ramps, legible signage, and responsive surface materials. Another vital finding is the importance of participatory planning. Effective pedestrian interventions must involve the people who use these spaces daily. Engaging residents, business owners, and local planners ensures that design solutions are not only technically sound but also socially accepted and sustainable. Despite the valuable insights gained, the study is not without limitations. It focuses solely on Al-Tahlia Street and may not capture the full diversity of pedestrian needs across Jeddah. The PCA provided useful data-driven insights but did not account for temporal variations such as changes in pedestrian behavior across different times of day or seasons—critical factors in a climate like Jeddah's. The qualitative portion, while rich in narrative depth, was limited in scale and would benefit from broader community representation, especially among older adults, people with disabilities, and lower-income groups. Future research should expand the scope to include different street types and neighborhoods in Jeddah or other Saudi cities. The integration of spatial tools such as GIS, thermal imaging, and behavioral observation can offer more comprehensive analyses. A more participatory approach—inviting citizens to co-create and evaluate public spaces—will also enhance the relevance and impact of future urban design strategies. Understanding walkability not just through structured data but through lived experiences is essential to crafting inclusive, climate-responsive, and people-centered cities.

References

1. Aparicio, et al.: Walkability defined neighborhoods for sustainable cities. Cities **149**, 104944 (2024). https://doi.org/10.1016/j.cities.2024.104944
2. Ewing, R., Handy, S.: Measuring the unmeasurable: urban design qualities related to walkability. J. Urban Des. **14**(1), 65–84 (2009). https://doi.org/10.1080/13574800802451155
3. Forsyth, A.: What is a walkable place? The walkability debate in urban design. Urban Design Int. **20**(4), 274–292 (2015). https://doi.org/10.1057/udi.2015.22
4. Fonseca, et al.: Walkable cities: using the smart pedestrian net method for evaluating a pedestrian network in Guimarães, Portugal. Sustainability **14**(16), 10306 (2022). https://doi.org/10.3390/su141610306
5. Gandawijaya, R. A., Kirana, D. C., Morika, D.: Study of public seating facilities with behavior and activity approaches on pedestrian street Ir. H. Djuanda Bandung. IOP Conf. Series Earth Environ. Sci. 794(1), 012078 (2021). https://doi.org/10.1088/1755-1315/794/1/012078
6. Ibrahim, S., Younes, A., Abdel-Razek, S.A.: Impact of neighborhood urban morphologies on walkability using spatial multi-criteria analysis. Urban Science **8**(2), 70 (2024). https://doi.org/10.3390/urbansci8020070
7. Kędra, A., Maleszyk, P., Visvizi, A.: Engaging citizens in land use policy in the smart city context. Land Use Policy **129**, 106649 (2023). https://doi.org/10.1016/j.landusepol.2023.106649
8. Kim, J.: The impact of sidewalk width on pedestrian comfort and urban social interaction. J. Urban Des. **26**(3), 445–460 (2021). https://doi.org/10.1080/13574809.2020.1865954
9. Kim, et al.: Sustainable, walkable cities for the elderly: Identification of the built environment for walkability by activity purpose. Sustain. Cities Soc. **100**, 105004 (2024). https://doi.org/10.1016/j.scs.2023.105004
10. Kraus, L., Rosenbloom, S.: Enhancing accessibility in urban spaces for individuals with disabilities. J. Disability Policy Stud. **29**(2), 108–119 (2019). https://doi.org/10.1177/1044207318757635
11. Lo, R.H.: Walkability: What is it? J. Urbanism: Int. Res. Placemaking Urban Sustain. **2**(2), 145–166 (2009). https://doi.org/10.1080/17549170903092867
12. Lo, R.H.: Walkability: what is it? J. Urbanism: Int. Res. Placemaking Urban Sustain. **2**(2), 145–166 (2009). https://doi.org/10.1080/17549170903092867
13. Martadinata, H., Permana, S.: Urban green space design for enhanced walkability and comfort. Buildings **14**(3), 651 (2023). https://doi.org/10.3390/buildings14030651
14. Nie, N. H., Hull, C. H., Jenkins, J. G., Steinbrenner, K., Bent, D. H.: Statistical package for the social sciences (2nd ed.) (1975). McGraw-Hill
15. Rosenfield, K.: Times square celebrates the grand opening of the shields (2017). R., Gomes da Silva, E. J., Lima e Lima, T., & Osorio, N.: Walkability: a review of trends. J Urbanism Int. Res. Placemaking Urban Sustain. 16(1), 19–41 (2021). https://doi.org/10.1080/17549175.2021.1936601
16. Shields, R., Gomes da Silva, E. J., Lima e Lima, T., Osorio, N.: Walkability: a review of trends. J. Urban.: Int. Res. Placemaking Urban Sustain. 16(1), 19–41 (2021). https://doi.org/10.1080/17549175.2021.1936601
17. Soleimani, A., Janani, H.: The effects of microclimatic conditions on pedestrian comfort in hot-humid climates. Sustain. Cities Soc. **47**, 101401 (2019). https://doi.org/10.1016/j.scs.2019.101401
18. Visvizi, A., Abdel-Razek, S. A., Wosiek, R., Malik, R.: Conceptualizing walking and walkability in the smart city through a model composite w2 Smart City Utility Index. Energies **14**(23), 8193 (2021). https://doi.org/10.3390/en14238193

19. Visvizi, A., Godlewska-Majkowska, H.: Not only technology: From smart city 1.0. through smart city 4.0 and beyond (an introduction), in: Visvizi, A., Godlewska-Majkowska, H. (eds) Smart Cities: Lock-in, Path-dependence and Non-linearity of Digitalization and Smartification. Routledge (in production), London & New York (2024). ISBN 9781032539508, https://doi.org/10.1201/9781003415930
20. Visvizi, A., Wosiek, R., Malik, R.: The smart city competitiveness index (SMCI): Conceptualization, modelling, application – An evidence-based insight. Land Use Policy **148**, 107408, (2025). https://doi.org/10.1016/j.landusepol.2024.107408
21. Whyte, W. H.: City: rediscovering the center. University of Pennsylvania Press (1988)

Towards Touristic-Based Development for Al-Tayseer Area in Makkah City

Aseel Fakieh and Asmaa Ibrahim

Abstract This paper investigates how regenerative tourism strategies can be designed for informal settlements in Makkah, with a focus on Al-Tayseer, to prevent spatial segregation and marginalization of residents while advancing Saudi Vision 2030's tourism objectives. Al-Tayseer is a strategically located district in Makkah, situated near the Grand Mosque and frequently utilized by pilgrims during Hajj and Umrah. The area features mid-range accommodations and essential infrastructure supporting religious tourism. Employing a mixed-methods approach, the study integrates quantitative surveys of Al-Tayseer residents and qualitative interviews with stakeholders. Sampling criteria, research instruments, and data analysis techniques are detailed to ensure methodological transparency and reliability. The research reveals that while tourism-led regeneration offers economic opportunities, it also risks exacerbating social exclusion, displacement, and the commodification of marginalized communities. Residents express concerns about housing security, cultural identity, and the lack of participatory planning. The paper contributes a critical, ethically grounded framework for regenerative urban tourism in sacred cities. It highlights the necessity of participatory mechanisms, legal safeguards, and context-sensitive planning to balance tourism growth with social equity.

Keywords Informal settlements · Urban fabrics · Land use · Context conservation · Urban historic landscape · Placemaking · Cultural tourism · 2030 Vision · SDG-11

A. Fakieh (✉)
Effat College of Architecture and Design, Effat University, Jeddah, Saudi Arabia
e-mail: eng.aseelfakieh@gmail.com

A. Ibrahim
Urban Design Lab, College of Architecture and Design, Effat University, Jeddah, Saudi Arabia
e-mail: asibrahim@effatuniversity.edu.sa

A. Visvizi et al. (eds.), *Research and Innovation Forum 2025 Volume 1*,
Springer Proceedings in Complexity, https://doi.org/10.1007/978-3-032-09066-9_22

1 Introduction

In recent years, tourism-led urban regeneration has become a central strategy for cities seeking to enhance their competitiveness and attract regional, national, and international visitors [1, 2]. This trend is evident in diverse urban contexts, where the rapid growth of tourism has reshaped cityscapes and stimulated the renovation of historic neighborhoods, often serving as a catalyst for broader urban transformation [3, 4]. However, alongside these opportunities, tourism-driven change has also intensified debates around gentrification, displacement, and inequality, particularly in areas where informal settlements and vulnerable communities are present [5, 6]. Recent integrative reviews highlight that while tourism gentrification and lifestyle migration can produce exclusionary outcomes, they also offer pathways for regeneration when guided by inclusive governance and robust local participation frameworks [7, 8]. Moreover, the imperative of universal accessibility in cultural heritage and tourist environments has gained traction, with researchers emphasizing the need for innovative technological solutions and strengthened governance to ensure that regeneration efforts do not marginalize existing residents or diminish social equity [9, 10].

Against this backdrop, Makkah's urban landscape is undergoing significant transformation under the ambitions of Saudi Vision 2030, which positions tourism as a cornerstone of economic diversification and cultural revitalization [11, 12]. The regeneration of informal settlements such as Al-Tayseer, situated at the intersection of religious heritage and major tourism corridors, encapsulates both the promise and the complexity of these changes. Al-Tayseer is a strategically located district in Makkah, situated near the Grand Mosque and frequently utilized by pilgrims during Hajj and Umrah. The area features mid-range accommodations and essential infrastructure supporting religious tourism. The present study critically examines how regenerative tourism strategies can be integrated in Al-Tayseer to avoid the marginalization or spatial segregation of its residents, while still achieving the objectives of tourism-based development.

The methodology employed in this research involves a mixed-methods approach. Quantitative data was collected through surveys administered to residents and stakeholders in Makkah, including those familiar with the Al-Tayseer area. These surveys aimed to capture perceptions regarding the current conditions, challenges, and potential for tourism development. The quantitative data was analyzed using descriptive statistics to identify key trends and patterns. This primary data was complemented by the qualitative method of using interviews with experts.

This paper is organized to provide a comprehensive and critical examination of regenerative tourism strategies in the context of informal settlements in Makkah, specifically focusing on Al-Tayseer. The study begins with a review of recent literature and theoretical frameworks that address the intersections of urban regeneration, tourism, gentrification, and social equity [1, 7]. This is followed by a detailed explanation of the research methodology, including the mixed-methods approach, sampling criteria, and data analysis techniques, in alignment with best practices for transparency and reliability [9, 10]. Empirical findings from fieldwork in Al-Tayseer are then presented, highlighting resident priorities, stakeholder perspectives, and spatial patterns of accessibility and risk. The discussion critically engages with the ethical and practical implications of tourism-driven regeneration, referencing current debates on poverty tourism and the challenges of implementing inclusive strategies in sacred urban contexts [13, 14]. The paper concludes by synthesizing the main contributions and outlining recommendations for future research that address the long-term impacts of participatory and ethical tourism interventions [8, 11].

2 Literature Review

The intersection of tourism-led urban regeneration and the fate of informal settlements has become a central concern in contemporary urban studies, especially in rapidly transforming cities of the Global South. Scholars widely acknowledge that while tourism can serve as a catalyst for urban revitalization and economic diversification, it often brings with it significant risks of social exclusion, displacement, and spatial segregation, particularly in areas where informal settlements are prevalent [1, 2]. In the context of Makkah and similar sacred cities, these dynamics are further complicated by the religious and cultural significance of place, raising unique ethical and governance challenges [11, 12].

Recent systematic reviews and empirical studies highlight that urban regeneration projects, when driven primarily by tourism and investment interests, frequently accelerate gentrification and the transformation of informal neighborhoods, often to the detriment of their original residents [5, 15]. The literature underscores that such processes are rarely neutral; rather, they tend to privilege certain social groups while marginalizing others, unless deliberate measures are taken to ensure equity and inclusion [6, 7]. In particular, insufficient community participation and the lack of robust, context-sensitive frameworks for local engagement are identified as persistent obstacles to inclusive regeneration [9, 15].

A growing body of research argues that participatory planning and the empowerment of informal settlement residents are critical for achieving socially just and sustainable regeneration outcomes [16]. Studies from diverse urban contexts demonstrate that when local communities are actively involved in the design, implementation, and governance of tourism-led regeneration, the risks of displacement and social fragmentation are significantly reduced [8, 10]. Such participatory approaches are not only ethically imperative but also enhance the legitimacy, effectiveness, and resilience of urban interventions [8, 10].

The ethical dimensions of tourism in informal and sacred urban contexts have also received increasing scholarly attention. The literature on "poverty tourism" and the commercialization of marginalized communities raises concerns about the commodification of poverty and the potential erosion of cultural and spiritual values [13, 14]. In cities like Makkah, where religious sanctity is paramount, these tensions are particularly acute, necessitating frameworks that prioritize dignity, agency, and cultural sensitivity in tourism development [13, 14].

Recent methodological advances, such as the use of geospatial technologies and participatory mapping, have improved the identification and analysis of informal settlements, enabling more targeted and equitable interventions [18]. These tools support the development of regeneration strategies that are responsive to the needs and aspirations of local residents, while also aligning with broader tourism and urban development goals.

Despite these advances, literature consistently identifies significant gaps. There remains a lack of integrated frameworks that combine the economic imperatives of tourism with the social imperatives of inclusion and spatial justice. Moreover, the long-term impacts of tourism-led regeneration on the well-being, identity, and cohesion of informal settlement residents are still insufficiently understood, particularly in the context of sacred cities [7, 15].

Accordingly, the existing literature strongly supports the hypothesis that regenerative tourism strategies in informal settlements like Al-Tayseer can only achieve both development and social equity objectives if they are grounded in participatory, ethical, and context-sensitive frameworks. Such approaches must foreground the voices and rights of residents, ensure robust mechanisms for community engagement, and critically engage with the ethical tensions inherent in the commodification of sacred and marginalized spaces. This review thus provides the conceptual and empirical foundation for investigating how these principles can be operationalized in the specific context of Al-Tayseer, informing the structure and methodology of the present study.

3 Methods and Materials

This research investigates the urban redevelopment potential of the Al-Tayseer area in Makkah City, focusing on tourism and economic growth. The study area, Al-Tayseer, is a northern residential district strategically located 1.15 km from Al-Haram (Holy Mosque), spanning 375,125.52 square meters. This area is characterized by its proximity to key pilgrimage routes and its existing infrastructure, including pilgrimage hotels and traditional residential structures, which attract substantial tourist traffic due to Makkah's religious significance. Al-Tayseer's accessibility is enhanced by major roadways connecting it to Jeddah and other urban centers, making it a crucial area for managing pilgrimage logistics and fostering tourism. The Royal Commission for Makkah City and Holy Sites assesses Al-Tayseer's redevelopment potential using elevational standards, and this research further focuses on a specific section covering 33,402.29 square meters.

The development of informal settlements, including Al-Tayseer, in Saudi Arabia has increased significantly in recent decades due to factors such as high population growth, unaffordable land prices, and inadequate urban planning. In Makkah, approximately one-third of the city's population resides in informal settlements, which constitute about half of the total urban area (Municipality of Makkah, 2008). The Royal Commission for Makkah City and holy sites categorizes these informal settlements into zones, with Al-Tayseer falling under the "upgraded zone." Understanding the historical context of informal settlement development in Makkah, including the pre-oil, pre-planning, and planning stages, is crucial for addressing the current situation.

To comprehensively assess the challenges and opportunities for regenerative tourism in Al-Tayseer, the study adopted a mixed-methods approach, integrating quantitative surveys, qualitative interviews, and spatial mapping. This methodological design was chosen to capture the multifaceted nature of urban regeneration in informal settlements, ensuring that the perspectives of residents, experts, and stakeholders were adequately represented and that findings aligned with the goals of Vision 2030.

For the quantitative component, a snowball sampling technique was employed. This approach was selected due to the social dynamics and accessibility challenges typical of informal settlements, where formal household lists are often unavailable and community trust is crucial for participation. The initial respondents were identified through local community leaders and urban planners familiar with Al-Tayseer, who then referred additional participants, allowing the sample to expand organically within the community. This method ensured the inclusion of diverse voices, particularly those who might otherwise be overlooked in more conventional sampling frameworks.

A total of 400 surveys were distributed, with 357 valid responses collected and analyzed. This sample size was calculated based on an estimated population of 5,000 households in Al-Tayseer, targeting a 95% confidence level and a 5% margin of error. The achieved response rate provided a robust quantitative foundation, with the margin of error for survey estimates maintained at ±5%, ensuring the reliability of the findings for the target population.

The survey instrument itself was meticulously designed, comprising 51 questions across eight critical dimensions, including demographics, perceptions of informal settlements, urban morphology, the impact of urbanization, tourism's role in poverty alleviation, infrastructure quality, tourism potential, and policy gaps. The questions utilized a mix of multiple-choice formats, rating scales, and open-ended prompts to facilitate both quantitative analysis and the collection of nuanced qualitative insights. Demographic analysis revealed that the sample was predominantly male (54.7%), with most respondents aged above 35 years (66.7%). Employment status indicated a majority of full-time or part-time workers (45.7%), and 81.2% identified as residents of Makkah, ensuring that the data reflected local perspectives.

In addition to the quantitative survey, a qualitative method was employed to provide a deeper, expert-driven understanding of the challenges and opportunities facing Al-Tayseer's urban regeneration. This phase consisted of structured interviews with three key participants, each selected for their expertise in urban planning, infrastructure, and cultural preservation. The purpose of these interviews was to explore, in greater depth, the themes and issues that emerged from the survey, enabling a more nuanced interpretation of the data and a richer contextualization of the findings.

The interviewees included an experienced urban planning engineer, a civil engineer involved in major urban projects, and an academic specializing in architecture and urban planning. Their combined perspectives offered a comprehensive view of Al-Tayseer's historical evolution, current policy environment, and future potential within the framework of Vision 2030.

4 Results

The survey results provide a comprehensive insight into the perceptions and challenges associated with informal settlements in Makkah, with a particular focus on Al-Tayseer. Regarding familiarity, 43.6% of respondents reported being very familiar with informal settlements in the city, while 36.8% were somewhat familiar and 19.6% were not familiar. When asked about the primary causes of informal settlements, 44.4% identified inadequate urban planning as the main factor, followed by 20.5% who cited economic migration.

A significant majority, 72.3%, agreed that informal settlements negatively affect Makkah's image, while only 10.6% disagreed and 17.1% remained neutral. Government efforts to improve conditions were rated as effective by just 18.8% of respondents, with 42.7% considering them ineffective and 38.5% neutral. The prioritization of informal settlement improvement was strongly supported, with 56.4% strongly agreeing and 34.2% agreeing that it should be a key urban development focus; only 9.4% were neutral or disagreed.

Assessments of living conditions in Al-Tayseer were largely negative: 52.1% rated them as poor, 27.4% as average, and only 20.5% as good. Infrastructure issues, including roads and sanitation, emerged as the most pressing problem (65.8%), followed by overcrowding (18.8%). Nearly half (48.7%) described the area's general infrastructure as inadequate, 31.6% as average, and 19.7% as sufficient. When comparing Al-Tayseer to other areas in Makkah, 59.8% rated its living conditions as worse, 23.9% as comparable, and 16.3% as better. The most urgent infrastructure needs identified were roads and sanitation facilities (71.2%), with electricity also noted (16.3%).

In terms of housing and building conditions, 64.1% rated Al-Tayseer as worse than the rest of Makkah, 25.6% as comparable, and only 10.3% as better. The main factors contributing to the development of informal settlements were lack of infrastructure (41.9%) and population growth (30.8%). The quality of life for residents was seen as negatively affected by local conditions by 68.4% of respondents, while 24.3% indicated a moderate impact and 7.3% saw no impact.

Government effectiveness in addressing Al-Tayseer's challenges was rated as effective by only 17.9%, with 49.6% neutral and 32.5% ineffective. Awareness of government plans or policies was limited, as 62.4% were unaware of any such initiatives. For policy improvement, 53.8% suggested better infrastructure planning and greater community engagement. Regarding socio-economic status, 61.5% noted a negative impact from the area's development, 29.1% saw little to no impact, and 9.4% observed some positive effects.

The survey also captured perceptions of the built environment. Only 15.4% agreed that buildings in Al-Tayseer are well-maintained and safe, while 57.3% disagreed. Streets were considered adequately wide by just 22.2%, with 61.1% disagreeing. Architectural harmony was noted by only 13.6% of respondents, and 68.2% disagreed. Adequate housing was recognized by 19.8%, while 58.1% disagreed. Opinions on building density were mixed, with 29.3% agreeing it contributed to community, 42.7% disagreeing, and 28% neutral.

Urbanization's effects were also mixed: 28.4% agreed it had improved infrastructure, but 52.3% disagreed. While 37.6% felt rapid growth made the area more vibrant, 41.2% disagreed. Only 24.7% agreed that urbanization increased job availability, while 51.8% disagreed. A large majority (64.1%) felt that population influx strained resources, and 56.4% said development had not kept pace with growth.

Tourism's role in local development was viewed with some optimism but also skepticism. While 39.3% agreed that tourism provides significant employment opportunities, 48.2% were neutral and 12.5% disagreed. Economic contributions from tourist activities were acknowledged by 45.3%, but only 11.1% strongly agreed that residents benefit directly. Infrastructure improvements due to tourism were noted by 31.5%, while 53.1% remained neutral. Tourism's role in promoting cultural exchange was recognized by 38.5% of respondents.

Infrastructure assessments revealed that only 19.7% felt the road network was sufficient, while 52.1% disagreed. Water and sanitation facilities were deemed adequate by 24.6%, with 51.3% disagreeing. Electrical services were considered reliable by 33.8% of respondents. Public spaces were well-maintained according to just 17.3%, and only 18.8% agreed that the infrastructure supports a healthy living environment.

Tourism potential was seen as underutilized: 29.9% agreed that Al-Tayseer has unique attractions, but only 16.8% felt that historical and cultural sites were well-promoted. High-quality accommodation was recognized by 22.5%, and recreational activities by 18.2%. The preservation of local culture through tourism was agreed upon by 26.5%.

Policy perceptions were generally critical. Only 18.8% agreed that current urban policies address residents' needs, and 21.6% felt there was a clear plan for sustainable development. Just 24.8% believed that policies support the integration of tourism with local development, and only 26.5% felt the government actively involves residents in planning. Safety and security were considered adequate by 32.4%.

Overall, the results above highlight significant dissatisfaction with infrastructure, policy, and government responsiveness in Al-Tayseer, as well as a cautious optimism about the potential for tourism to contribute to local development—provided that planning and community engagement are meaningfully improved. The survey also explored the role of tourism, revealing mixed perceptions. While some residents acknowledged the potential for tourism to create jobs and stimulate the local economy, most expressed skepticism about the direct benefits to the community. Infrastructure improvements and cultural exchange were noted, but concerns persisted regarding whether tourism development genuinely addressed local needs.

Analysis of the interviews revealed several key themes. First, the historical context of Al-Tayseer was highlighted as a critical factor shaping its current challenges. Experts traced the origins of informal settlement growth to the area's strategic proximity to Al-Haram, noting that decades of neglect and fragmented governance had resulted in a lack of cohesive urban planning, inadequate infrastructure, and diminished quality of life. The lack of open public spaces and recreational amenities was repeatedly identified as a factor driving resident dissatisfaction and outward migration.

Second, tourism and economic growth were discussed as both opportunities and risks. All experts agreed that tourism could serve as a catalyst for local economic revitalization, but emphasized that its benefits would only be realized if development was community-centered and culturally sensitive. Suggestions included the creation of artisan markets, seasonal bazaars, and the integration of cultural attractions into the local economy. However, concerns were raised about the potential for excessive commercialization to erode Al-Tayseer's unique historical and cultural identity.

Third, the governance framework was critiqued for its lack of coherence and limited resident participation. The experts advocated for a participatory governance model, recommending the establishment of a dedicated management authority for Al-Tayseer to improve policy coordination and accountability. They stressed that residents must be actively involved in planning processes to ensure that regeneration aligns with local needs and aspirations.

Fourth, the alignment of Al-Tayseer's regeneration with Vision 2030 was affirmed, particularly in terms of enhancing urban environments, improving quality of life, and promoting cultural and religious tourism. The experts proposed that infrastructure upgrades, increased connectivity to larger urban projects (such as the Masar Project), and the preservation of cultural heritage should be core components of the regeneration strategy.

Finally, the interplay between tourism and cultural preservation emerged as a central theme. The experts recommended the development of dedicated cultural zones, the promotion of heritage tourism, and the adoption of architectural guidelines that respect traditional styles. They also emphasized the importance of empowering the local community through skills development and direct involvement in tourism initiatives, as well as implementing sustainable tourism practices to prevent resource strain and cultural commodification.

Accordingly, the qualitative interviews provided critical insights that complemented the quantitative survey findings. While the survey revealed widespread dissatisfaction with infrastructure, policy, and government responsiveness, the interviews illuminated the underlying causes and offered concrete strategies for integrated, sustainable, and culturally sensitive regeneration. Together, these methods underscore the necessity of participatory planning, community empowerment, and the careful balancing of tourism development with heritage preservation in the future transformation of Al-Tayseer.

5 Discussion of Findings

The findings of this study offer direct insight into the research question of how regenerative tourism strategies in Al-Tayseer can be designed to prevent marginalization or spatial segregation of residents while advancing tourism-based development. The quantitative survey revealed a pervasive dissatisfaction among residents regarding infrastructure, living conditions, and the effectiveness of government interventions. More than half of respondents rated living conditions as poor, with infrastructure issues—such as inadequate roads, sanitation, and electricity—identified as the most urgent needs. These results are consistent with broader regional research, which highlights the persistent challenges of unplanned urban growth and limited basic services in informal settlements [11, 15].

The qualitative interviews provided further context, emphasizing that historical neglect and fragmented governance have contributed to Al-Tayseer's current difficulties. Experts noted that while tourism presents economic opportunities, its benefits are unlikely to reach residents unless development is participatory and culturally sensitive. This aligns with findings from Jordan and Tanzania, where inclusive, community-driven approaches to urban regeneration have shown greater potential for sustainable outcomes and social cohesion [17, 18]. Both the survey and interviews indicated that improved infrastructure planning and greater community engagement are critical for addressing local challenges, echoing the importance of co-production and participatory governance found in recent studies [17].

The complex relationship between tourism and local development was also evident. While some optimism exists about tourism's potential to generate employment and stimulate the local economy, both quantitative and qualitative data suggest that direct benefits to residents remain limited, and concerns about cultural commodification and uneven distribution of gains persist. This tension reflects global debates on tourism-led regeneration in marginalized urban areas [6, 7].

In summary, the results support the hypothesis that regenerative tourism in Al-Tayseer can only succeed if grounded in participatory, ethical, and context-sensitive strategies. Addressing infrastructure deficits, enhancing policy transparency, and fostering meaningful community involvement are essential steps. These findings reinforce evidence from other contexts that inclusive, co-produced urban regeneration is not only more equitable but also more effective in achieving sustainable development and social cohesion in informal settlements [17, 18].

6 Conclusion

In conclusion, this research has demonstrated that regenerative tourism in Al-Tayseer can only succeed if it is rooted in participatory, context-sensitive, and inclusive approaches that prioritize the needs and aspirations of residents. The findings reveal significant dissatisfaction with current infrastructure, policy, and government responsiveness, while also highlighting the underutilized potential of tourism as a tool for local development. Both quantitative and qualitative results indicate that without meaningful community engagement and targeted investment in basic services, tourism-led regeneration risks perpetuating marginalization and spatial segregation.

Based on these findings, several precise recommendations are proposed. Local government and urban planners should prioritize comprehensive infrastructure upgrades—particularly in roads, sanitation, and electricity—and ensure transparent, participatory planning processes that actively involve Al-Tayseer residents at every stage. The establishment of a dedicated local management authority is recommended to coordinate regeneration efforts and monitor progress. Tourism authorities and the private sector are encouraged to develop tourism products and services that directly benefit the local community, such as artisan markets and cultural festivals, and to invest in capacity-building programs that enable residents to participate in and profit from tourism activities. Community organizations and NGOs should facilitate community mobilization, advocate for residents' rights, and provide platforms for dialogue between stakeholders to ensure that local voices are central to regeneration strategies. Academics and researchers should continue to provide evidence-based policy advice and conduct rigorous impact assessments to guide the adaptive management of regeneration projects.

Future research should focus on longitudinal studies to evaluate the enduring effects of participatory regeneration and tourism development in informal settlements. Comparative research across different urban contexts in Saudi Arabia and the broader region would also provide valuable insights for policy transfer and adaptation.

References

1. García-Hernández, M., de la Calle-Vaquero, M., Yubero, C.: Cultural heritage and urban tourism: historic city centres under pressure. Sustainability **9**(8), 1346 (2017)
2. Yrigoy, I.: Urban tourism, housing and gentrification: the case of Airbnb in Barcelona. Urban Studies **59**(5), 1017–1035 (2022)
3. Füller, H., Michel, B.: 'Stop being a tourist!' New dynamics of urban tourism in Berlin-Kreuzberg. Int. J. Urban Reg. Res. **38**(4), 1304–1318 (2014)
4. Gravari-Barbas, M., Guinand, S.: Tourism and gentrification in contemporary metropolises: international perspectives. Routledge (2017)
5. Gant, A.C.: Tourism gentrification: new perspectives and challenges. Tour. Geogr. **24**(1–2), 1–10 (2022)

6. Colomb, C., Novy, J.: Urban tourism and its discontents: an introduction. In: Protest and resistance in the tourist city (pp. 1–30). Routledge (2017)
7. Ioannides, D., Röslmaier, M., van der Zee, E.: Tourism gentrification and social sustainability: a review. Curr. Issue Tour. **26**(7), 1020–1034 (2023)
8. Sequera, J., Nofre, J.: Touristification, transnational gentrification and urban change in Lisbon: the neighbourhood of Alfama. Urban Studies **57**(15), 3169–3189 (2020)
9. Pinto, P., Guerreiro, J., Silva, F.: Universal accessibility in cultural tourism: challenges and opportunities. J. Tourism Futures **9**(1), 89–104 (2023)
10. Buhalis, D., Darcy, S.: Accessible tourism: concepts and issues. Channel View Publications (2011)
11. Alhowaish, A.K.: Tourism development and Vision 2030 in the Kingdom of Saudi Arabia: challenges and opportunities. Tourism Plan. Develop. **18**(3), 315–331 (2021)
12. Ministry of Tourism. (2022). Saudi Vision 2030 and the future of tourism. [Official Report]
13. Frenzel, F., Koens, K., Steinbrink, M.: Slum tourism: poverty, power and ethics. Routledge (2012)
14. Jack, G., Phipps, A.: Tourism and intercultural exchange: why tourism matters. Channel View Publications (2005)
15. Abdelfattah, R. S., Maghelal, P. K.: The anatomy of urban regeneration-led gentrification in the Middle East and North Africa: a systematic literature review. J. Infrastruct. Policy Develop. 8(8), Article 5993. EnPress Publisher (2024)
16. Kędra, A., Maleszyk, P., Visvizi, A.: Engaging citizens in land use policy in the smart city context, Land Use Policy, **129**, 106649 (2023). https://doi.org/10.1016/j.landusepol.2023.106649
17. OHI-02-2021-0023: (2021). The social role of open spaces in informal settlements in Jordan. Open House International, 46(2), 232–250
18. Frediani, A. A., Cociña, C., & Boano, C.f.: Pathways to the 'good life': Co-producing prosperity research in informal settlements in Tanzania. Urban Planning 5(3), 191–204 (2020). https://doi.org/10.17645/up.v5i3.3177

Humanizing Cities Through Urban Design: Urban Design Principles and Policymaking Considerations

Asmaa Hussain Alganawi, Hossny Azizalrahman, and Anna Visvizi

Abstract The pace urbanization over the past decades altered the fabric of cities, often prompting the emergence of built environment seemingly disconnected from the needs of city residents. As a way of alleviating this trend, modern urban planning prioritizes social well-being, accessibility, and sustainability. Given the interconnected nature of urban planning, policymaking, funding requirements and investors' expectations, the question is how to measure the degree to which urban environments are human centered. Referencing the case of Jeddah, the objective of this paper is to develop a framework that will serve as a tool to assess the degree to which a city, and so urban design, embrace human-centered principles. Taking into account tangible factors, such as design characteristics and public space availability, and non-tangible factors, such as citizen/community engagement, a conceptual framework developed is developed. It is then applied to evaluate the city of Jeddah. The findings indicate that Jeddah is well positioned to support future strategies to improve living standards across physical, social, and economic dimensions. The study highlights as well in which domains urban inclusivity and a city's resilience may still be advanced in view of contributing to a more cohesive and engaged society.

Keywords Quality of life · Urban design · Urban design principles · Wellbeing · Jeddah · Policy-making · SDG-11

A. H. Alganawi · H. Azizalrahman
King Abdulaziz University, Al Jamiaa District, Abdullah Sulayman Street, Jeddah 21589, Saudi Arabia

A. Visvizi (✉)
SGH Warsaw School of Economics, Al. Niepodległości 162, 02-554 Warsaw, Poland
e-mail: avisvi@sgh.waw.pl

Effat University, 8482 Qasr Khouzam, Al-Nazlah Al-Yamaniyah, Jeddah 22332, Saudi Arabia

A. Visvizi et al. (eds.), *Research and Innovation Forum 2025 Volume 1*,
Springer Proceedings in Complexity, https://doi.org/10.1007/978-3-032-09066-9_23

1 Introduction

Over the past decades, the increasing pace of urbanization has substantially altered urban landscapes, in a number of cases leading to a deterioration of quality of life and well-being of cities' inhabitants. Although the imperative of "humanizing cities" has been voiced already in the 1960s [1, 2], challenges remain and a lot needs to be done if cities are to regain their capacity to accommodate the individuals' and the society's needs. Initially, the imperative of "humanizing cities" focused on reducing vehicle-dominated infrastructure. It has since expanded to include broader urban design principles centered on walkability, inclusivity, and citizens' well-being [3]. In this context, the need to restructure urban design around a holistic human-centered agenda geared toward ensuring that people remain the primary agents of their environment has been stressed [4–6]. In this view, a humane city prioritizes well-being of its inhabitants by encouraging social interaction, enhancing accessibility, and fostering a sense of belonging [5].

A 'humanized city' can be defined as an urban environment designed to enhance residents' quality of life by integrating human-centric elements. These elements include (i) built environment (BE), characterized by walkability, efficient transportation systems, green spaces; (ii) infrastructure necessary to foster citizen/community engagement, and (iii) the resulting improved quality of life and well-being [1–8]. Collectively, these factors contribute to creating sustainable, inclusive, and livable urban spaces that cater to the population's diverse needs [9]. In this sense a humanized city is a concept that embodies tangible and intangible components. The tangible components include built environment, such as buildings, infrastructure, like roads and utilities, and public space, like places to gather, e.g. squares, and recreational spaces, e.g. parks. The intangible aspects, directly and indirectly related to the tangible factors, include representations of cultural identity, social stability, accessibility, resilience, and community engagement. These intangible elements represent non-material characteristics that influence residents' interactions and sense of connection to the city. Still, it is necessary to recognizing both dimensions if the human aspect of a city is to be understood. Thus, a humanized city (see Figure 1 for details) is a city where residents' well-being, city infrastructure, and natural elements are effectively integrated in a fashion that fosters social stability, citizen engagement, community building and thus social and economic development [10, 11]. Figure 1 presents a generalized framework for humanized cities based on these tangible and intangible variables and their interrelated components.

Despite the several claims, raised by academia and in public discourse, on the need to humanize urban areas, very little has been done to evaluate how the imperative of humanizing cities is implemented around the globe [12]. For this reason, in this paper an evaluative framework designed to examine how much Jeddah has been humanized is developed. Consequently, gaps and areas for improvement through a comparative analysis with other global cities are identified. By exploring the city's urban fabric through both tangible and intangible dimensions, this paper offers valuable

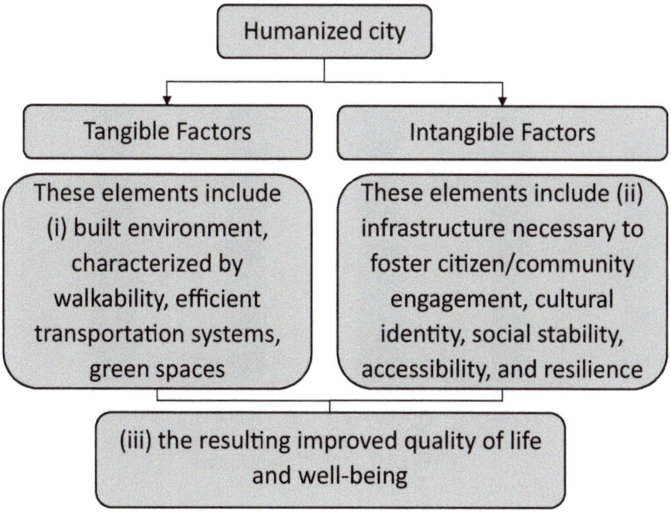

Fig. 1 Tangible and intangible indicators for city evaluation. *Source* Authors

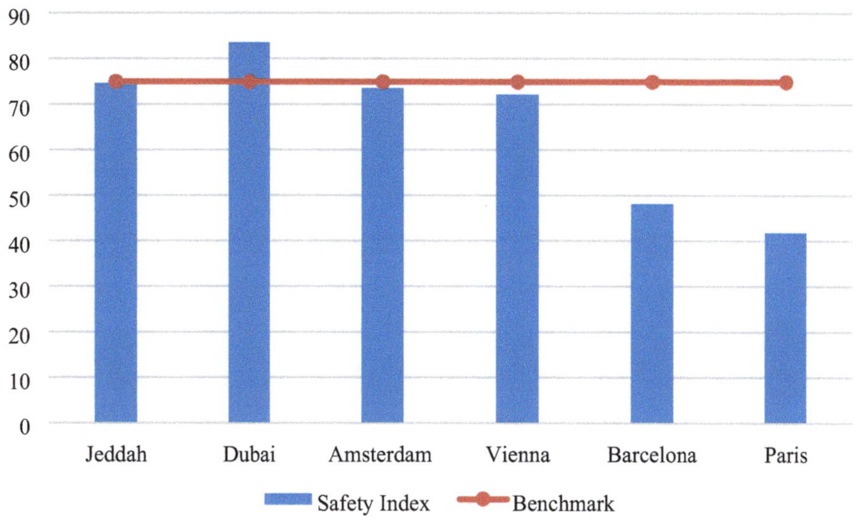

Fig. 2 The safety index. *Source* Authors

and actionable insights into strategies for enhancing social cohesion, accessibility, and overall quality of life for Jeddah's residents.

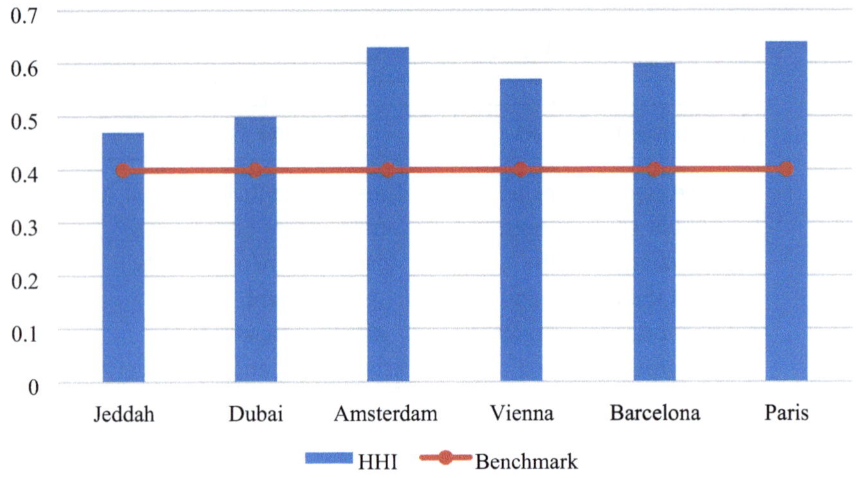

Fig. 3 Economic diversity (Herfindahl–Hirschman Index). *Source* Authors

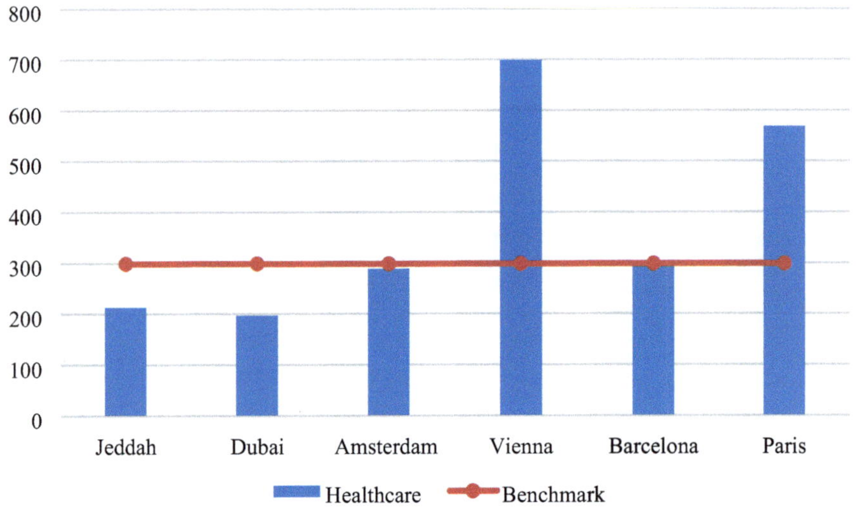

Fig. 4 Graph of healthcare access. *Source* Authors

2 Literature Review

This section explores the key themes, methodologies, and debates within humanizing cities to offer insights into the subject and lay down a foundation for appreciating the research and its potential implications. The concept of "humanizing cities" has evolved significantly over the past several decades, driven by the increasing need to prioritize human experience and well-being in urban environments. Initially rooted

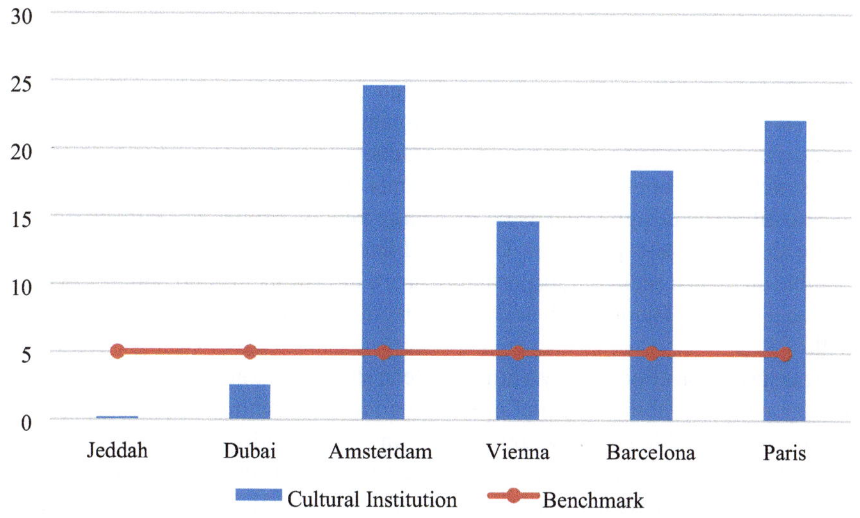

Fig. 5 Graph of cultural institution/gallery. *Source* Authors

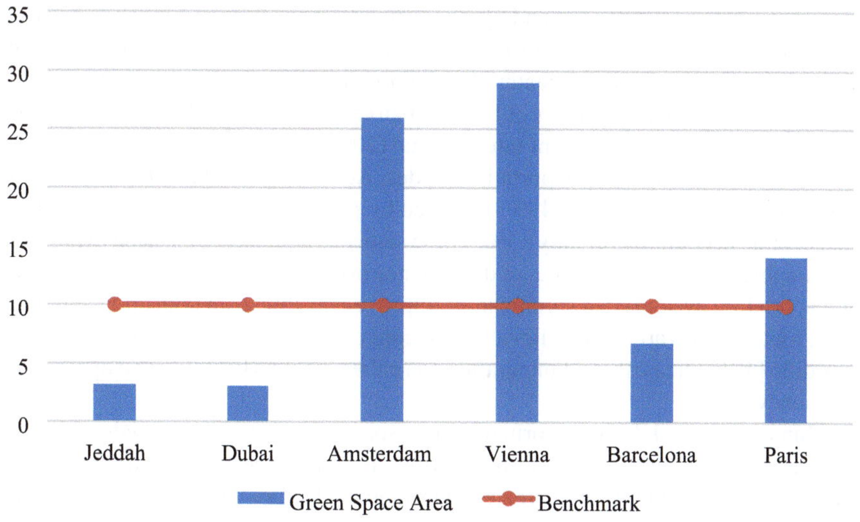

Fig. 6 Green space area. *Source* Authors

in critiques of car-centric urban planning in the 1960s, this idea has expanded to encompass broader urban design principles focused on walkability, inclusivity, and quality of life [1, 13]. Jacobs [7] was among the first urban scholars to challenge modernist urban planning, arguing that cities were designed primarily for automobiles rather than people. Her advocacy for mixed-use neighborhoods, vibrant public spaces, and pedestrian-friendly streets laid the foundation for humanizing cities.

She emphasized that fostering social interactions in dynamic urban environments is essential for building livable communities. Building on Jacobs' work, Speck [14] identified walkability as a core urban design principle rather than a luxury. His analysis highlighted the economic, environmental, and health benefits of designing pedestrian-friendly cities, urging urban planners to shift from vehicle-dominated infrastructure to people-centered environments. Similarly, Gehl [6] reinforced that successful urban design depends on balancing people, nature, and infrastructure to create shared urban experiences. He argued that cities should be structured around tangible elements, such as transportation, green infrastructure, and public spaces, as well as intangible factors, such as cultural identity, social cohesion, and accessibility [10].

Transportation systems play a crucial role in shaping urban experiences and determining a city's livability. Several urban environments continue to prioritize private cars over sustainable mobility options, limiting accessibility and social interaction. Jordan Lewis asserts that streets should be multifunctional public spaces where pedestrians, cyclists, and vehicles coexist harmoniously [6, 15]. This perspective aligns with contemporary urban planning strategies, which advocate for wide sidewalks, street furniture, natural landscapes, and mixed-use transit corridors to enhance human experience [16]. Alternative mobility models, such as the Dutch "Woonerf" street design, integrate pedestrians and cyclists into shared urban spaces while controlling vehicle speeds through surface treatments and traffic-calming measures. Studies have shown that such human-centered transportation strategies improve safety, social engagement, and overall quality of life in urban environments [17]. A global mobility study published in Environment International (2024) surveyed over 800 cities across 60 countries to analyses transportation modes. According to GTU (2024), car usage dominates urban transport at 43%, followed by public transit at 43%, with walking and cycling accounting for only 13%. These statistics underscore the need to promote sustainable mobility and reduce car dependence in cities worldwide [18]. IoT-enabled transportation systems can significantly improve real-time data collection and analysis, route optimization, and alleviation of congestion [9]. Moreover, the findings of Gordillo [19] indicate that AI integration can enhance urban livability and operational efficiency.

Human-centered design prioritizes urban spaces that cater to the needs and experiences of individuals. Abras [20] describes it as a process where users actively influence design outcomes to create environments that are inclusive, walkable, and socially connected. The imperative of human-scale urban planning that fosters accessibility, livability, and equitable development is frequently voiced in the literature [19, 21]. Community participation is critical to ensuring urban sustainability and strengthening social integration. Community engagement involves collaboration with local groups, incorporating their perspectives into urban planning decisions [22]. Auernhammer [23] argues that participatory urbanism enables residents to shape their environments through active involvement in decision-making. Recent urban initiatives, such as the Project for Public Spaces [5] highlight innovative approaches to humanizing cities through placemaking and tactical urbanism. Montgomery [24] argues that successful urban design must accommodate all demographics, including children, the elderly, and underrepresented communities, to create equitable spaces.

Integrating green infrastructure is a vital component of humanized urban environments. Research shows that green spaces—including parks, green paths, and urban vegetation—enhance air quality, mitigate pollution, and regulate temperature fluctuations [25, 26]. Urban areas should allocate at least 15% of their space to green infrastructure, with an additional 30% dedicated to pedestrian-friendly streets [27]. Studies indicate that access to nature improves mental well-being, reduces stress, and fosters physical activity. Tsui [28] asserts that urban green spaces contribute to increased satisfaction among city dwellers by reinforcing ecological and social connectivity.

From a different angle, placemaking emphasizes the creation of safe, culturally rich public spaces that reflect a community's identity. Myers [29] describes placemaking as promoting social engagement, cultural exchange, and collective well-being through interactive urban environments. Chakravarthy [30] further underscores the importance of integrating mixed-use developments that encourage walking, cycling, and public gathering, enhancing the vibrancy of city life. UN-SDG [26] highlights placemaking as an integral component of sustainable development, fostering strong local communities and ensuring urban spaces remain accessible to all. Alanbar and Badawy [31] have supplemented these studies by addressing the role of the park cooling island in mitigating urban heat island (UHI). Despite extensive studies on humanizing cities, several research gaps remain: empirical assessments, interdisciplinary approaches, cultural contexts, and technological integration.

This study draws upon several theoretical frameworks to address these gaps. Applying these frameworks will provide a deeper understanding of how human-centric planning and design can improve Jeddah's livability, sustainability, and social well-being. The following frameworks are thus referenced:

- Human-Centered Design: Prioritizing user experiences in urban spaces [32, 33].
- Social Capital Theory: Exploring how humanized cities foster social networks and community cohesion [34–36].
- Place Attachment Theory: Examining emotional connections between individuals and urban spaces [37].
- Participatory Urbanism: Advocating community-driven urban planning [38, 39].

3 Methods and Materials

This study adopts a mixed-method approach, integrating desk research, conceptual framework building, modelling, and empirical analysis. Mixed-method research is widely used in urban studies due to its ability to combine quantitative and qualitative assessments for spatial planning and livability [40]. The methodology follows a structured framework incorporating:

- Development of a Generic Evaluative Framework—Establishing an analytical framework based on tangible and intangible urban design principles [6, 7, 36, 41].
- Indicator Selection, Validation, and Weighting—Defining and assessing measurable benchmarks for human-centered urban development using established methodologies [41].
- Data Collection and Comparative Analysis—Gathering data from selected cities and validating humanization indices using global standards [1, 42].
- Statistical Methods for Evaluation—Employing normalization techniques, and comparative statistical modeling to ensure reliable findings [43].

The components of the framework are elaborated below.

3.1 A Humanized City: A Framework of Analysis. Dimensions, Benchmarks, Indicators

A multi-criteria evaluation framework for humanizing cities is proposed here, addressing basic dimensions of social, economic and education, healthcare, cultural and environmental, and the quality of life and recreational.

3.2 Applying the Humanized City Framework: Indicator Selection, Normalization, and Weighting

Indicators selected for evaluating urban humanization were drawn from urban planning theories and validated through empirical studies. The selection followed a multi-stage validation process, ensuring relevance across different city contexts [41, 46]. Indicators were derived from urban planning theories, including walkability, accessibility, mixed-use development, and social cohesion [6–8]. Experts from urban planning institutes validated the selection through structured Delphi method consultations [45], Selected indicators were compared against global indices, such as the Global Livability Index and Oxford Economics Global Cities [27, 42, 48]. The data normalization formulae, scoring, and weighting are set out below.

Data Normalization

a. Positive-impact indicators (higher = better)

$$y_{ij} = \frac{x_{ij} - b_i}{b_i}$$

b. Negative-impact indicators (lower = better)

$$y_{ij} = \frac{b_i - x_{ij}}{b_i}$$

where y_{ij} is normalized data of assessed object on ij indicator, x_{ij} is original value of the object on ij indicator, b_i is the benchmark value of the ij indicator.

Scoring

a. Dimension Score for city j

$$S_{cj} = \sum_{i \in c} w_i y_{ij}$$

b. Humanization Index for city j

$$H_j = \sum_{c} \left(w_c \times S_{cj}\right)$$

where H_j is humanization index assessed on city j, w_c is the dimension weight of c, and S_{cj} is a dimension score

3.3 Data Collection and Benchmarking

Data was sourced from reputable urban indices and databases, including Oxford Economics Global Cities Index—A dataset assessing economic and urban development metrics [11]. Global Livability Index—Evaluates infrastructure, healthcare, environmental sustainability, and social cohesion [50]. Local Urban Planning Reports—Provides city-specific datasets for Jeddah and comparison cities [51]. Moreover, the study conducts a comparative benchmarking process, assessing Jeddah's performance relative to cities with strong urban humanization strategies. Selected benchmark cities— Paris, Amsterdam, and Zurich—were chosen based on high rankings in livability indices [42], geographic diversity of urban planning models [1], and urban scale and primary function [46].

3.4 Statistical Methods

To ensure data consistency and comparability, the following statistical techniques were employed: (1) data were normalized using Min-Max Normalization to ensure uniform comparisons [43]. By integrating validated indicators, weighting techniques, and statistical analysis, this methodology establishes a structured, adaptable, and data-driven approach to assessing city humanization.

4 Results and Analysis

The researchers tested the proposed evaluative framework in selected cities: Jeddah, Dubai, Amsterdam, Vienna, Barcelona, and Paris, because of their distinctive characteristics. They compared the city's data with each indicator's reference value to determine whether the city's performance was better, identical, or worse than the benchmark of humanization. First, Raw data are populated in the evaluative framework. Second, normalized values are worked out, followed by weighted score as shown below. A score closer to 1 indicates a better performance model or a higher level of humanization in that specific dimension. A score closer to 0 suggests a lower performance in that dimension. The performance of Jeddah is also graphed to afford a visual perspective on its humanization performance.

Considering the data presented in figures (Figs. 1–7) and tables (1–4), drawing on these, Table 5 (below) offers a comparative evaluation of Jeddah, Dubai, Amsterdam, Vienna, Barcelona, and Paris. Five dimensions of comparison are applied, i.e. social, economic and education, healthcare, cultural and environmental, and quality of life and recreational aspects, of five cities, i.e. those dimensions that were identified as key for the attainment of a humanized city through urban design. Each of these dimensions is assessed on a scale from 0 to 1, whereby the closer the score to 1 the better the performance of a city in a given dimension.

The findings thus presented demonstrate that Jeddah performs relatively well in the economic and education dimension, scoring 0.20, the highest among all five cities included in the analysis, thus suggesting a comparatively strong foundations in economic access and educational provision. In the social dimension, Jeddah

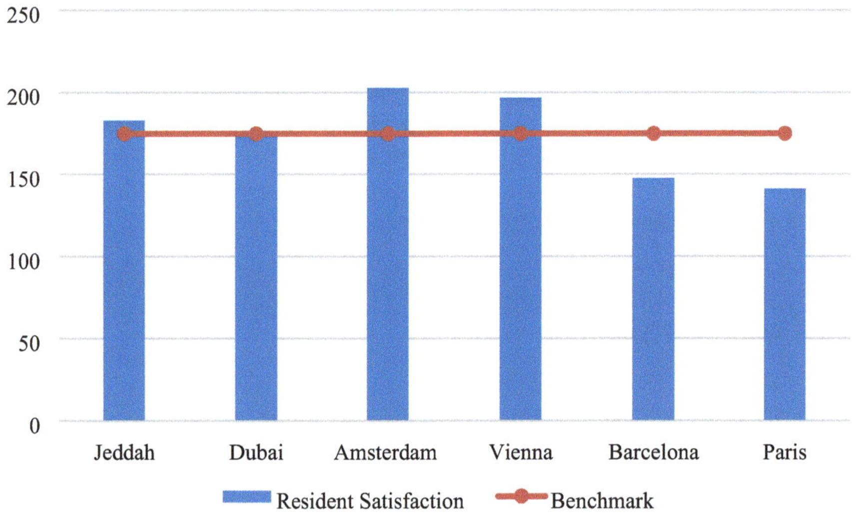

Fig. 7 Resident satisfaction (Quality of Life). *Source* Authors

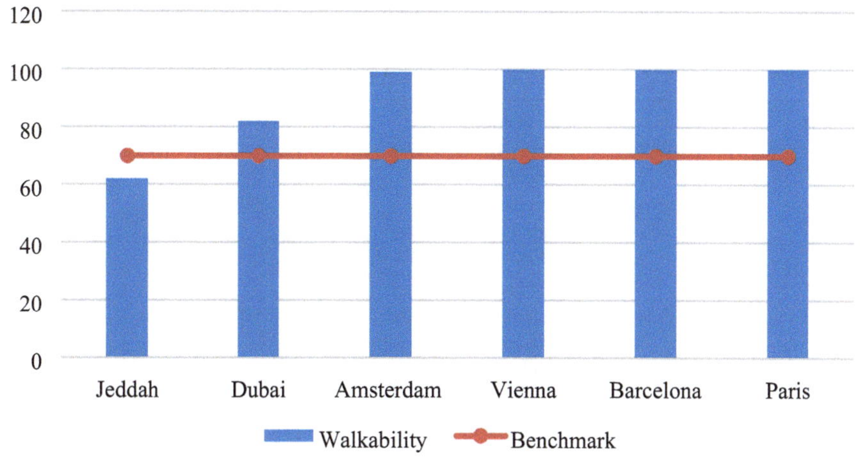

Fig. 8 Walkability. *Source* Authors

also scores 0.15, matching Amsterdam and slightly surpassing Vienna, indicating a moderate degree of social cohesion and inclusion.

However, the data also highlights areas where improvement is needed. That is, Jeddah scores just 0.01 in healthcare and only 0.0004 in the cultural and environmental dimension. These extremely low values signal a significant shortfall in both the provision of health services and the availability or integration of cultural assets and environmental features; all critical aspects of a humanized urban experience. In the domain of quality of life and recreation, Jeddah's score of 0.068 remains well below the benchmark. The latter is set by cities like Amsterdam (0.197) and Vienna (0.19). Jeddah's low score in this domain confirms that there is limited access to recreational spaces and lifestyle-enhancing amenities.

As regards other cities, Amsterdam and Vienna demonstrate consistently high values across most dimensions. Amsterdam, for instance, scores 0.1884 in cultural and environmental, and 0.197 in quality of life and recreation, respectively. This suggests the existence of a robust urban fabric that supports residents well-being in a comprehensive way. Vienna's performance is similarly strong, particularly in healthcare (0.2) and cultural and environmental (0.1592) domains, showcasing the city's long-standing commitment to healthcare provision and, of course, culture. As for Dubai, in some ways, it resembles the case of Jeddah. That is, it has a strong social dimension score (0.2), and decent performance in economic and education (0.16) domain. Yet, similarly as Jeddah, it performs below the benchmark in healthcare and cultural infrastructure. Meanwhile, Paris and Barcelona present mixed profiles. Paris performs well in healthcare (0.14) and cultural and environmental (0.1318) domains. Barcelona's strengths are more moderate but consistent across dimensions, with particularly low social (0.03) and economic (0.04) scores; which is surprising.

Taken together, this analysis illustrates that while Jeddah is positioned to build on its economic and social foundations, its trajectory toward a more human-centered

Table 1 A humanized city: an evaluative framework: dimensions, benchmarks and indicators

Dimension	Indicator	Definition	Effect	Unit	Benchmark	Data source
Social	Safety perception index	Percentage of respondents who rate local authorities	+	0–100%	75	Numbeo (2025a)
	Unemployment rate	Percentage of the working population that is unemployed and actively seeking work	–	0–100%	≤5	World Bank (2024)
	Gini index	Percentage of income inequality within a population	–	0–100%	≤30	World Bank (2019)
Economic and education	Economic diversity	Herfindahl–Hirschman Index of employment across sectors in three schemas	–	0–1	≤0.40	World Integrated Trade Solution (2022)
	Government expenditure on education	Percentage of GDP for expenditure on education	+	0–100%	≥4%	World Bank (2022)
Healthcare	Healthcare access	Number of beds per 100,000 population	+	Count/100,000	≥300 per 100,000	Ministry of Health (2023), World Bank (2020b)
	Life expectancy at birth	Total (years) of life expectancy	+	Years	≥64 years	World Bank (2020c)
Cultural and environmental	Cultural institution/gallery	Number of museums, galleries, theatres per 100,000	+	Count/100,000	≥5 per 100,000	Statista (2021)
	Green space	Area of public parks & pedestrian zones per capita	+	m^2/person	≥10 m^2/person	Ragab Khalil (2014), Russo & Cirella (2018), World Bank (2020a)

(continued)

Table 1 (continued)

Dimension	Indicator	Definition	Effect	Unit	Benchmark	Data source
The quality of life & recreational	Resident satisfaction	Score from annual quality-of-life survey	+	10–230	≥175	Numbeo (2025b)
	Walkability	Score distance to nearest pedestrian/bike network	+	0–100	≥70	Walk Score Survey (2021)

Source Authors, based on Britton et al. (2024) for dimensions of the matrix, and other sources, as indicated. [11, 44, 45]

Table 2 Weighting of indicators

Dimension	Weight (w_c)	Source
Social	20%	The Global Liveability Index 2024
Economic and education	20%	
Healthcare	20%	
Cultural and environmental	20%	
The quality of life and recreational	20%	

Source * Principle of assigning equal weight to each category and The Global Liveability Index 2024

urban environment will require targeted improvements in healthcare, cultural integration, and investment in built environment, including recreational infrastructure. These findings emphasize the necessity for holistic urban planning approaches that not only address physical design but also embrace community engagement and inclusivity, thereby fostering a resilient and livable city for all residents.

5 Discussion and Conclusions

This paper has attempted to establish an evaluative framework for assessing the level to which the city of Jeddah may be considered humanized. The results indicate that while some aspects of Jeddah's urban design lend themselves toward facilitating the humanization of the city, to attain a more human-centered urban environment it is necessary to introduce improvements in healthcare, cultural integration, and built environment. For instance, Jeddah falls below the benchmark in walkability. The proposed humanization framework can significantly enhance urban design by fostering stronger bonds between residents and their surroundings. By prioritizing walkability, inclusive public spaces, and culturally resonant urban interventions, the framework aligns with global best practices that encourage deeper communal ties. Social capital plays a significant role, as humanized cities inherently promote social

Table 3 Result of raw data in the evaluative framework

Dimension	Indicator	Effect	Unit	Benchmark	Formula	y_{ij}	S_{cj}	$H_j = \sum_c (w_c \times S_{cj})$
Social	Safety perception index	+	0–100%	75	$y_{ij} = \frac{x_{ij}-75}{75}$	…	$S_{cj} = \sum_{i \in c} w_i y_{ij}$	*Social* × *20%*
	Unemployment rate	–	0–100%	≤5	$y_{ij} = \frac{5-x_{ij}}{5}$	…		
	Gini index	–	0–100%	≤30	$y_{ij} = \frac{30-x_{ij}}{30}$	…		
Economic and education	Economic diversity	–	0–1	≤0.40	$y_{ij} = \frac{0.40-x_{ij}}{0.40}$	…	$S_{cj} = \sum_{i \in c} w_i y_{ij}$	*Economic & Ed.* × *20%*
	Government expenditure on education	+	0–100%	≥4%	$y_{ij} = \frac{x_{ij}-75}{75}$	…		
Healthcare	Healthcare access	+	Count/100,000	≥300 per 100,000	$y_{ij} = \frac{x_{ij}-300}{300}$	…	$S_{cj} = \sum_{i \in c} w_i y_{ij}$	*Healthcare* × *20%*

(continued)

Table 3 (continued)

Dimension	Indicator	Effect	Unit	Benchmark	Formula	y_{ij}	S_{cj}	$H_j = \sum_c (w_c \times S_{cj})$
	Life expectancy at birth	+	Years	≥64 years	$y_{ij} = \frac{x_{ij}-64}{64}$	…		
Cultural and environmental	Cultural institution/ gallery	+	Count/100,000	≥5 per 100,000	$y_{ij} = \frac{x_{ij}-5}{5}$	…	$S_{cj} = \sum_{i \in c} w_i y_{ij}$	*Cul. & Env.* × *20%*
	Green space	+	m^2/ person	≥10 m^2/ person	$y_{ij} = \frac{x_{ij}-10}{10}$	…		
The quality of life & recreational	Resident satisfaction	+	10–230	≥175	$y_{ij} = \frac{x_{ij}-175}{175}$	…	$S_{cj} = \sum_{i \in c} w_i y_{ij}$	*QuL & Rec.* × *20%*
	Walkability	+	0–100	≥70	$y_{ij} = \frac{x_{ij}-70}{70}$	…		

Source Authors, based on Azizalrahman & Hasyimi, 2018 https://www.sciencedirect.com/science/article/abs/pii/S2210670718307480; data drawn form the following databases

Current Safety Index by City, 2024. https://www.numbeo.com/crime/region_rankings_current (Accessed on May 31, 2025)

Ministry of Health Saudi Arabia, 2023. https://www.moh. gov.sa/en/Ministry/Statistics/book/ Documents/Statistical-Yearbook-2023.pdf (Accessed on May 31, 2025)

Ministry of Health UEA, 2021. https://mohap.gov.ae/documents/20117/1212145/Healthcare (Accessed on May 31, 2025)

https://www.iamsterdam.com/en/see-and-do/museums-and-galleries (Accessed on May 31, 2025)

https://www.barcelona.de/en/barcelona-museums.html (Accessed on May 31, 2025)

https://worldcitiescultureforum.com/city/vienna/ (Accessed on May 31, 2025)

Mostafa and Alshahrani [57]. https://gulfnews.com/uae/government/dubai-municipality-to-create-31-square-metres-of-green-space-per-resident (Accessed on May 31, 2025). https://www.numbeo.com/quality-of-life/in/cities (Accessed on May 31, 2025).https://www.walkscore.com/score (Accessed on May 31, 2025).

Table 4 Normalized results

Dimension	Unit	Jed dah	Dubai	Am sterdam	Vienna	Barcelona	Paris
Social	0–1	0.7 8	1.0 0	0.76	0.7 2	0.15	0.00
Economic and Education	0–1	1.0 0	0.8 2	0.05	0.4 1	0.23	0.00
Healthcare	0–1	0.0 3	0.0 0	0.18	1.0 0	0.20	0.74
Cultural and environmental	0–1	0.0 0	0.0 9	1.00	0.5 9	0.74	0.89
	0–1	0.0 1	0.0 0	0.88	1.0 0	0.14	0.42
The quality of life & recreational	0–1	0.6 7	0.5 6	1.00	0.9 0	0.10	0.00
	0–1	0.0 0	0.5 2	0.97	1.0 0	1.00	1.00

Source Authors

Table 5 The humanized city framework applied: rankings and comparisons

Dimension	Scale*	Jeddah	Dubai	Amsterdam	Vienna	Barcelona	Paris
Social	0–1	0.15	0.20	0.15	0.14	0.03	0
Economic and Education	0–1	0.20	0.16	0.01	0.08	0.04	0
Healthc are	0–1	0.01	0	0.04	0.20	0.04	0.14
Cultural and Environmental	0–1	0.00 04	0.00 97	0.18 84	0.15 92	0.08 92	0.13 18
The quality of life & recreational	0–1	0.06 8	0.10 9	0.19 7	0.19 0	0.11 0	0.10 0

Source Authors; * A score closer to 1 indicates a better performance model or a higher level of humanization in that specific dimension. A score closer to 0 suggests a lower performance in that dimension

interaction, trust, and collective well-being. The framework recognizes the importance of designing urban areas that encourage engagement, collaboration, and accessibility, facilitating interactions that strengthen social networks and contribute to overall urban livability. Beyond theoretical frameworks, this study's practical implications are equally significant for urban planners and policymakers. It highlights how such efforts can bridge the existing gap in prioritizing human needs, particularly in public spaces, streetscapes, service provisions, and community areas. The framework not only establishes design principles but also emphasizes strategies for sustainable implementation, ensuring that urban interventions are both adaptable and responsive to Jeddah's evolving demographic and economic landscape.

Furthermore, while humanization is a universal goal, its execution in Jeddah must consider local governance structures and cultural dynamics. Municipal regulations, decision-making processes, and community engagement play crucial roles in determining the feasibility of urban interventions. Acknowledging these factors provides a clearer understanding of the implementation challenges and ensures that proposed solutions are tailored to Jeddah's socio-political and cultural context. By integrating these considerations, the framework evolves beyond theoretical abstraction and offers a practical, locally relevant model for fostering a human-centric urban environment.

This paper has discussed the ongoing commitment to humanizing urban spaces, emphasizing the need to place people at the center of urban development. It highlights the necessity of balancing physical infrastructure, such as mobility systems and green spaces, with social dynamics like participation and belonging. Further, the study presents an evaluative framework for Jeddah that incorporates the connection of people to their environment and promotes social interactions. It stresses the importance of adapting global humanization models to local contexts, as well as addressing challenges like resource inequity and governance limitations. It advocates for systemic interventions in urban planning that align with social and economic realities, calling for refined policies and community engagement to improve livability in cities like Jeddah. Moreover, it highlights the need for data-driven, adaptable strategies to effectively implement urban humanization. The following recommendations are suggested for Jeddah.

First, enhancing walkability and public spaces, i.e. Jeddah's rapid urban expansion has prioritized vehicular infrastructure at the expense of pedestrian-friendly environments. The city could learn from the 15-minute city concept, propounded in Paris, which prioritizes walkability by ensuring essential services are within a short walking distance. Second, strengthening community engagement in city governance, i.e. while civic participation is gaining momentum in Jeddah, challenges persist regarding inclusive decision-making. A structured framework for participatory governance, drawing inspiration from successful local initiatives like community-driven beautification projects. The superblocks initiative in Barcelona was developed with community input, transforming streets into pedestrian-friendly zones. Third, utilizing evidence-based metrics for evaluating the degree to which cities are humanized, i.e. policymakers should adopt data-driven assessments, such as pedestrian density mapping, resident satisfaction surveys, and environmental impact studies. By referencing past redevelopment projects, including the Jeddah Corniche waterfront enhancements, urban planners can measure successes and refine future humanization interventions. Fourth, prioritizing sustainability and green infrastructure building. Specifically, given Jeddah's arid climate, sustainable urban development must integrate heat-mitigating design principles, such as green rooftops, water-efficient landscaping, and shaded public spaces. Encouraging the expansion of public transportation networks and reducing car dependence. Amsterdam is a catalyst from which Jeddah can benefit to improve a sophisticated cycling infrastructure for a livable and eco-friendly city. Fifth, leveraging technology in an effort to humanize cities, i.e. smart city solutions can improve both public space accessibility and resident engagement [56]. Technologies such as interactive digital kiosks in key districts and real-time air quality monitoring can help communities feel more connected to their environment. Partnerships with local universities and tech startups could foster innovation in adaptive infrastructure tailored to Jeddah's unique urban challenges.

References

1. UN-Habitat: World Cities Report 2016: Urbanization and Development: Emerging Futures. UN-Habitat (2016). https://unhabitat.org/sites/default/files/download-manager-files/WCR-2016-WEB.pdf
2. Visvizi, A., Pérez del Hoyo, R. (Eds.). (2021). Smart Cities and the UN SDGs. Elsevier. https://doi.org/10.1016/C2020-0-01556-2
3. Urban Land Institute: Building healthy places toolkit: strategies for enhancing health in the built environment. Urban Land Institute, Washington, DC (2015). https://uli.org/wp-content/uploads/ULIDocuments/Building-Healthy-Places-Toolkit.pdf
4. Visvizi, A., Lytras, M.D.: Rescaling and refocusing smart cities research: from megacities to smart villages. J. Sci. Technol. Policy Manage. **9**(2), 134–145 (2018). https://doi.org/10.1108/JSTPM-02-2018-0020
5. Project for Public Spaces: How to turn a place around. project for public spaces. https://www.pps.org/product/how-to-turn-a-place-around2#:~:text=How%20to%20Turn%20a%20Place%20Around%20is%20a%20user%2Dfriendly,and%20improve%20their%20public%20spaces
6. Gehl, J.: Cities for people. Island Press (2010). https://www.academia.edu/36365278/CITIES_FOR_PEOPLE?auto=download
7. Jacobs, J.: The death and life of great American cities. Random House (1961). https://www.petkovstudio.com/bg/wp-content/uploads/2017/03/The-Death-and-Life-of-Great-American-Cities_Jane-Jacobs-Complete-book.pdf
8. Beatley, T. What are biophilic cities [PDF]. ResearchGate (2019). https://www.researchgate.net/publication/302586347_Biophilic_Cities_What_Are_They
9. Ushakov, D., Dudukalov, E., Mironenko, E., Shatila, K.: Big data analytics in smart cities' transportation infrastructure modernization. Transport. Res. Proc. **63**, 2385–2391 (2022)
10. Jacobsen, K. S., et al.: The importance of tangible and intangible factors in human–carnivore coexistence. Conserv. Biol. 35(4), 1222–1233. https://www.researchgate.net/publication/347540545_The_importance_of_tangible_and_intangible_factors_in_human-carnivore_coexistence.
11. Oxford Economics: Global cities performance review. Global Cities Index 2024. Oxford Economics (2022). file://C:/Users/vip/Downloads/Oxford_Economics_Global_Cities_Index_2024.pdf
12. Visvizi, A., Wosiek, R., Malik, R.: The smart city competitiveness index (SMCI): conceptualization, modelling, application—An evidence based insight. Land Use Policy 148, 107408 (2025). https://doi.org/10.1016/j.landusepol.2024.107408
13. Visvizi, A., Abdel-Razek, S.A., Wosiek, R., Malik, R.: Conceptualizing walking and walkability in the smart city through a model composite w^2 smart city utility index. Energies **14**(23), 8193 (2021). https://doi.org/10.3390/en14238193
14. Speck, J.: Walkable city rules: 101 steps to making better places. Island Press (2018). https://www.researchgate.net/publication/331761355_Walkable_City_Rules_101_Steps_to_Making_Better_Plac
15. Ivanova, D., Yeralina, E., Shatila, K.: Strategies for sustainable transportation in roadway system in urban areas. In: E3S Web of Conferences (Vol. 389, p. 05001). EDP Sciences (2023)
16. Larice, M., Stein, A. (Eds.). The urban design reader. Routledge (2013). https://citygreen.com/woonerf-street-concept-for-shared-cityspaces/
17. Citygreen: Woonerf: street concept for shared city spaces (2024). Retrieved December 12, 2024, from https://citygreen.com/woonerf-street-concept-for-shared-city-spaces/
18. Global transportation usage: In the ABC of mobility. Environ. Int. (2024). https://www.sciencedirect.com/science/article/pii/S0160412024001272
19. Ramírez-Gordillo, T., Mora, H., Maciá-Lillo, A., Amador, S., Gil, D.: Human-centric solutions and ai in the smart city context: the industry 5.0 perspective. In: Visvizi, A., Troisi, O., Corvello, V. (eds) Research and Innovation Forum 2023. RIIFORUM 2023. Springer Proceedings in Complexity. Springer, Cham. https://doi.org/10.1007/978-3-031-44721-1_16

20. Abras, C., Maloney-Krichmar, D., Preece, J.: User-Centered Design. In Bainbridge, W. Encyclopedia of Human-Computer Interaction. Thousand Oaks: Sage Publications (2004). https://aim.johnkeston.com/wp-content/uploads/2012/01/User-centered_design_encyclopedia_chapter.pdf
21. Almahmood, M., et al.: Human-centred public urban space: exploring how the 're-humanisation' of cities as a universal concept has been adopted and is experienced within the socio-cultural context of Riyadh. Urban Res. Pract. (2018). https://doi.org/10.1080/17535069.2018.1539512file:///C:/Users/vip/Downloads/Humancentredpublicurbanspaceexploringhowtherehumanisation-ofcitiesasauniversalconcepthasbeenadoptedandisexperiencedwithinthe.pdf
22. Clinical and Translational Science Awards Consortium, Community Engagement Key Function Committee Task Force on the Principles of Community Engagement. Principles of community engagement (2nd ed.) (2011). Centers for Disease Control and Prevention. https://stacks.cdc.gov/view/cdc/11699
23. Auernhammer, J.: Human-centered AI: the role of human-centered design research in the development of AI. Proceedings of DRS2020: Synergy 5, 2173–2185 (2020). https://www.researchgate.net/publication/343588407_Human-centered_AI_The_role_of_Humancentered_Design_Research_in_the_development_of_AI
24. Montgomery, C.: Happy city: transforming our lives through urban design (2013). https://happycities.com/the-book
25. Haq, S. M. A.: Urban green spaces and an integrative approach to sustainable environment. J. Environ. Protect. 2, 601–608 (2011). https://www.researchgate.net/publication/276488760_Urban_Green_Spaces_and_an_Integrative_Approach_to_Sustainable_Environment
26. UN-Habitat: Sustainable development goals (2015). [Online] Available at: https://www.undp.org/content/undp/en/home/sustainable-development-goals.html
27. UN-Habitat: Streets as public spaces and drivers of urban prosperity (2013). Nairobi. https://indicators.report/indicators/i-70/
28. Tsui, J.: What will the future of sustainable cities look like. environmental protection 2020. [online] Available at: https://eponline.com/articles/2020/08/12/what-will-the-future-of-sustainable-cities-look-like.aspx. Accessed 31 March 2021
29. Myers, F. R.: Ways of place-making. La Ricerca Folklorica (45), 101–122 (2002). https://www.researchgate.net/publica-tion/271815526_Ways_of_Place-Making
30. Chakravarthy, M.: Cultural Placemaking (2019). https://medium.com/@meerachakravarthy/culturalplacemaking-5ef10b89ebf9
31. Alanbar, M., Badawy, N.: The role of park cooling island (PCI) in mitigating urban heat island (UHI). In: Visvizi, A., Troisi, O., Corvello, V. (eds) Research and innovation forum 2023. RIIFORUM 2023. Springer proceedings in complexity. Springer, Cham (2024). https://doi.org/10.1007/978-3-031-44721-1_18
32. Soni, S., Mason, J., & Sherman, J.: Beyond human-centered design. Child Welfare, 100(1), 81–110 (2022). https://www.jstor.org/stable/48663139
33. Shehayeb, D., Tawfik, M., ElSayed, M., Emad, S., Halawa, E.: Humanizing smart cities: a preconception to a better life for all. In: 2023 2nd international conference on smart cities 4.0 (pp. 218–222). IEEE (2023)
34. Putnam, R. D. (Ed.): Democracies in flux: the evolution of social capital in contemporary society. Oxford University Press (2002)
35. Khandokar, F., Paranagamage, P., Price, A.D.F., Austin, S.A.: Briefing: user-perspectives on walkable neighbourhoods. Urban Design and Planning **162**(4), 155–158 (2009). https://doi.org/10.1680/udap.2009.162.4.155
36. Guan, S., Wang, J.: Research on the optimal design of community public space from the perspective of social capital. Sustainability **15**(12), 9767 (2022). https://doi.org/10.3390/su15129767
37. Cerina, V., Fornara, F., Manca, S. (2017). Architectural style and green spaces predict older adults' evaluations of residential facilities. Eur. J. Ageing 14, 207–217. Morgan (2010)

38. Dyer, M., Corsini, F., Certomà, C.: Making urban design a public participatory goal: toward evidence-based urbanism. Proc. Instit. Civil Eng.-Urban Design Plan. **170**(4), 173–186 (2017)
39. Stein, J., Fisher, S.: Smart citizens, participatory urbanism, and the future of city design. In the New Companion to Urban Design (pp. 549–560). Routledge (2019)
40. Creswell, J. W., Plano Clark, V. L. (2017). Designing and conducting mixed methods research. SAGE Publications
41. Ahas, R., Silm, S., Tiru, M.: Urban livability indicators: assessing quality of life. J Urban Stud. **56**(2), 134–150 (2019)
42. OECD: Better life index: measuring well-being in cities. OECD Publishing (2022)
43. Field, A.: Discovering statistics using SPSS. SAGE Publications (2018)
44. Gehl, J.: Cities for people. Island Press (2010)
45. The Global Liveability Index: Summary report. (2024). Retrieved November 15, 2024, from https://pages.eiu.com/rs/753-RIQ-438/images/TheGlobal-Liveability-Index-Summary-Report-final.pdf
46. World Economic Forum: Global social mobility index 2020: Why economies benefit from fixing inequality [Report]. World Economic Forum 2020. Retrieved from https://www.weforum.org/publications/global-social-mobility-index-2020-why-economies-benefit-from-fixing-inequality
47. Carmona, M., de Magalhaes, C., Natarajan, L.: Public places, urban spaces: the dimensions of urban design. Routledge (2021)
48. Rowe, G., Wright, G.: The Delphi technique as a forecastingtool. Int. J. Forecast. **15**(4), 353–375 (1999). https://doi.org/10.1016/S0169-2070(99)00018-7
49. Saaty, T.L.: The analytic hierarchy process: planning, priority setting. McGraw-Hill, Resource Allocation (1980)
50. Evans, J., Karvonen, A., Luque-Ayala, A., Martin, C., Mccormick, K., Raven, R., Voytenko Palgan, Y.: Smart and sustainable cities? Pipedreams, practicalities and possibilities. Local Environ. **24**(7), 557564 (2019). https://doi.org/10.1080/13549839.2019.1624701
51. Economist Intelligence Unit: Global Livability Index (2021). https://www.eiu.com/n/campaigns/global-liveability-index2024/#:~:text=The%20Global%20Liveability%20Index%202021,ability%20to%20contain%20covid%2D19
52. Mandeli, K.N. (2008). The realities of integrating physical planning and local management into urban development: a case study of Jeddah, Saudi Arabia, Habitat International, Volume 32, Issue 4, 2008, Pages 512–533. https://doi.org/10.1016/j.habitatint.2008.02.005
53. Current Safety Index by City (2024). https://www.numbeo.com/crime/region_rankings_current. Accessed 31 May 2025
54. Ministry of Health Saudi Arabia (2023). https://www.moh.gov.sa/en/Ministry/Statistics/book/Documents/Statistical-Yearbook2023.pdf. Accessed 31 May 2025
55. Ministry of Health UEA, 2021. https://mohap.gov.ae/documents/20117/1212145/Healthcare. Accessed 31 May 2025
56. https://www.iamsterdam.com/en/see-and-do/museums-and-galleries
57. Mostafa, A. M., Alshahrani, A.: Front. Sustain. Cities 6, 1416983 (2024). https://doi.org/10.3389/frsc.2024.1416983
58. Bosco, G., Riccardi, V., Sciarrone, A., D'Amore, R., Visvizi, A.: AI-driven innovation in smart city governance: achieving human-centric and sustainable outcomes. Transform. Govern.: People, Process Policy (2024). https://doi.org/10.1108/TG-04-2024-0096

Medicine and Healthcare Innovation in Increasingly Complex Contexts

Private Equity in US Healthcare: Issues and Developments

Andrew G. Buks and Anna Visvizi

Abstract Private equity (PE) investment in healthcare has surged around the world over the past years. Nevertheless mixed opinions exists as to its value and implications for the healthcare system and the patients. In this context, the case of PE investment in the healthcare sector in the United States (US) has been largely criticized in the literature, whereby the key lines of concern have revolved around such issues as care delivery, provider ownership, and patient experience. Indeed, the US private equity investors spent more than $200 billion on health care acquisitions in 2021; with over $1 trillion invested over the decade 2010–2020. Investment was concentrated in high-margin specialist practices, hospitals, nursing homes, physician groups, and emergency care services. Admittedly, the PE business model is based on maximizing financial returns through leveraged buyouts, cost containment, and short-term asset management. While this model offers potential operational efficiencies, evidence links PE ownership to worsened patient care outcomes, increased healthcare costs, reduced access for vulnerable populations, and operational instability. By focusing on the case of the US, this paper reviews PE's business structure, its sectoral spread, impact on outcomes and costs, and the evolving regulatory landscape. By referencing the case of the European Union (EU) and the EU member-states, this paper seeks to navigate the contingencies of PE investing in the US healthcare, thus adding to the still nascent debate on how to tame the PE potential to the benefit of the society.

Keywords Private equity · Healthcare financing · Health system consolidation · Quality of care · Healthcare costs · Patient outcomes · Health policy reform · SDG-3

A. G. Buks (✉)
Black Hills State University, Spearfish, SD, USA
e-mail: andrew.buks@bhsu.edu

A. Visvizi
SGH Warsaw School of Economics, Warsaw, Poland

Effat University, Jeddah, Saudi Arabia

A. Visvizi
e-mail: avisvi@sgh.waw.pl

A. Visvizi et al. (eds.), *Research and Innovation Forum 2025 Volume 1*,
Springer Proceedings in Complexity, https://doi.org/10.1007/978-3-032-09066-9_24

1 Introduction

Private equity (PE) investment in healthcare has surged around the world over the past years [1, 2]. While a connection between the global financial crisis and the inability of the state to finance healthcare provision may be hypothesized, research highlights substantial variability in this respect around the world [3, 4]. The role of PE in healthcare varies internationally, and thus PE investment in healthcare sector received mixed evaluations as to its value and implications for the healthcare system and the patients [5, 6]. Given the scope of PE investment in the healthcare sector in the United States (US) criticism has become more prominent there [7], rather than in relation to other cases internationally. It the former case, the key lines of criticism have revolved around such issues as care delivery, provider ownership, accessibility, patient experience, as well as ethical considerations [8–11]. Once limited to fringe areas of elective and outpatient care, today PE investment extends to primary care, hospital systems, behavioral health, and hospice. PE firms are drawn to healthcare by its reliable revenue streams, inelastic demand, and fragmented provider markets ready for consolidation [12], as well as tangible assets, e.g., real estate. Of course, other factors weigh in the analysis too. Increased PE activity in the healthcare sector provoked vibrant public and academic debate. Opponents argue that PE firms are driven by short-term financial imperatives, thus neglecting the specificity of an industry traditionally driven by professional ethics and long-term patient care goals, as well as by the imperative of universality. Associated with the latter are the principles of equity, inclusivity, and public responsibility. An apparent tension thus emerges between investors' focus on return and patient care. This tension is central to understanding the broader effects of PE ownership in the healthcare sector. More balanced voices in the debate suggest however that PE investment in the healthcare sector bears the promise of improved efficiency, knowledge sharing, organizational learning, and innovation [10, 13, 14].

Telling in this context is the 2018 Hahnemann University Hospital, a 496-bed safety-net facility in central Philadelphia acquisition [15, 16]. It was acquired by Paladin Healthcare (PH), a PE firm known for distressed hospital investments. At the time of acquisition, the hospital served approximately 150,000 patients annually, many of whom were uninsured or on Medicaid. The hospital operated a top-ranked residency training program affiliated with Drexel University College of Medicine. PH structured the acquisition by separating the hospital's real estate from its operations, placing the physical property into a separately controlled real estate investment vehicle. Shortly after the deal closed, PH sought to lease the facility back to the hospital at higher rates. When operational revenues could not support either the inflated lease payments or the high debt service imposed through the acquisition, PH announced that the hospital would close due to financial un-sustainability [17–19]. Despite public protests, union resistance, and political scrutiny, including a Federal review, the hospital officially ceased operations in September 2019. As a result, tens of thousands of low-income residents were left without a primary emergency and specialty care facility. The closure also displaced nearly 600 medical residents and 2,500 employees, creating ripples throughout Pennsylvania's healthcare labor market.

The case of the US is very specific in that it is underpinned by the historically-rooted (limited) role of the government in the economy and the role of the state in the provision of basic services to its citizens. In the European Union (EU) and the EU member-states, the welfare state started consolidating following the end of the WWI, but the first signs of its emergence may be traced to the last decades of the 19th century [20]. In the US, questions and matters of welfare were, by necessity, taken over by charitable societies [21]. In this sense, substantial differences will exist between PE activity in the healthcare sectors in the US and in the EU. PE is present in the EU and European healthcare markets, particularly in countries like Germany, France, and the Netherlands, as well as in England [22]. Still, its expansion has been limited, as compared to the US. This divergence is largely a function of structural, i.e., fragmentation, and regulatory, i.e., financing models, differences between the EU and the US. Referencing the case of the EU though may offer some useful insights into ways current PE activity in the healthcare sector may be navigated. As the debate on PE in healthcare gathers pace, and the number of publications increases, some first trends and patterns may be identified. Therefore, rather than resorting to an outright critique of PE firm investments in the US healthcare, this paper aligns itself with those authors who argue for in-depth research, data collection, reconsideration of some managerial practices, as well as research-informed dialogue with regulators [23–26]. To this end, the remainder of this paper is structured as follows. Section 2 offers a brief insight into the methods employed and materials used. Section 3 elaborates on the PE business strategies. Section 4 elaborates on societal and cost considerations, while also adding a few points on positive sides of PE. Discussion and conclusions follow.

2 Methods and Materials

This study employs a qualitative, critical literature review methodology, complemented by case study and nested comparative analysis. The aim is to explore how private equity (PE) investment has reshaped the US healthcare sector and what comparative insights may be drawn from the EU context. The research is grounded in a critical review of scholarly and policy literature published between 2010 and 2024. Peer-reviewed articles, empirical studies, and policy reports were sourced from the mainstream data-bases, whereby literature selection was guided by relevance to the structural, financial, and ethical impacts of PE in healthcare. Specifically, it focused on mechanisms of PE investment and financial engineering; impact of PE investment on healthcare access, quality, and workforce; US regulatory gaps and challenges; comparative EU-US healthcare structures and governance.

A nested analytical framework was employed to draw insight from select EU healthcare systems, including Germany, France, and the Netherlands. The value of nested comparative analysis, and so its relevance in the context of this study, is that it combines elements of case study research and cross-case comparison to understand complex social, political, or institutional phenomena [27]. Nested analysis seeks to combine large-N (quantitative or broad comparative) and small-N (in-depth case study) analysis in a structured way to enhance generalizability (via broad comparisons or patterns) and causal depth (via deep case-level exploration). In this way, context-sensitive comparison, focusing not on equivalency but on structurally informed divergence in regulation, public–private balance, and market entry [28] is feasible.

Against critique of PE activity in the US healthcare, in this paper, the comparative lens creates an opportunity to provide a more objective perspective on PE activity in the US, and thus explain the mechanisms underpinning its dynamic growth. In this way, it also creates an opportunity to suggest ways of navigating challenges that it triggers. Limitations of this study include a reliance on secondary English-language data, potentially omitting perspectives from the EU member states with limited English-language reporting. Moreover, the interpretive nature of critical analysis introduces subjectivity. These constraints are mitigated by triangulating multiple sources, employing transparent criteria, and situating findings within their institutional context.

3 A Critical Insight in to the PE Business Model in the Healthcare Sector

PE firms have become dominant actors in healthcare investment, particularly in the US. The PE business model, tends to be defined by its pursuit of short- to medium-term financial returns via leveraged buyouts, aggressive consolidation, and strategic asset extraction [13]. Here, it role in introducing commercial imperatives into a field governed historically by professional ethics and social mission is highlighted. It is also argued that to maximize short-term profitability, PE firms typically implement a series of aggressive cost-cutting strategies: these may include reducing staff (especially registered nurses), limiting capital expenditures (e.g., infrastructure upgrades), and stripping assets by selling real estate and leasing it back at inflated rates.

This tactic, known as asset extraction, prioritizes returns for investors while compromising long-term sustainability and care quality. This section examines the structure and strategies of the PE model in healthcare, with specific attention to three dimensions: financial engineering (including leveraged buyout and extractive strategies), sectoral penetration, and consolidation through platform roll-ups.

The core mechanism of the PE business model in US healthcare is the leveraged buyout (LBO), whereby a firm acquires a healthcare provider primarily with borrowed capital, the repayment of which is typically assigned to the acquired entity itself. This debt-financed growth strategy enables PE firms to achieve high returns on equity while minimizing their own capital exposure. However, it often destabilizes the financial foundations of the provider organization, particularly when revenue projections fail to materialize. In the US this model has been used to restructure hospitals, emergency physician groups, hospice providers, and nursing homes. Notably, the 2006 $33 billion buyout of HCA Healthcare by a consortium including Bain Capital and KKR was, at the time, the largest LBO in history and a paradigmatic example of PE financialization in hospital systems [12].

Extractive strategies such as dividend recapitalizations (where acquired companies take on additional debt to pay investor dividends), management fees, and real estate spinoffs have become common features of PE ownership. These tactics may erode financial resiliency and detract from investments in patient safety, staffing, or infrastructure [29, 30].

What distinguishes the US experience is not merely the volume of PE investment but its comprehensive reach across the care continuum. From emergency medicine and primary care to behavioral health and end-of-life services, PE firms have sought to integrate high-margin and high-volume specialties into consolidated portfolios. PE-backed firms have established dominance in several high-margin, high-volume specialties. In anesthesiology, for example, national groups, such as Envision, manage more than 30% of all hospital anesthesia contracts in the US. Similarly, in radiology, large-scale consolidations by firms like Radiology Partners have resulted in the creation of some of the country's largest physician networks, allowing for centralized billing and broader contracting leverage with insurers [31].

The urgent care sector has also witnessed rapid expansion under PE sponsorship, with firms such as MedExpress and FastMed, leveraging large capital infusions to scale operations across multiple states. This strategy capitalizes on predictable reimbursement streams, demographic trends (e.g., aging population), and gaps in public oversight. Hospice care and behavioral health have become especially attractive to PE due to rising demand and under-regulated billing environments. In 2021 alone, over 30 PE transactions were recorded in the hospice sector, marking a sharp increase in investor interest [32]. Similarly, in behavioral health, firms have targeted addiction treatment centers and youth residential facilities, often with minimal public scrutiny [26, 33].

Another mechanisms PE firms employ for healthcare consolidation is the platform roll-up. This strategy involves acquiring a mid-sized practice or firm, such as a regional dermatology chain, to serve as a platform, followed by the sequential acquisition of smaller entities that are absorbed into the platform structure. The goal is to increase market share, exploit economies of scale, and raise exit valuations for resale or public offering. Roll-up strategies have been particularly aggressive in specialties like dermatology, ophthalmology, fertility, and dentistry. While this strategy supports brand consolidation, administrative unification, and centralized contracting, it can also undermine physician autonomy and standardize care protocols in ways that prioritize profitability over patient needs. Furthermore, the resulting market consolidation can reduce regional competition, driving up prices and limiting consumer choice in both urban and suburban care environments. By bundling multiple practices into a single corporate entity, PE firms may gain regional monopoly power, reduce competition, and engage in price maximization tactics, often at the expense

Table 1 Private equity ownership by healthcare sector [37]

Healthcare sector	Estimated PE ownership (%)
Urgent care	17
Behavioral health	25 (In some states)
Nursing homes	11
Hospitals	10
Dental chains	13
Anesthesia	19
Radiology	6
Fertility clinics	8
Digital health	15

Source Authors, based on Appelbaum and Batt [12], Hoffman and Jackson [38] and Private Equity Stakeholder Project [37]

of patient choice and provider autonomy [34]. While PE advocates argue that roll-ups offer operational efficiency, critics suggest they erode clinical independence and may lead to care standardization driven by cost rather than evidence.

Furthermore, PE firms generally operate on a three-seven-year exit timeline, during which they attempt to grow revenue aggressively through market consolidation, higher pricing, and increased patient volume. Once they achieve their target financial metrics, the firm typically sells the company to another buyer or takes it public, often without regard for the long-term health of the institution [35].

Finally, digital health represents another frontier of PE activity. Companies like One Medical, Carbon Health, and Kindbody, have attracted substantial investment to grow their tech-enabled, app-based care platforms. These ventures promise convenience and broadened access for select populations, especially in metropolitan areas. However, critics argue that the intense focus on user experience, rapid scaling, and market valuation may divert attention from clinical oversight, evidence-based care protocols, and patient data security [36] (Table 1).

As for the EU, the EU member-states operate under healthcare systems that are publicly funded and centrally regulated, with limited room for for-profit delivery in core services. In Germany, for instance, legal restrictions limit hospital ownership by non-medical investors, while in France, medical practice ownership requires direct physician participation [28]. Consequently, PE firms in Europe have concentrated their investments in private-pay niches, such as fertility treatment, dental chains, diagnostics, and outpatient surgery, as well as care-homes. These sectors operate at the margins of national health systems and offer fewer barriers to entry. Furthermore, the European Commission and national competition authorities maintain stronger enforcement mechanisms to prevent excessive consolidation and monopolistic behavior. As mentioned earlier, the EU healthcare market is highly fragmented, with the lines of divisions along national borders, and historically and culturally entrenched financing models. For this reason, in place of platform roll-ups, European PE activity often emphasizes specialist-focused growth strategies, targeting high-income urban markets with relatively stable out-of-pocket payments [39]. While financial engineering is still present, debt levels tend to be lower and recapitalizations rarer, in part due to stronger financial disclosure and solvency rules.

4 Safety, Quality, Cost and Equitable Access Considerations

A central concern in academic debate surrounding PE ownership in healthcare is its impact on the quality of patient care and clinical outcomes. As PE firms aim to maximize returns within a short holding period, the structural emphasis on profitability frequently leads to operational changes that compromise the integrity and safety of care delivery. Recent empirical studies and case reports suggest a consistent pattern of declining patient outcomes following PE acquisition, particularly in high-acuity settings such as hospitals and skilled nursing facilities.

The analysis of over 5 million Medicare hospitalizations across 310 facilities before and after PE acquisition, suggests that there was a 25% increase in hospital-acquired conditions, including infections, falls, and pressure ulcers [40]. The same study documented a 38% rise in central line-associated bloodstream infections and a 100% increase in surgical site infections at PE-owned hospitals compared to non-PE institutions. These outcomes were attributed to staffing reductions, rushed throughput, and degraded infection control protocols, i.e., strategies linked to cost-cutting priorities under PE ownership. Parallel findings have been observed in nursing homes, where the elderly and medically vulnerable are particularly at risk. The mechanisms driving these declines are multifaceted. PE-owned facilities often reduce registered nurse hours, replacing skilled staff with lower-paid aides to save on labor costs [12]. They may also delay or eliminate capital investments in equipment and infrastructure, which can deteriorate over time and increase risks of infection and injury. These operational shifts, though financially logical from an investor standpoint, create clinical environments less equipped to provide safe, consistent, and responsive care. To be just, however, research suggests that PE activity may be associated with reducing the dynamics of full-time employment, followed by partial substitution toward part-time workers. It is also argued that while technology adoption may be restrained, the number of vendors expands, thus motivating the emergence of a cluster. Thus, a case for improved operating margins can be made [41].

As for the negative developments, research highlights the case of behavioral health, where continuity of care and qualified clinicians are critical. Here, PE-backed addiction treatment centers often emphasize volume over therapeutic quality, leading to high readmission rates and inconsistent treatment efficacy [36]. In cosmetic and elective surgery, rapid consolidation of practices by PE firms, with early indications of shortened consultations, reduced aftercare, and increased revision surgeries has been observed [42].

These studies collectively reveal the core tension pertaining to PE investment in healthcare, i.e., the financial incentives that drive PE success often undermine the systems of accountability, patient focus, and long-term investment that define high-quality healthcare delivery. While not all PE-owned entities perform poorly and some may indeed innovate or improve logistics, the general picture is negative, especially when compared with non-PE counterparts [see also: 22].

Given the magnitude of these findings, experts have called for the inclusion of ownership data in national hospital quality dashboards, as well as the creation of post-acquisition monitoring frameworks by public health agencies. Without these interventions, the market is left to govern itself, often to the detriment of patients, especially those who are most vulnerable (Table 2).

One of the most consistent findings across the healthcare economics literature is that PE ownership is associated with increased healthcare costs for both patients and payers. This rise in expenditure is not incidental; it is a structural consequence of the PE business model, which relies on maximizing revenue during short ownership periods. PE firms typically implement strategies such as up-coding, in which services are billed as more intensive than they were in reality; surprise billing, whereby patients are charged out-of-network rates in emergency settings; and the imposition of administrative overhead, often in the form of management service fees paid to entities affiliated with the owning PE firm. These mechanisms extract financial value from the healthcare system without adding measurable improvements to care.

Empirical studies corroborate this trend. For instance, it was found that prices at ambulatory surgical centers rose by approximately 50% within two years of PE acquisition [43]. Likewise, it was reported that outpatient visit costs increased significantly following PE buyouts, even though the quality of care, measured by clinical outcomes and consultation length, remained unchanged [44]. These findings strongly suggest that cost escalation is driven not by innovation or patient-centered enhancement, but by revenue optimization tactics embedded in the PE playbook. These higher charges often result in elevated insurance premiums, greater out-of-pocket spending, and increased financial strain for patients, particularly those with chronic conditions or limited coverage. Importantly, the increase in cost does not correspond to gains in

Table 2 Clinical outcomes before versus after PE acquisition

Metric	Before PE acquisition (%)	After PE acquisition (%)	Change (%)
Hospital-acquired infections	1.5.0	1.6.9	↑ 38
Surgical site infections	1.3.0	1.6.0	↑ 100
Patient falls	1.10.0	1.12.7	↑ 27
Untreated pain (nursing homes)	1.20.0	1.21.6	↑ 8
Mortality rate (nursing homes)	1.12.0	1.13.3	↑ 11

Source Authors, based on Gupta et al. [29] and StepStone Group [40]

Table 3 Cost effects of PE ownership

Service type	Observed change post-PE	Sources
Ambulatory surgery centers	↑ 50% In patient charges	Lin et al. [43]
Physician outpatient visits	↑ Spending due to upcoding and billing intensity	Singh et al. [44]
Overall expenditures	↑ In 9 of 12 studies	Borsa et al. [45]
Medicaid acceptance	↓ Acceptance in multiple specialties	Nie et al. [46]
Surprise billing incidents	↑ In emergency and anesthesia care	Bruch et al. [31], GAO [47]

Source Authors, based on Lin et al. [43], Singh et al. [44], Nie et al. [46] and Borsa et al. [45]

care quality—an imbalance that underscores the extractive nature of PE's approach to healthcare investment (Table 3).

PE's impact on access is shaped by its business model: expand where profitable, exit where not. PE firms tend to open or acquire facilities in commercially dense urban areas, where insurance reimbursement rates are high. Conversely, they are more likely to reduce services or close facilities in rural, underserved, or Medicaid-heavy regions. Found that PE-acquired hospitals reduced their proportion of Medicaid patients over time and closed or outsourced high-cost departments like obstetrics and emergency medicine [37]. PE-owned urology and specialty clinics were significantly less likely to accept Medicaid than non-PE counterparts, restricting access for publicly insured patients [46]. While some PE-backed platforms boast expanded appointment availability and shorter wait times, these improvements are selective, i.e., targeting privately insured, higher-margin patients. As such, PE tends to create parallel systems, i.e., one optimized for billing and profits, the other shrinking in reach, scope, and affordability.

5 Discussion and Conclusions

Private equity (PE) has emerged as a powerful force in accelerating innovation and modernization in healthcare technology. The intersection of rising healthcare costs, demographic shifts, and the digital transformation of care delivery has created fertile ground for PE investment [48]. PE firms are uniquely positioned to provide healthcare tech companies with long-term capital and strategic guidance, helping them scale rapidly in a fragmented and highly regulated sector. These investors apply proven investment strategies—such as buy-and-build models and operational improvement frameworks including revenue cycle management, population health analytics, and AI-powered diagnostics. PE also enables health tech firms to digitize clinical workflows, automate administrative burdens, and deploy predictive analytics to improve patient outcomes. In the U.S., PE investment has had a significant, though nuanced, impact on healthcare delivery across multiple domains. For instance, in the hospice sector, investor-backed acquisitions have contributed to improved palliative care delivery and enhanced digital infrastructure [49]. Similarly, consolidation among plastic surgery practices has led to more standardized care models and better patient access through coordinated administrative systems [42]. In the digital health space, PE has accelerated the adoption of AI tools that support remote diagnostics and telemedicine integration [50]. While concerns about profit motives and quality erosion persist, especially in acute care environments, evidence exists that in targeted, tech-enabled sectors, PE investment fosters modernization and contributes to the delivery of more efficient, accessible, and data-driven healthcare.

Efforts to address private equity's role in healthcare remain fragmented, inconsistent, and largely reactive. Although some states have taken steps to improve transparency and impose oversight, these measures vary significantly in scope and enforcement, creating gaps in protection and accountability. At the federal level, no comprehensive legislation currently exists to track or regulate private equity acquisitions in the health sector. The US Government Accountability Office [47] has reported that existing data systems maintained by the Centers for Medicare & Medicaid Services (CMS) are inadequate to detect and respond to changes in healthcare ownership. The GAO found that many healthcare providers—particularly nursing homes and hospitals acquired by PE firms—use complex and opaque corporate structures that hinder effective oversight. As a result, regulators often do not know who truly owns or controls a facility, making it difficult to link ownership with patient outcomes or financial performance. The GAO recommended the creation of a national ownership registry, which would collect data on mergers, lease agreements, and the financial relationships that structure provider operations.

Several states have moved ahead independently. For instance, Minnesota enacted legislation in 2024 banning private equity firms from owning hospitals and nursing homes within its borders [51]. This law followed a series of closures that occurred within 24 months of PE acquisitions, leaving communities without access to emergency and specialty care. In Indiana, new requirements mandate that the Attorney General review all significant healthcare mergers, especially when PE involvement is suspected. Meanwhile, Oregon has proposed enhancements to its certificate-of-need laws, which would give the state greater authority to scrutinize acquisitions involving financial sponsors with short-term investment strategies [52].

The experience of California illustrates both political opportunity and resistance. In 2023, Assembly Bill 3129 sought to require disclosure of all healthcare transactions involving private equity and to trigger automatic reviews when care institutions changed hands. Despite support from consumer advocacy organizations and physician groups, the bill was vetoed, reportedly after significant lobbying from business interests, including the California Chamber of Commerce and PE industry associations. Critics argued that the bill would deter investment and innovation, while proponents cited growing evidence of quality deterioration in PE-owned entities [36, 41].

In the US, the current landscape leaves patients, providers, and communities vulnerable to the consequences of financial engineering. Sudden facility closures, staffing collapses, and long-term disinvestment are often downstream effects of ownership structures that regulators neither review nor monitor systematically. To counteract this risk, scholars have called for federal legislation that mandates public ownership reporting, establishes national quality audits tied to ownership changes, and includes claw-back provisions for PE firms that shutter facilities soon after acquisition [12, 38, 47] (Table 4).

The rise of PE in healthcare not only raises operational and regulatory concerns, but also challenges the core assumptions about what healthcare is and who it should serve. At the center of this concern lies the seeming incompatibility between the profit-maximization imperative of private investors and the social contract underpinning the provision of healthcare, where the latter prioritizes equity, continuity, and patient dignity. Indeed, PE-owned firms are legally bound to prioritize investor returns. This fiduciary duty often leads to decisions that compromise care quality, including premature discharges, overbooking, staff under-resourcing, and service closures in low-reimbursement areas. In this context, patients become financial units, and clinical staff turn into tools to fulfill performance targets. The structural design of private equity ownership often conflicts with the ethical foundations of healthcare, prioritizing investor returns in ways that can undermine patient-centered care and professional integrity [51].

These dynamics contribute to growing moral distress among clinicians, many of whom report feeling pressured to meet financial quotas that interfere with clinical judgment. Nurses, in particular, report increased workloads and reduced autonomy, leading to burnout and higher turnover in PE-owned facilities [12]. Furthermore, PE ownership introduces new forms of opacity and conflict of interest. Management service agreements between PE firms and provider groups may hide financial relationships, while fragmented corporate structures allow firms to shield assets from liability in the event of malpractice or facility collapse. The broader implications are systemic. That is, if allowed to proceed unchecked, the expansion of PE in healthcare threatens to erode public trust, widen disparities, and create a dual-track system where care is optimized for profit rather than need. It risks converting healthcare from a profession to a portfolio commodity, thus severing the relational and ethical foundations on which effective healthcare depends.

Table 4 Regulatory and legislative response by state

State	Policy action	Status (as of 2024)
Minnesota	Full ban on PE acquisitions in healthcare	Enacted
Indiana	AG review of healthcare mergers	Active
California	Ownership disclosure law (AB 3129)	Vetoed
Oregon	Certificate-of-need + review for hospital buyouts	Proposed (pending)
National (US)	GAO review + proposed federal registry for PE owners	Under review

Source Authors, based on Kannan et al. [41], U.S. Government Accountability Office [47] and Fuse-Brown and Hall [51]

The ethical dilemma is not simply that PE firms prioritize profit. It is also that they do so with limited transparency and minimal accountability. PE's growing footprint in healthcare reflects a broader financialization of essential services. While proponents argue that PE firms inject needed capital and introduce operational discipline, the empirical record tells a more cautionary tale: decreased care quality, rising costs, widened disparities, and increased instability.

Healthcare is a domain where efficiency cannot come at the cost of equity and safety. The PE model's time-limited horizon, i.e., often measured in quarters, not decades, it encourages decisions that may optimize revenue but degrade the infrastructure and ethics of care delivery. Particularly concerning is the erosion of safety-net institutions, which often serve as the first and last resort for the poor, the elderly, and the uninsured. To protect patients, governments must establish stronger regulatory frameworks, including mandatory reporting of ownership changes, pre-acquisition review panels, and post-acquisition audits that assess staffing levels, patient outcomes, and service continuity. Some states have taken meaningful steps, but comprehensive federal action remains essential. Ultimately, healthcare should serve patients, not investors. A system driven by fiduciary duty alone cannot fulfill the moral, professional, and communal responsibilities of medicine. If private equity is to play a role in healthcare's future, it must be reformed, regulated, and held accountable to standards befitting the sector it seeks to profit from. Clearly, more insight and more research are needed to evaluate the implications comprehensively. Also, ways of navigating the challenges need to be discussed very carefully. The discussion need to involve the healthcare professionals, the regulators, as well as patients themselves. Best practices and instances of failure need to be considered and lessons need to be drawn. The EU, or more broadly, European experience, as well as developments elsewhere in the world, may offer an insight too. Future research needs to chart and address these issues.

References

1. Zhang, X., Zimmerman, A., Zhang, Y., Ogbuoji, O., Tang, S.: Rapid growth of private hospitals in China: emerging challenges and opportunities to health sector management. Health Policy **44**, (2024). https://doi.org/10.1016/j.lanwpc.2023.100991
2. Vural, I.E.: Financialisation in health care: an analysis of private equity fund investments in Turkey. Soc. Sci. Med. **187**, 276–286 (2017). https://doi.org/10.1016/j.socscimed.2017.06.008
3. Nguyen, T.C., Castro, V., Wood, J.: The impact of financial crises on social spending: delving into the effects in developed and developing countries. Int. J. Financ. Econ. **29**(4), 4721–4741 (2024). https://doi.org/10.1002/ijfe.2901
4. Piscopo, J., Groot, W., Pavlova, M.: Determinants of public health expenditure in the EU. PLoS ONE **19**(3), e0299359 (2024). https://doi.org/10.1371/journal.pone.0299359
5. Moise, A.D., Popic, T.: Political determinants of health: health care privatization and population health in Europe. J. Health Polit. Policy Law **49**(5), 769–782 (2024). https://doi.org/10.1215/03616878-11257024
6. Kumar, A.: Opportunities in healthcare private equity in India. In: Burns, L.R. (ed.) India's Healthcare Industry: Innovation in Delivery, Financing, and Manufacturing, pp. 424–438. Cambridge University Press (2014)

7. Geyman J.: Private equity looting of U.S. health care: an under-recognized and uncontrolled scourge. Int. J. Soc. Determ. Health Health Serv. **53**(2), 233–238 (2022). https://doi.org/10.1177/00207314221134041
8. Hoffman, S., Ganguli, I.: Access to Primary Care and Health Care Fragmentation, p 2327. Faculty Publications (2026). https://scholarlycommons.law.case.edu/faculty_publications/2327
9. Cai C, Song Z.: A policy framework for the growing influence of private equity in health care delivery. JAMA **329**(18), 1545–1546 (2023). https://doi.org/10.1001/jama.2023.2801. PMID: 37052901; PMCID: PMC10699936
10. Barrett, J., Kernodle, Z.: Private Equity in Healthcare: Exploring the Challenges and Opportunities (2023). https://doi.org/10.17615/bwhw-8r52
11. Katz Olson, L.: Ethically Challenged. Private Equity Storms US Health Care, Johns Hopkins University Press (2022). https://doi.org/10.1353/book.100172
12. Appelbaum, E., Batt, R.: Private Equity Buyouts in Healthcare: Who Wins, Who Loses? Institute for New Economic Thinking (2020). https://doi.org/10.36687/inetwp118
13. Bangerter, L.R.: The role of private equity in health care for older adults. Public Policy & Aging Rep. **33**(2), 41–43 (2023). https://doi.org/10.1093/ppar/prad002
14. Hussain, M., Ahmad, S.Z., Visvizi, A.: Government regulation and organizational effectiveness in the health-care supply chain. Transform. Gov. People Process Policy **16**(4), 649–662 (2022). https://doi.org/10.1108/TG-06-2022-0090
15. Pomorski, C.: The death of Hahnemann Hospital. The New Yorker. https://www.newyorker.com/magazine/2021/06/07/the-death-of-hahnemann-hospital (2021)
16. Lexa, F.J.: Hahnemann—the fast death of an academic medical center and its implications for US radiology. J. Am. Coll.E Radiol. **17**(1, Part A), 82–85 (2020). https://doi.org/10.1016/j.jacr.2019.08.001
17. D'Mello, K.: Hahnemann's closure as a lesson in private equity healthcare. J. Hosp. Med. **15**, 318–320 (2020). https://doi.org/10.12788/jhm.3378
18. Orlowski, J.M., Thompson, T.: Lessons to learn from Hahnemann University Hospital's closure. Acad. Med. **95**(4), 509–511 (2020). https://doi.org/10.1097/ACM.0000000000003170
19. Roberts, L.W.: The closure of Hahnemann University Hospital and the experience of moral injury in academic medicine. Acad. Med. **95**(4), 485–487 (2020). https://doi.org/10.1097/ACM.0000000000003151
20. Stein, K., Sander, A.: The emergence of the western welfare state. In: Castles, F.G., Leibfried, S., Lewis, J., Obinger, H., Pierson, C. (eds.) The Oxford Handbook of the Welfare State, pp. 61–82. Oxford University Press, Oxford (2012). ISBN 978019965051
21. Hall, P.D.: A historical overview of philanthropy, voluntary associations, and NonProfit Organizations in the United States, 1600–2000. In: Powell, W.W., Steinberg, R. (eds.) The NonProfit Sector: A Research Handbook, 2nd edn., pp. 32–65. Yale University Press, New Haven (2006)
22. Bach-Mortensen, A., Goodair, B., Degli Esposti, M.: Involuntary closures of for-profit care homes in England by the Care Quality Commission. Health Policy **5**(4), e297-e302 (2024). https://doi.org/10.1016/S2666-7568(24)00008-4
23. Unruh, L., Rice, T.: Private equity expansion and impacts in United States healthcare. Health Policy **155**, 105266 (2025). https://doi.org/10.1016/j.healthpol.2025.105266
24. Richards, M.R., Shi, M., Whaley, C.M.: Managing margins: PE effects on financial, physical, and human capital working paper 32840. NBER Working Paper Series National Bureau of Economic Research (2024). http://www.nber.org/papers/w32840
25. Gracias, S.: Private equity investment in health care and ineffective antitrust regulations. Univ. Chic. Bus. Law Rev. **3**(2), Article 10 (2024). https://chicagounbound.uchicago.edu/ucblr/vol3/iss2/10
26. Karamardian, M., Jagtiani, E., Chawla, A., Nembhard, I.M.: An update on impacts of private equity ownership in health care: extending a systematic review. In: Health Management, Policy and Innovation (www.HMPI.org), vol. 9, issue 2. https://hmpi.org/2024/06/19/an-update-on-impacts-of-private-equity-ownership-in-health-care-extending-a-systematic-review/ (2024)

27. Lieberman, E.S.: Nested analysis as a mixed-method strategy for comparative research. Am. Polit. Sci. Rev. **99**(3), 435–452 (2005). https://doi.org/10.1017/S0003055405051762
28. Wendt, C., Frisina, L., Rothgang, H.: Healthcare system types: a conceptual framework for comparison. Health Policy **93**(2–3), 85–93 (2009). https://doi.org/10.1111/j.1467-9515.2008.00647.x
29. Gupta, A., Howell, S.T., Yannelis, C., Gupta, R.: Owner incentives and performance in healthcare: private equity investment in nursing homes. Rev. Financ. Stud. **37**(4), 1029–1077 (2024). https://doi.org/10.1093/rfs/hhad032
30. Orewa, G.N., Karabukayeva, A., Pradhan, R., Jimoh, I., Weech-Maldonado, R.: The effects of private equity ownership in U.S. nursing homes quality and financial performance: a systematic review. Health Policy **161**, 105388 (2025). https://doi.org/10.1016/j.healthpol.2025.105388
31. Borsa, A., Bejarano, G., Ellen, M., Bruch, J.D.: Evaluating trends in private equity ownership and impacts on health outcomes, costs, and quality: systematic review. BMJ **382**, e075244 (2023). https://doi.org/10.1136/bmj-2023-075244
32. Braun, R.T., Bond, A.M., Qian, Y., Zhang, M., Casalino, L.P.: Private equity in dermatology: effect on price, utilization, and spending. Health Aff. **40**(5), 719–726 (2021). https://doi.org/10.1377/hlthaff.2020.02062
33. Swenson, P.A.: Ethically challenged: private equity storms health care. New Polit. Sci. (2023). https://doi.org/10.1080/07393148.2023.2181543
34. Ishtiaque Rahman, M.: Understanding private equity-owned HHAs in the U.S.: a performance comparison between PE-owned and non-PE-owned agencies. Health Policy **153**, 105250 (2025). https://doi.org/10.1016/j.healthpol.2025.105250
35. Braun, R.T., McBride, T.D., Pollack, H.A.: Private equity and physician practices: profit maximization and impacts on patient care. JAMA Health Forum **3**(8), e222082 (2022)
36. Singh, Y., Reddy, M., Zhu, J.M.: Life cycle of private equity investments in physician practices: an overview of private equity exits. Health Aff. Sch. **2**(4), qxae047 (2024). https://doi.org/10.1093/haschl/qxae047
37. Private Equity Stakeholder Project.: Private Equity Healthcare Deals: 2024 in Review. https://pestakeholder.org/reports/healthcare-deals-2024-in-review/ (2024)
38. Hoffman, A.K., Jackson, A.: Behavioral health at a crossroads: private equity, addiction treatment, and regulatory gaps. J. Health Polit. Policy Law **47**(6), 787–809 (2022). https://doi.org/10.1215/03616878-9681641
39. Bruch, J.D., Gondi, S., Song, Z.: Changes in hospital income, use, and quality associated with private equity acquisition. JAMA Intern. Med. **180**(11), 1428–1435 (2020). https://doi.org/10.1001/jamainternmed.2020.3552
40. StepStone Group.: European Healthcare Private Equity: Strategic Pathways and Outperformance. https://www.stepstonegroup.com/news-insights/european-healthcare-private-equitys-outperformance-and-strategic-pathways/ (2024)
41. Kannan, S., Bruch, J.D., Song, Z.: Changes in hospital adverse events associated with private equity acquisition. JAMA **330**(24), 2365–2375 (2023). https://doi.org/10.1001/jama.2023.22095
42. Dhawan, R., Brooks, K., Shauly, O., Shay, D.: Increasing private equity investments in plastic surgery practices in the United States: analysis of trends and consolidation. Aesthetic Surg. J. **45**(7), 715–723 (2025). https://academic.oup.com/asj/article-abstract/45/7/715/8066290
43. Lin, H., Munnich, E.L., Richards, M.R., Whaley, C.M., Zhao, X.: Private equity and healthcare firm behavior: evidence from ambulatory surgery centers. J. Health Econ. **91**, 102801 (2023). https://doi.org/10.1016/j.jhealeco.2023.102801
44. Singh, Y., Song, Z., Polsky, D., Bruch, J.D., Zhu, J.M.: Association of private equity acquisition of physician practices with changes in health care spending and utilization. JAMA Health Forum **3**(9), e222886 (2022). https://doi.org/10.1001/jamahealthforum.2022.2886
45. Borsa, A., Bejarano, G., Ellen, M., Bruch, J.D.: Evaluating trends in private equity ownership and impacts on health outcomes, costs, and quality: systematic review. BMJ **382**, e075244 (2023). https://doi.org/10.1136/bmj-2023-075244

46. Nie, J., Hsiang, W., Lokeshwar, S.D., McMahon, G., Demkowicz, P.C., Kenney, P.A., Breyer, B.N., Leapman, M.S.: Association between private equity acquisition of urology practices and physician medicare payments. Urology **174**, 75–81 (2023). https://doi.org/10.1016/j.urology.2022.12.025
47. U.S. Government Accountability Office.: Private Equity in Healthcare: Preliminary Findings on Ownership and Oversight. GAO-23-105563 (2023). https://www.gao.gov. Last accessed 1 July 2025
48. Champagne, D., Davidson, A., Littlejohns, J., Podpolny, D.: Private Equity Opportunities in Healthcare Tech. McKinsey & Company. https://www.mckinsey.de/~/media/McKinsey/Industries/Private%20Equity%20and%20Principal%20Investors/Our%20Insights/Private%20equity%20opportunities%20in%20healthcare%20tech/Private-equity-opportunities-in-healthcare-tech-vF.pdf (2019)
49. Soltoff, A., Williams, D., Braun, R.: Financial and clinical impact of investor-backed hospice acquisitions. J. Pain Symptom Manag. https://www.jpsmjournal.com/article/S0885-3924(25)00161-7/pdf (2025)
50. Karami, M., Madlool, H.: AI and digital health integration in primary health care systems. Front. Digit. Health. https://www.frontiersin.org/journals/digital-health/articles/ (2025). https://doi.org/10.3389/fdgth.2025.1532361/pdf
51. Fuse-Brown, E.C., Hall, M.A.: (2024) Private equity and the corporatization of healthcare. Stanf. Law Rev. **76**(3), 527–596. https://review.law.stanford.edu/wp-content/uploads/sites/3/2024/03/Fuse-Brown-Hall-76-Stan.-L.-Rev.-527.pdf (2024)
52. Debevoise & Plimpton LLP.: Oregon Imposes Significant Restrictions on Private Equity In Healthcare. https://www.debevoise.com/insights/publications/2025/06/oregon-imposes-significant-restrictions-on-private (2025)

Digital Revolution in Pharmaceutical Industry: A Network Analysis of Industry 5.0 Technologies

Mert Mecit Erim, Carlotta D'Alessandro, Katarzyna Szopik-Depczyńska, and Giuseppe Ioppolo

Abstract The pharmaceutical industry is experiencing significant transformation through digital technologies (DT), with Industry 5.0 representing a paradigm shift toward human-centric and sustainable manufacturing. This study presents a network analysis of scientific literature on DT in pharmaceuticals, examining the implementation of Industry 5.0 concepts and identifying research gaps. Through systematic review of 1,410 articles from Scopus (2018–2025), the analysis shows that research on Industry 5.0 in pharmaceuticals remains in early stages. The network analysis identifies "Digitalization," "COVID-19," "Nanotechnology," and "Artificial Intelligence" as dominant concepts. Research gaps exist in the social, psychological, economic, and regulatory domains of Industry 5.0 implementation. This implies certain opportunities for scholars to contribute to understanding the application and impact of Industry 5.0 technologies in this vital sector, particularly in the areas of technological integration, regulatory compliance, and operational optimization.

Keywords Industry 5.0 · Digital technologies · Pharmaceutical industry · SDG-3

M. M. Erim (✉) · C. D'Alessandro · G. Ioppolo
Department of Economics, University of Messina, Messina, Italy
e-mail: mert.erim@studenti.unime.it

C. D'Alessandro
e-mail: carlotta.dalessandro@unime.it

G. Ioppolo
e-mail: giuseppe.ioppolo@unime.it

K. Szopik-Depczyńska
Department of Corporate Management, Institute of Management, University of Szczecin, Szczecin, Poland
e-mail: katarzyna.szopik-depczynska@usz.edu.pl

A. Visvizi et al. (eds.), *Research and Innovation Forum 2025 Volume 1*,
Springer Proceedings in Complexity, https://doi.org/10.1007/978-3-032-09066-9_25

1 Introduction

The world population has grown dramatically, reaching approximately 8.2 billion in 2024 [1], creating mounting challenges for the pharmaceutical sector, particularly in ensuring widespread access to essential medicines, as it strains production capacity, increases demand for diverse therapeutic solutions, and puts pressure on supply chain efficiency. WHO estimates that at least one-third of the world's population lacks access to essential medicines, with this figure reaching up to one-half in poorer regions of Asia and Africa [2]. This access crisis is compounded by inefficiencies in traditional production and supply chain processes, including delays, high costs, and inadequate real-time data integration [3].

Digital Technologies (DTs) have emerged as potential solutions to these challenges, offering ways to increase efficiency, enhance quality, and improve accessibility. Industry 4.0 technologies including artificial intelligence, Internet of Things (IoT), and blockchain are already transforming pharmaceutical operations through automation, interconnectivity, and real-time data processing [4, 5]. However, these technologies face important implementation challenges, including standardization gaps, integration issues, and data security risks [6].

Industry 5.0 represents a paradigm shift beyond Industry 4.0's automation focus, emphasizing human-centric approaches and sustainable practices. It optimizes human–machine interaction for more flexible, personalized production systems [7], though implementation faces challenges including Information Technology (IT) infrastructure limitations and data sharing risks [8]. The COVID-19 pandemic accelerated adoption of these technologies while highlighting their role in supply chain resilience [9]. Comprehensive reviews of Industry 5.0 in pharmaceuticals remain limited, with existing studies either providing minimal coverage of Industry 5.0 implications [9] or focusing on limited geographical contexts [8]. This scarcity of comprehensive analysis, combined with the rapid evolution of DTs and their varying impacts across different aspects of pharmaceutical operations, creates an important need for systematic investigation.

Considering these gaps, this research addresses three fundamental questions:

How are Industry 5.0 DTs currently being implemented across different aspects of pharmaceutical operations?

What are the key challenges and potential benefits of their implementation in the pharmaceutical sector?

Which aspects of this digital transformation remain understudied in the current literature?

Such findings can be used for guiding future technological investments, improving implementation strategies, and ensuring that the pharmaceutical sector's digital transformation effectively addresses the growing global healthcare demands while maintaining focus on human-centric and sustainable approaches.

To address these questions, this study employs a network analysis of scientific literature focused on DT in the pharmaceutical sector, with the aim to systematically map technological trends, applications, and knowledge gaps. Both practitioners and researchers can benefit from the findings of this study and in order to understand the trajectory of pharmaceutical sector digitalization and identify overlooked areas for future investigation.

The paper is structured as follows: Sect. 2 presents the methodology and data collection process. Section 3 analyzes the resulting technology, application networks and discusses the implications of the findings for the sector's ongoing digital transformation. Finally, Sect. 4 concludes with recommendations for future research and practice.

2 Methodology

The research strategy for this study employs a two-step methodological framework designed to systematically investigate DTs in the pharmaceutical sector. The first step involves a systematic literature review to identify and visualize prominent concepts and their basic relationships in the literature. This research was conducted at the end of February 2025. Figure 1 provides a visual representation of the entire methodological process.

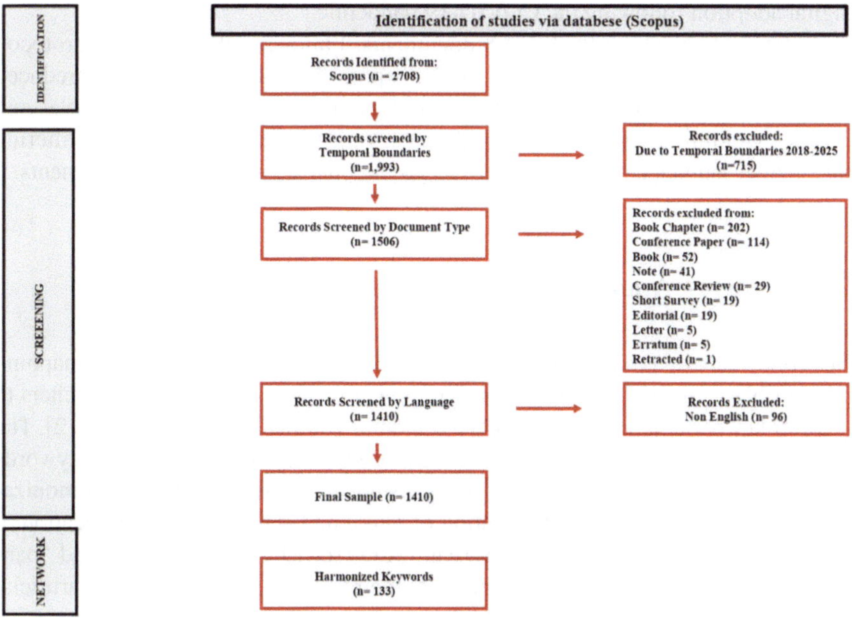

Fig. 1 Visual summary representing of the PRISMA protocol. *Source* Authors

2.1 Search Strategy and Selection Process

When seeking an answer to a specific research question or hypothesis through a systematic search, the aim is to minimize bias and achieve reliable findings by using clear and systematic methods [10]. This process adheres to the Preferred Reporting Items for Systematic Reviews and Meta-Analyses (PRISMA) protocol, which provides a standardized framework for conducting systematic reviews [11]. The PRISMA protocol comprehensive approach favors methodological transparency and reproducibility in literature reviews, which is particularly appropriate for emerging technological fields where research patterns may not yet be well-established.

The search strategy was designed to capture the broad spectrum of digital technology applications in the pharmaceutical sector. The search string combined terms related to DT and pharmaceutical applications: ("digital tech*" OR "smart tech*" OR "emerging tech*" OR "digital transform*") AND ("Pharma*"). The selection of these specific terms was based on preliminary exploratory searches that identified key terminology used in relevant literature.

Scopus was selected due to its comprehensive coverage of peer-reviewed literature and consistent indexing quality in technological and pharmaceutical fields. The timeframe was set to 2018–2025 to capture recent developments in digital transformation, particularly given the emergence of Industry 5.0 concepts and accelerated digital adoption following the COVID-19 pandemic.

The initial search yielded 2,708 documents. Following the PRISMA protocol, several filtering steps were applied. The temporal boundary (2018–2025) reduced the sample to 1,993 documents. Document type filtering retained only articles and reviews, yielding 1,506 documents (868 articles and 638 reviews). Language filtering for English-language publications produced the final sample of 1,410 documents.

2.2 Network Analysis

Network analysis is an approach used to explore complex systems by mapping relationships (edges) between various components (nodes), allowing researchers to uncover key patterns, clusters, and central themes within research literature [12]. The keyword network analysis was conducted using VOSviewer software [13]. Keywords with similar meanings were harmonized to ensure accurate analysis. This harmonization process implied the standardization of plural and singular forms, unification of terms referring to the same concept (e.g., "COVID-19", "coronavirus", and "pandemic") and consolidation of acronyms with their full terms (e.g., "AI" with "artificial intelligence").

A minimum threshold of 4 keyword occurrences was established for inclusion in the network analysis. This threshold was selected based on the final sample size (1,410 documents) to balance between capturing emerging concepts and maintaining network clarity. This approach is in line with similar bibliometric studies in emerging technological fields, where lower thresholds help identify nascent research directions while filtering out incidental terms.

3 Results and Discussions

The network analysis (Fig. 2) identified 133 harmonized keywords, with the most prominent keywords based on node size in the visualization are "Digitalization", "Covid-19", "Nanotechnology", and "Artificial Intelligence".

Digitalization emerges as the dominant concept, connecting with virtually all other nodes in the network. This centrality reflects the ongoing digital transformation across pharmaceutical research and development, highlighting how digital processes now permeate pharmaceutical operations. Industry 5.0 states that the need for digitalization and artificial intelligence-supported technologies is critical to increase production efficiency and flexibility [14].

COVID-19 represents another significant node in the network, interacting with keywords from sectors such as health, social media, and artificial intelligence. COVID 19 has had a devastating impact across the world [15]. During this period, AI applications emerged as crucial tools, helping detect case clusters, predict disease spread, and accelerate drug discovery and vaccine development processes [16].

Nanotechnology is a new application defined as the understanding and control of matter with dimensions between 1 and 100 nm [17]. Nanotechnology has enabled operations at the nanometer scale with pharmaceutical applications including cancer treatment [18], AIDS treatment [19], and nuclear medicine [20]. Applications of nanotechnology in pharmacology are now undeniably linked to the potential of drug targeting [21].

Artificial Intelligence (AI) represents another dominant concept. AI is becoming an integral part of drug discovery, playing a key role in the drug discovery and development value chain from target identification to clinical development [22]. In cancer diagnosis, deep learning has demonstrated its clinical potential by enabling highly accurate tumor classification using microRNA stem-loop cohorts [23].

Drug development emerges as another important concept because it is notably costly, lengthy, and high-risk [24]. One key challenge facing the pharmaceutical industry is maintaining cost-effective drug development programs amid rising costs and efficiency concerns [25]. Artificial intelligence contributes significantly to drug development and repurposing of existing drugs, especially for emerging diseases such as COVID-19 [26].

Industry 5.0 technologies offer promising solutions to these challenges. Industry 5.0 integrates organizational strategies and advanced technologies to create and manage operations and supply chains that are resilient, sustainable, and centered around human needs [27]. The implementation of Industry 5.0 technologies aims to ensure and enhance sustainability in the pharmaceutical supply chain [8].

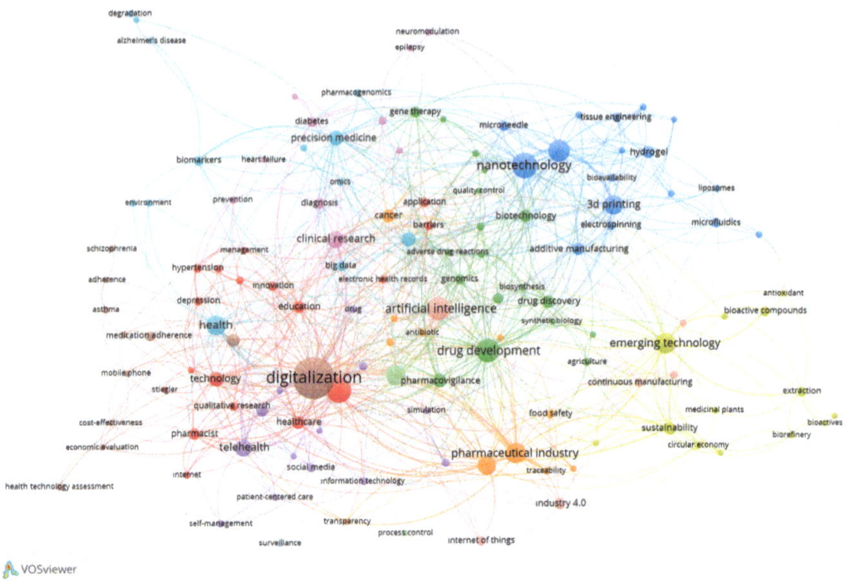

Fig. 2 Visual network map. *Source* Authors

Human-Centric AI (HCAI) emerges as one of the most significant technologies in Industry 5.0, aiming to support human self-sufficiency, enhance creativity, and facilitate social participation [28]. Although the medical field generates vast amounts of data, challenges remain in effectively utilizing this information to identify stakeholder needs. HCAI addresses this gap by employing machine learning-powered natural language processing that works backward from team-generated questions to extract insights from multiple datasets [29].

Personalized medicine represents another key Industry 5.0 technology, tailoring treatment based on individual clinical, genetic, genomic, and environmental factors [30]. This approach enables the linking of molecular and clinical profiles, allowing physicians to make more accurate patient care decisions and empowering patients to make informed lifestyle choices [31]. Essentially, personalized medicine involves administering the right drug at the right dose at the right time [32].

Biomanufacturing leverages biological systems to generate commercially valuable biomolecules for pharmaceutical applications [33]. Its products can also be isolated from natural sources like blood or cultured cells and microbes, using either natural or genetically modified cells, tissues, or enzymes through techniques such as genetic, metabolic, protein, or synthetic engineering [33]. This approach has created a revolution in more efficient, scalable, and sustainable production methods.

Several important Industry 5.0 technologies that did not appear prominently in the network analysis deserve consideration for their potential pharmaceutical applications.

Collaborative robotics involves close human–machine interaction in shared workspaces [34]. Robots are safely delivering laundry, meals, and medications through hospital corridors shared with staff, patients, and visitors [35].

Human digital twin is an emerging technology that could revolutionize the healthcare system with its ability to enable personalized healthcare using various tools such as artificial intelligence and blockchain [36].

Neuromorphic computing mimics the brain's structure to develop energy-efficient hardware that performs complex tasks by leveraging the distributed design of traditional electronics [37]. This technology offers advanced approaches to understanding and treating complex neurological diseases [38].

Nonetheless, there is little attention to social, psychological, economic, and regulatory perspectives on Industry 5.0 implementation in pharmaceuticals.

From a social-psychological perspective, the integration of pharmaceutical variables with social and cultural factors remains understudied. Social psychology explains the mental characteristics of the individual that are directly involved in their social adaptation and then shows how the group itself consists of independent personalities, how it functions as a whole, how it controls the individual [39], and this approach has potential applications in improving drug treatment outcomes [40]. Research and improvement of this aspect is very important because social and psychological factors can significantly influence patient behavior, treatment adherence, and illness perceptions. By considering this together with pharmaceutical variables, healthcare providers can develop more personalized and effective treatment strategies, which in turn improves patient experience and treatment quality.

On the economic side of the pharmaceutical market, drug discovery and development have become more difficult due to conditions such as stricter regulations, evolving treatment standards, and more challenging reimbursement [41]. Stakeholders in the pharmaceutical industry, who see creating well-being in terms of health and quality of life as one of the main goals, state that the process needs to be improved in order for this to be sustainable [42]. In this context, the adoption of Industry 5.0 is of critical importance; because it promises to make great contributions to the sector economically by reducing costs, improving sustainability, improving production processes and increasing operational efficiency [43].

With the recent withdrawal of high-profile drugs from the market, it has become critical to reexamine the current process of drug regulation to protect public health [44]. Research on how emerging technologies like blockchain and AI-supported clinical trials will be evaluated by regulatory authorities is notably lacking, creating uncertainty regarding implementation feasibility.

This study offers important implications for policy makers and sectoral stakeholders. As seen in the existing literature, the applicability of Industry 5.0 technologies is faced with regulatory uncertainties. Policy makers need to develop regulators that will ensure the integration of these technologies and the development of the sector. New technologies that are preparing to enter our lives with Industry 5.0 require new regulators.

In order for Industry 5.0 to be successfully implemented, it is important to include all stakeholders in society and the sector in the process, this will ensure that the decisions to be taken are more inclusive, ethical and applicable.

While the existing literature primarily focuses on Industry 4.0 technologies, research on Industry 5.0 is limited, especially on human-centered approaches, biomanufacturing and human–machine interaction.

These findings contribute to the ongoing discussion on Industry 5.0 and its implications for the pharmaceutical sector. In line with Horvath [45] who noted that the pharmaceutical sector is one of the most innovation-driven sectors in the context of Industry 5.0, our research strengthens the firm independence and responsiveness of digital technologies. Additionally, our study complements the broader discourse on the applications of Blockchain and Blockchain-based applications in the smart city context of Industry 5.0, as discussed by Visvizi et al. [46], by providing a sector-specific perspective grounded in SDG 3.

4 Conclusions

This study analyzes DT in the pharmaceutical sector with a particular focus on Industry 5.0, revealing significant literature gaps and future research opportunities. While Industry 4.0 technologies such as artificial intelligence, big data analytics, and IoT are increasingly adopted in pharmaceutical production and supply chain management [6], Research on Industry 5.0's human-centered approaches, biomanufacturing methods remain under-researched.

With Industry 5.0, technology no longer merely assists in diagnosis or treatment but enables truly personalized approaches. Where traditional methods accepted an "average patient phenotype" for drug production, personalized medicine, a key Industry 5.0 innovation, now enables customized treatment methodologies tailored to individual patient needs.

The investigation of the gaps identified in the current study is extremely important for literature. Investigating how and in which aspects the DT of Industry 5.0 are currently used in the sector plays a critical role in revealing the current status of the sector, how far it has advanced and in which areas improvements are needed. Without investigating this issue, problems arise in completing the transition to Industry 5.0 and successfully implementing strategic planning.

Knowing the basic challenges and potential benefits that Industry 5.0 applications will bring will enable us to determine a realistic and balanced strategy and will help the process to be carried out more healthily. By specifying the current challenges, solutions can be sought, and precautions can be taken in this regard.

It is also extremely important to reveal the areas that are less examined in the current literature. First of all, in order for the integration of this process with society to be successfully ensured, these areas should be examined and the concerns in society should be resolved. It has a very critical place in terms of how the research should continue, where it should focus and guiding the policies to be prepared in this regard.

Industry 5.0 technologies offer promising solutions for these processes while necessitating examination not only at technical and operational levels but also from social, psychological, economic, and regulatory perspectives.

Future research should aim to address the psychological, economic and regulatory dimensions required for the sustainable and human-centered digital transformation of the sector, while also addressing the impact of Industry 5.0 technologies on pharmaceutical manufacturing, patient experiences and global healthcare accessibility, thus closing the current knowledge gaps. How Industry 5.0 can contribute to improving patient experience, increasing drug accessibility and ensuring ethical data management requires further research. Indeed, these aspects need to be investigated and concerns addressed so that the industry and the technologies can integrate successfully with society. It is important to investigate the potential benefits of this transformation for companies in future research, and the importance of companies adopting Industry 5.0 technologies for sustainable and profitable economic growth should be mentioned in the literature.

Acknowledgements Co-financed by Borsa DM 630/2024 Laboratorio farmaceutico Erfo s.p.a. Co-financed by the Minister of Science under the "Regional Excellence Initiative" Program.

References

1. Ritchie, H., Mathieu, E.: How many people die and how many are born each year? Our World in Data. https://ourworldindata.org/births-and-deaths (2023)
2. Goodwin, P.E.: Right idea, wrong result—Canada's access to medicines regime. Am. J. Law Med. **34**(4), 567–584 (2008). https://doi.org/10.1177/009885880803400404
3. Schuhmacher, A., Gassmann, O., Hinder, M.: Changing R&D models in research-based pharmaceutical companies. J. Transl. Med. **14**(1), (2016). https://doi.org/10.1186/s12967-016-0838-4
4. Xu, M., David, J.M., Kim, S.H.: The fourth industrial revolution: opportunities and challenges. Int. J. Financ. Res. **9**(2), 90 (2018). https://doi.org/10.5430/ijfr.v9n2p90
5. Chang, A., El-Rayes, N., Shi, J.: Blockchain technology for supply chain management: a comprehensive review. FinTech **1**(2), 191–205 (2022). https://doi.org/10.3390/fintech1020015
6. Soni, S.J., Patel, A.N.: Digital transformation and Industry 4.0 in pharma manufacturing: the role of IOT, AI, and Big data. J. Integr. Sci. 1–8 (2024). https://doi.org/10.37022/jis.v7i4.92
7. Nashath, Z., Tang, D.Y.Y., Chew, K.W., Show, P.L.: Medicine and pharmaceuticals biomanufacturing–Industry 5.0. Prospect. Ind. 5 Biomanufacturing 135–160 (2021). https://doi.org/10.1201/9781003080671-7-7
8. Agrawal, A., Sharma, A., Srivastava, P.K., Devi, M.S.: Moving Towards Industry 5.0 for Sustainable Supply Chain in the Pharmaceutical Industry. In: 2024 11th International Conference on Computing for Sustainable Global Development (INDIACom), pp. 820–826 (2024). https://doi.org/10.23919/indiacom61295.2024.10498268
9. Mastrantonas, A., Kokkas, P., Chatzopoulos, A., Papoutsidakis, M., Stergiou, C., Vairis, A., Kanetaki, Z.: Identifying the effects of Industry 4.0 in the pharmaceutical sector: achieving the sustainable development goals. Discov. Sustain. **5**(1), (2024). https://doi.org/10.1007/s43621-024-00716-2

10. Moher, D., Liberati, A., Tetzlaff, J., Altman, D.G.: Preferred reporting items for systematic reviews and meta-analyses: the PRISMA statement. Ann. Intern. Med. **151**(4), 264–269 (2009). https://doi.org/10.7326/0003-4819-151-4-200908180-00135
11. Page, M.J., McKenzie, J.E., Bossuyt, P.M., Boutron, I., Hoffmann, T.C., Mulrow, C.D., Shamseer, L., Tetzlaff, J.M., Akl, E.A., Brennan, S.E., Chou, R., Glanville, J., Grimshaw, J.M., Hróbjartsson, A., Lalu, M.M., Li, T., Loder, E.W., Mayo-Wilson, E., McDonald, S., McGuinness, L.A., Stewart, L.A., Thomas, J., Tricco, A.C., Welch, V.A., Whiting, P., Moher, D.: The PRISMA 2020 statement: an updated guideline for reporting systematic reviews. BMJ, n71 (2021). https://doi.org/10.1136/bmj.n71
12. Chiesi, A.M.: Network analysis. International encyclopedia of the social & behavioral sciences, pp. 10499–10502. Elsevier (2001). https://doi.org/10.1016/B0-08-043076-7/04211-X
13. Van Eck, N.J., Waltman, L.: Software survey: VOSviewer, a computer program for bibliometric mapping. Scientometrics **84**(2), 523–538 (2009). https://doi.org/10.1007/s11192-009-0146-3
14. Xu, X., Lu, Y., Vogel-Heuser, B., Wang, L.: Industry 4.0 and Industry 5.0—inception, conception and perception. J. Manuf. Syst. **61**, 530–535 (2021). https://doi.org/10.1016/j.jmsy.2021.10.006
15. Rahman, M.M., Bodrud-Doza, M., Shammi, M., Md Towfiqul Islam, A.R., Moniruzzaman Khan, A.S.: COVID-19 pandemic, dengue epidemic, and climate change vulnerability in Bangladesh: Scenario assessment for strategic management and policy implications. Environ. Res. **192**, 110303 (2021). https://doi.org/10.1016/j.envres.2020.110303
16. Malik, Y.S., Sircar, S., Bhat, S., Ansari, M.I., Pande, T., Kumar, P., Mathapati, B., Balasubramanian, G., Kaushik, R., Natesan, S., Ezzikouri, S., El Zowalaty, M.E., Dhama, K.: How artificial intelligence may help the Covid-19 pandemic: pitfalls and lessons for the future. Rev. Med. Virol. **31**(5), 1–11 (2020). https://doi.org/10.1002/rmv.2205. (Portico)
17. Hulla, J., Sahu, S., Hayes, A.: Nanotechnology. Hum. Exp. Toxicol. **34**(12), 1318–1321 (2015). https://doi.org/10.1177/0960327115603588
18. Cho, K., Wang, X., Nie, S., Chen, Z. (Georgia), Shin, D.M.: Therapeutic nanoparticles for drug delivery in cancer. Clin. Cancer Res. **14**(5), 1310–1316 (2008). https://doi.org/10.1158/1078-0432.ccr-07-1441
19. Rudolph, C., Schillinger, U., Ortiz, A., Tabatt, K., Plank, C., Müller, R.H., Rosenecker, J.: Application of novel solid lipid nanoparticle (SLN)-gene vector formulations based on a dimeric HIV-1 TAT-peptide in vitro and in vivo. Pharm. Res. **21**(9), 1662–1669 (2004). https://doi.org/10.1023/b:pham.0000041463.56768.ec
20. Reilly, R.M.: Carbon nanotubes: potential benefits and risks of nanotechnology in nuclear medicine. J. Nucl. Med. **48**(7), 1039–1042 (2007). https://doi.org/10.2967/jnumed.107.041723
21. Couvreur, P., Vauthier, C.: Nanotechnology: intelligent design to treat complex disease. Pharm. Res. **23**(7), 1417–1450 (2006). https://doi.org/10.1007/s11095-006-0284-8
22. Vijayan, R.S.K., Kihlberg, J., Cross, J.B., Poongavanam, V.: Enhancing preclinical drug discovery with artificial intelligence. Drug Discovery Today **27**(4), 967–984 (2022). https://doi.org/10.1016/j.drudis.2021.11.023
23. Tran, K.A., Kondrashova, O., Bradley, A., Williams, E.D., Pearson, J.V., Waddell, N.: Deep learning in cancer diagnosis, prognosis and treatment selection. Genome Med. 13(1), (2021). https://doi.org/10.1186/s13073-021-00968-x
24. Hinkson, I.V., Madej, B., Stahlberg, E.A.: Accelerating therapeutics for opportunities in medicine: a paradigm shift in drug discovery. Front. Pharmacol. **11**, (2020). https://doi.org/10.3389/fphar.2020.00770
25. Mak, K.-K., Pichika, M.R.: Artificial intelligence in drug development: present status and future prospects. Drug Discov. Today **24**(3), 773–780 (2019). https://doi.org/10.1016/j.drudis.2018.11.014
26. Zhou, Y., Wang, F., Tang, J., Nussinov, R., Cheng, F.: Artificial intelligence in COVID-19 drug repurposing. Lancet Digit. Health **2**(12), e667–e676 (2020). https://doi.org/10.1016/s2589-7500(20)30192-8
27. Ivanov, D.: The Industry 5.0 framework: viability-based integration of the resilience, sustainability, and human-centricity perspectives. Int. J. Prod. Res. **61**(5), 1683–1695 (2022). https://doi.org/10.1080/00207543.2022.2118892

28. Shneiderman, B.: Human-centered artificial intelligence: three fresh ideas. AIS Trans. Hum.-Comput. Interact. 109–124 (2020). https://doi.org/10.17705/1thci.00131
29. Bedenkov, A., Moreno, C., Agustin, L., Jain, N., Newman, A., Feng, L., Kostello, G.: Customer centricity in medical affairs needs human-centric artificial intelligence. Pharm. Med. **35**(1), 21–29 (2021). https://doi.org/10.1007/s40290-020-00378-1
30. Chan, I.S., Ginsburg, G.S.: Personalized medicine: progress and promise. Annu. Rev. Genomics Hum. Genet. **12**(1), 217–244 (2011). https://doi.org/10.1146/annurev-genom-082410-101446
31. Ginsburg, G.: Personalized medicine: revolutionizing drug discovery and patient care. Trends Biotechnol. **19**(12), 491–496 (2001). https://doi.org/10.1016/s0167-7799(01)01814-5
32. Annadurai, K., Danasekaran, R., Mani, G.: Personalized medicine: a paradigm shift towards promising health care. J. Pharm. Bioallied Sci. **8**(1), 77 (2016). https://doi.org/10.4103/0975-7406.171732
33. Zhang, Y.-H.P., Sun, J., Ma, Y.: Biomanufacturing: history and perspective. J. Ind. Microbiol. Biotechnol. **44**(4–5), 773–784 (2017). https://doi.org/10.1007/s10295-016-1863-2
34. Vicentini, F.: Collaborative robotics: a survey. J. Mech. Des. **143**(4), (2020). https://doi.org/10.1115/1.4046238
35. Bloss, R.: Collaborative robots are rapidly providing major improvements in productivity, safety, programing ease, portability and cost while addressing many new applications. Ind. Robot. Int. J. **43**(5), 463–468 (2016). https://doi.org/10.1108/ir-05-2016-0148
36. Okegbile, S.D., Cai, J., Niyato, D., Yi, C.: Human digital twin for personalized healthcare: vision, architecture and future directions. IEEE Netw. **37**(2), 262–269 (2023). https://doi.org/10.1109/mnet.118.2200071
37. Marković, D., Mizrahi, A., Querlioz, D., Grollier, J.: Physics for neuromorphic computing. Nat. Rev. Phys. **2**(9), 499–510 (2020). https://doi.org/10.1038/s42254-020-0208-2
38. Raikar, A.S., Andrew, J., Dessai, P.P., Prabhu, S.M., Jathar, S., Prabhu, A., Naik, M.B., Raikar, G.V.S.: Neuromorphic computing for modeling neurological and psychiatric disorders: implications for drug development. Artif. Intell. Rev. **57**(12), (2024). https://doi.org/10.1007/s10462-024-10948-3
39. Review of Essentials of Social Psychology.: J. Abnorm. Psychol. Soc. Psychol. **17**(1), 104–106 (1922). https://doi.org/10.1037/h0068645
40. Montagne, M.: Social pharmacology: integrating pharmaceutical and social science research on drug effects. Drug Inf. J. **38**(4), 315–320 (2004). https://doi.org/10.1177/009286150403800402
41. Schuhmacher, A.: Pharma innovation: how evolutionary economics is shaping the future of pharma R&D. Drug Discovery Today **29**(12), 104222 (2024). https://doi.org/10.1016/j.drudis.2024.104222
42. Milanesi, M., Runfola, A., Guercini, S.: Pharmaceutical industry riding the wave of sustainability: review and opportunities for future research. J. Clean. Prod. **261**, 121204 (2020). https://doi.org/10.1016/j.jclepro.2020.121204
43. Mukherjee, A.A., Raj, A., Aggarwal, S.: Identification of barriers and their mitigation strategies for industry 5.0 implementation in emerging economies. Int. J. Prod. Econ. **257**, 108770 (2023). https://doi.org/10.1016/j.ijpe.2023.108770
44. Ray, W.A., Stein, C.M.: Reform of drug regulation—beyond an independent drug-safety board. N. Engl. J. Med. **354**(2), 194–201 (2006). https://doi.org/10.1056/nejmsb053432
45. Horvath, K.G.: Riding the innovation wave solo—does Industry 5.0 in the pharmaceutical sector make companies more independent than ever? Res. Innov. Forum **2023**, 489–499 (2024). https://doi.org/10.1007/978-3-031-44721-1_37
46. Visvizi, A., Malik, R., Guazzo, G.M., Çekani, V.: The Industry 5.0 (I50) paradigm, blockchain-based applications and the smart city. Eur. J. Innov. Manag. **28**(1), 5–26 (2024). https://doi.org/10.1108/ejim-09-2023-0826

Interpretable Machine Learning for Metastasis Prediction in Paragangliomas and Pheochromocytomas

Carmen García-Barceló, David Bernabeu Ferrer, David Gil, Higinio Mora, Samanta Ortuño-Miquel, and Pablo Herrero Gómez

Abstract Paragangliomas and pheochromocytomas (PPGLs) are rare neuroendocrine tumors whose metastatic behavior remains difficult to predict using conventional clinical markers. Recent advances in machine learning (ML) offer new opportunities for improving diagnostic support, but their adoption in healthcare is often limited by the opacity of model decision-making. In this work, we present a machine learning framework designed to predict metastatic potential in PPGLs, integrating multiple classification algorithms with explainable artificial intelligence (XAI) techniques. The proposed approach emphasizes model interpretability, combining data preprocessing strategies, classifier evaluation, and SHAP-based analysis to uncover the most influential features driving predictions. Our results confirm that ensemble-based models, particularly Random Forest, provide high accuracy and interpretable outputs, which are essential for clinical acceptance. This study highlights the importance of transparent ML models in supporting oncological decision-making and outlines future directions for enhancing trust and usability in medical applications.

Keywords Machine learning · Explainability · Data science · Tumor · Metastasis · Paragangliomas

C. García-Barceló (✉) · D. B. Ferrer · D. Gil · H. Mora · P. H. Gómez
University Institute for Computer Research, University of Alicante, San Vicente del Raspeig, Spain
e-mail: cgb72@gcloud.ua.es

D. Gil
e-mail: david.gil@ua.es

H. Mora
e-mail: hmora@ua.es

S. Ortuño-Miquel
Alicante Institute for Health and Biomedical Research (ISABIAL), Alicante, Spain
e-mail: ortuno_sam@isabial.es

A. Visvizi et al. (eds.), *Research and Innovation Forum 2025 Volume 1*,
Springer Proceedings in Complexity, https://doi.org/10.1007/978-3-032-09066-9_26

1 Introduction

Neuroendocrine tumors such as paragangliomas and pheochromocytomas (PPGLs) pose significant challenges in clinical oncology due to their unpredictable behavior and relatively low incidence. Despite being uncommon, these tumors can exhibit aggressive progression, including metastasis, which occurs in approximately 20% of cases and is a major contributor to cancer-related mortality [1]. Early and reliable identification of metastatic risk is thus crucial for tailoring treatment strategies and improving prognosis.

Traditional risk assessment methods often rely on anatomical and histopathological features, which are difficult to standardize in clinical practice and frequently lack sufficient predictive accuracy [2, 3]. In recent years, machine learning (ML) has emerged as a powerful alternative, offering data-driven insights into tumor behavior through the analysis of complex and high-dimensional clinical datasets [4]. Nevertheless, the limited interpretability of many ML algorithms—commonly described as "black-box" models—remains a significant barrier to their adoption in critical fields such as oncology.

To overcome this limitation, explainable artificial intelligence (XAI) techniques aim to provide transparency into the inner workings of ML models, enabling healthcare professionals to understand and trust automated predictions. Among the most prominent XAI tools is SHapley Additive exPlanations (SHAP), which attributes model predictions to input features in a consistent and interpretable manner [5].

This paper presents a machine learning pipeline for predicting metastatic risk in PPGLs, integrating SHAP-based explainability to enhance clinical usability. The proposed framework combines preprocessing strategies, multiple classification models, and interpretability techniques to identify the most relevant clinical features and assess their influence on model outputs.

The remainder of this paper is structured as follows: Sect. 2 reviews related work on metastasis prediction and the role of explainability in healthcare ML. Section 3 outlines the dataset, preprocessing procedures, and machine learning methodology. Section 4 presents the experimental results and SHAP-based analysis. Finally, Sect. 5 discusses the conclusions and future research directions.

2 Related Work

The integration of machine learning (ML) in healthcare has led to notable advancements in disease diagnosis, prognosis, and clinical decision support. In oncology, ML techniques are particularly valuable for analyzing heterogeneous and high-dimensional datasets, where traditional statistical methods may fall short [6]. Classification algorithms such as support vector machines, decision trees, and ensemble models have demonstrated strong performance in a variety of cancer prediction tasks.

Previous studies on renal and neuroendocrine tumors have explored ML applications for identifying tumor stages, progression risk, and metastasis potential. For instance, Jagga and Gupta [7] applied supervised learning to RNA sequencing data to classify stages of clear cell renal carcinoma, leveraging The Cancer Genome Atlas (TCGA) as a data source. In the specific case of paragangliomas and pheochromocytomas (PPGLs), most existing prognostic models are based on clinical or histopathological variables such as tumor size, location, and cellular morphology [8, 9]. While useful, these indicators often lack consistency and predictive reliability when applied across diverse clinical settings [10].

To overcome these limitations, Pamporaki et al. [11] proposed a machine learning approach for predicting metastatic risk in PPGLs using a multicenter dataset. Their results highlighted the value of ensemble classifiers—particularly AdaBoost—in improving prediction accuracy. However, the study did not incorporate explainability techniques, leaving open questions about how models arrived at their decisions.

More broadly, the increasing complexity of ML models in medical contexts has driven interest in explainable AI (XAI), which seeks to improve transparency and interpretability. Recent literature reviews confirm that many healthcare applications still lack adequate explainability, despite growing awareness of its importance [12]. Among the available methods, SHAP and Grad-CAM have emerged as widely used tools, offering both global and local insights into model behaviour.

3 Materials and Methods

3.1 *Overview of the Approach*

The methodology proposed in this study follows a modular pipeline designed to predict metastatic risk in PPGLs while ensuring the interpretability of the results. The process consists of four main stages: data preparation, model development, performance evaluation, and application of explainability techniques. A variety of supervised classification algorithms were implemented to explore the behavior of different modeling paradigms, ranging from linear classifiers to ensemble and neural network models. A diagram summarizing the overall framework is presented in Fig. 1.

3.2 *Dataset Description*

The dataset employed in this study was obtained from Pamporaki et al. [12] and is publicly available through *zenodo.org*: https://doi.org/10.5281/zenodo.7749613.

It contains anonymized clinical and biochemical data from 788 patients diagnosed with paragangliomas and pheochromocytomas across seven hospitals in Europe and

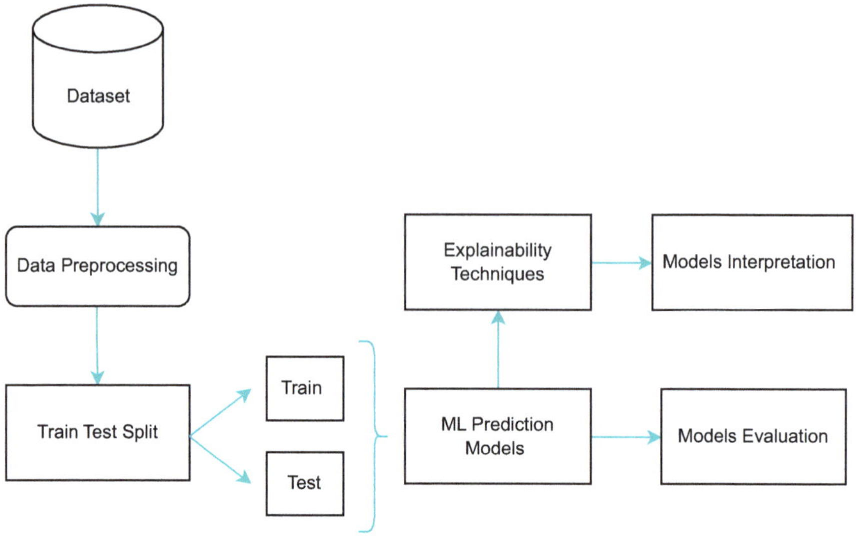

Fig. 1 Overview of the proposed methodology

the United States. Among these, 223 cases were metastatic (28.3%) and 565 non-metastatic (71.7%).

The dataset includes both demographic and clinical variables, such as sex, age, tumor location (adrenal or extra-adrenal), presence of SDHB mutation, tumor category, and prior disease history. It also incorporates biochemical markers—plasma levels of normetanephrine (NMN), metanephrine (MN), and methoxytyramine (MTY)—and a calculated tumor volume. An additional variable was created to group patients into pediatric (<18 years) and adult (≥ 18 years) categories. The target variable is a binary label indicating the presence or absence of metastasis.

3.3 Data Preprocessing

To prepare the data for modeling, several preprocessing steps were carried out:

- **Class imbalance correction**: Random OverSampling was applied to increase the representation of the minority class during training.
- **Missing value imputation**: Numerical variables with missing values were imputed using their mean values.
- **Outlier detection**: An extreme outlier was found in the NMN variable and removed based on physiological plausibility.
- **Categorical encoding**: Categorical variables were transformed using label or one-hot encoding depending on model requirements.

- **Feature scaling**: StandardScaler, MinMaxScaler, and RobustScaler were tested to normalize feature distributions.

3.4 Model Training and Evaluation

Ten classification algorithms were evaluated: Logistic Regression, Support Vector Classifier (linear, RBF, and polynomial kernels), Decision Tree, Random Forest, XGBoost, k-Nearest Neighbors (KNN), Naive Bayes, and Multilayer Perceptron (MLP). Grid search combined with 5-fold cross-validation was used to tune hyperparameters and select the optimal model configurations.

Models were evaluated using the following metrics:

- **Accuracy**: Proportion of correctly classified samples.
- **Balanced Accuracy**: Mean of sensitivity and specificity.
- **Precision**: Ratio of true positives to all predicted positives.
- **Recall (Sensitivity)**: Ratio of true positives to all actual positives.
- **F1-Score**: Harmonic mean of precision and recall.

3.5 Explainability Strategy

To interpret the predictions of the best-performing model, SHAP (SHapley Additive exPlanations) was used. SHAP provides both global and local explanations by assigning contribution values to each input feature. Summary plots and waterfall diagrams were generated to visualize how features influenced the prediction of metastasis risk, enabling a more transparent understanding of the model behavior.

4 Results

4.1 Evaluation of the Models

The predictive performance of each classification model was evaluated using the full feature set described in Sect. 3. Table 1 summarizes the metrics obtained for the ten algorithms tested. As shown, the **Random Forest classifier** consistently achieved the highest scores across all evaluation criteria, with an accuracy of 98.2%, balanced accuracy of 98.3%, and F1-score of 98.2%. It was closely followed by **XGBoost** and **Decision Tree**, which also demonstrated strong results.

Figure 2 shows a visual comparison of model performance in terms of accuracy and balanced accuracy.

Table 1 Results of the models with all available features

Algorithm	Accuracy	Balanced	F1-score	Precision	Recall
RandomForest	0.9823	0.9832	0.9823	0.9829	0.9823
XGBoost	0.9602	0.9603	0.9602	0.9602	0.9602
DecisionTree	0.9425	0.9440	0.9425	0.9444	0.9425
SVC_rbf	0.9336	0.9337	0.9336	0.9337	0.9336
SVC_poly	0.9159	0.9164	0.9160	0.9163	0.9159
LogisticRegression	0.9027	0.9033	0.9027	0.9033	0.9027
SVC_linear	0.8982	0.8986	0.8983	0.8987	0.8982
MLP	0.8982	0.8996	0.8983	0.9001	0.8982
KNN	0.8805	0.8837	0.8804	0.8881	0.8805
BernoulliNB	0.8717	0.8711	0.8716	0.8717	0.8717

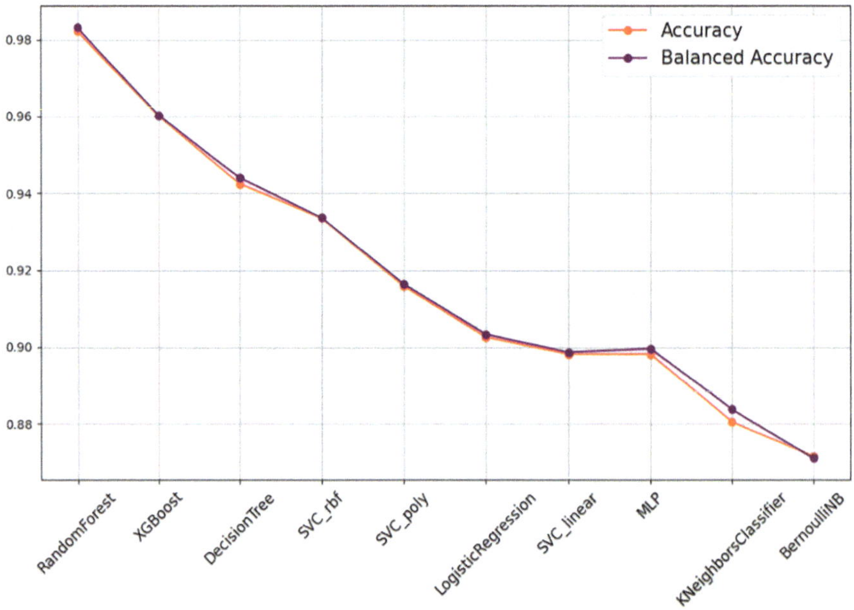

Fig. 2 Comparison of model performance in terms of accuracy and balanced accuracy

4.2 Explainability of the Random Forest Model

To interpret the behavior of the Random Forest classifier, SHAP (SHapley Additive exPlanations) was applied to quantify the influence of each input feature on the model's predictions. This analysis was conducted at both global and individual levels, enabling a comprehensive understanding of the model's decision logic.

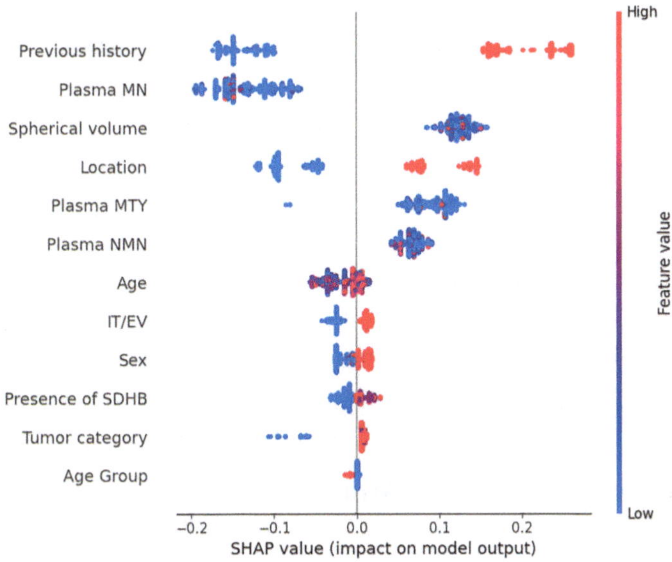

Fig. 3 SHAP summary plot for Random Forest predictions

At the global level, the SHAP summary plot (Fig. 3) reveals the most influential variables across all predictions. The variable *Previous History*, indicating whether the patient had a prior diagnosis of PPGLs, emerged as the strongest predictor of metastatic risk. High values of *Spherical Volume*, consistent with tumor growth beyond 4 cm in diameter, were also associated with an increased likelihood of metastasis. Additional variables contributing positively included extra-adrenal tumor location, presence of SDHB mutations, and elevated levels of plasma methoxytyramine (MTY).

Conversely, low values of MN and NMN, along with adrenal location and absence of prior disease, typically reduced the predicted risk. These findings align with clinical knowledge, reinforcing confidence in the model's learned associations.

To complement the global analysis, SHAP waterfall plots were generated for selected patients. These visualizations illustrate how each feature pushed the prediction above or below the base value (expected model output across the dataset). In one example, the patient's history of previous PPGL, a tumor volume exceeding 400 cm^3 and an extra-adrenal tumor location contributed cumulatively to a high metastasis probability (Fig. 4a). In contrast, a second patient exhibited low volume, no prior disease, and adrenal location—factors that significantly decreased the model's predicted probability (Fig. 4b).

Overall, the SHAP-based analysis confirms that the Random Forest model relies on clinically relevant features to make its predictions. This alignment between model logic and expert understanding enhances the trustworthiness and interpretability of the system in potential medical applications.

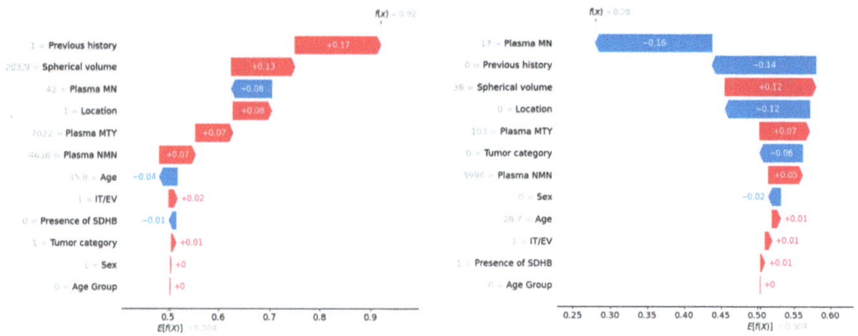

(a) Patient with high metastasis risk. (b) Patient with low metastasis risk.

Fig. 4 SHAP waterfall plots for two different patients

5 Conclusions and Future Work

This study presented a machine learning framework for predicting metastatic risk in paragangliomas and pheochromocytomas (PPGLs), integrating model explainability as a core component of the pipeline. By combining supervised learning algorithms with SHAP-based interpretability, the proposed system not only achieved strong predictive performance—particularly through the Random Forest model—but also provided insights into the clinical relevance of the most influential variables.

The results indicate that a small set of features, such as previous disease history, tumor volume, tumor location, and biochemical markers, carry significant predictive power. The use of SHAP facilitated both global and local interpretation of predictions, revealing that the model's decision-making process aligns with known clinical patterns. This transparency supports the potential adoption of the model in medical decision support scenarios, where interpretability is essential.

Looking ahead, several directions can be explored to enhance the robustness and applicability of this work. First, the evaluation of feature selection techniques could reduce the dimensionality of the input space, potentially improving generalization and reducing data collection costs. Second, validating the model on external datasets from diverse clinical environments would help assess its transferability and reliability. Finally, the integration of additional data modalities—such as genetic profiles or imaging biomarkers—could further enrich the predictive capabilities and support more comprehensive patient assessments.

By advancing both accuracy and interpretability, this research contributes to the broader goal of making artificial intelligence tools more transparent, accountable, and clinically useful in oncology.

References

1. Hanahan, D., Weinberg, R.A.: The hallmarks of cancer. Cell **100**(1), 57–70 (2000)
2. Miranda-Aguirre, A.P., Hernández-García, S., Tenorio-Torres, J.A., Farias-Alarcón, M.A., Erazo-Valle, S.A.: Experiencia en el manejo de los tumores neuroendocrinos en el Servicio de Oncología del Centro Médico Nacional "20 de noviembre", ISSSTE, en los últimos 10 años. Elsevier **11**(1), 18–28 (2012)
3. Neumann, H.P., Young, W.F., Eng, C.: Pheochromocytoma and paraganglioma. N. Engl. J. Med. **381**(6), 552–565 (2019)
4. Von Eschenbach, W.J.: Transparency and the black box problem: why we do not trust AI. Philos. & Technol. **34**(4), 1607–1622 (2021)
5. Bharati, S., Mondal, M.R.H., Podder, P.: A review on explainable artificial intelligence for healthcare: why, how, and when? IEEE Trans. Artif. Intell. **4**(4), 426–442 (2023)
6. Bera, K., Schalper, K.A., Rimm, D.L., Velcheti, V., Madabhushi, A.: Artificial intelligence in digital pathology—new tools for diagnosis and precision oncology. Nature Rev. Clin. Oncol. **16**(11), 703–715 (2019)
7. Jagga, Z., Gupta, D.: Classification models for clear cell renal carcinoma stage progression, based on tumor RNAseq expression trained supervised machine learning algorithms. In: BMC Proceedings vol. 8, pp. 1–7. BioMed Central (2014)
8. Thompson, L.D.: Pheochromocytoma of the Adrenal gland Scaled Score (PASS) to separate benign from malignant neoplasms: a clinicopathologic and immunophenotypic study of 100 cases. Am. J. Surg. Pathol. **26**(5), 551–566 (2002)
9. Kimura, N., Takayanagi, R., Takizawa, N., Itagaki, E., Katabami, T., Kakoi, N., Naruse, M.: Pathological grading for predicting metastasis in phaeochromocytoma and paraganglioma. Endocr. Relat. Cancer **21**(3), 405–414 (2014)
10. Wu, D., Tischler, A.S., Lloyd, R.V., DeLellis, R.A., de Krijger, R., van Nederveen, F., Nosé, V.: Observer variation in the application of the pheochromocytoma of the adrenal gland scaled score. Am. J. Surg. Pathol. **33**(4), 599–608 (2009)
11. Pamporaki, C., Berends, A.M., Filippatos, A., Prodanov, T., Meuter, L., Prejbisz, A., Eisenhofer, G.: Prediction of metastatic pheochromocytoma and paraganglioma: a machine learning modelling study using data from a cross-sectional cohort. Lancet Digit. Health **5**(9), e551–e559 (2023)
12. Allgaier, J., Mulansky, L., Draelos, R.L., Pryss, R.: How does the model make predictions? A systematic literature review on the explainability power of machine learning in healthcare. Artif. Intell. Med. **143**, 102616 (2023)

Enhancing Precision Medicine in Non-small Cell Lung Cancer Diagnosis Through Artificial Intelligence

Samanta Ortuño-Miquel, Cristina Alenda, Claudia Carratalá Lillo, Carmen García-Barceló, Ana Martínez García, and David Gil

Abstract Oncology diagnosis, particularly for Non-Small Cell Lung Cancer (NSCLC), plays a vital role in hospital procedures. However, despite the promising accuracy of existing algorithms, they are still not widely applied in clinical practice. This study aims to address the limitations of current approaches, focusing on NSCLC, a significant global health issue. We emphasize the role of normalization techniques, such as StainGAN and the Ruifrok and Johnston methods, which help reduce variability in histological staining and enhance data consistency for analysis. To improve model performance, we propose utilizing transfer learning with Hematoxylin and Eosin (H&E) stained images, a common diagnostic tool in pathology. Our methodology involves two strategies: analyzing whole-slide images to capture a broad view of molecular and environmental features, and focusing on smaller image patches for more detailed, localized analysis. We will employ well- established models such as ResNet and U-Net based models, chosen for their strengths in feature extraction and image segmentation. Both strategies will undergo thorough evaluation to ensure clinical applicability and robust performance. Ultimately, this

S. Ortuño-Miquel (✉)
Alicante Institute for Health and Biomedical Research (ISABIAL), Alicante, Spain
e-mail: samanta.ortuno@gmail.com

C. Alenda
Department of Pathology, Dr. Balmis General University Hospital, Alicante Institute for Health and Biomedical Research (ISABIAL), Alicante, Spain
e-mail: alenda_cri@gva.es

Present Address:
S. Ortuño-Miquel · C. C. Lillo · C. García-Barceló · A. M. García · D. Gil
Department of Computer Science Technology and Computation, University of Alicante, Alicante, Spain
e-mail: claudiacarratala@gmail.com

C. García-Barceló
e-mail: carmen.gbarcelo@gcloud.ua.es

A. M. García
e-mail: ana.martinezg@ua.es

D. Gil
e-mail: david.gil@ua.es

A. Visvizi et al. (eds.), *Research and Innovation Forum 2025 Volume 1*,
Springer Proceedings in Complexity, https://doi.org/10.1007/978-3-032-09066-9_27

research aims to advance AI-driven pathology for carcinoma to improve diagnostic accuracy, support more informed treatment decisions, and enhance patient outcomes in the context of precision medicine.

Keywords Lung cancer · Image normalization · Transfer learning · ResNet · U-Net · SDG-3 · SDG-9

1 Introduction

Lung cancer is the most prevalent cancer in men and the second most common in women, ranking as the leading cause of cancer-related mortality worldwide for both sexes [1]. It is mainly classified into two main histological types: Non-Small Cell Lung Cancer (NSCLC), which accounts for approximately 85% of the cases, and Small Cell Lung Cancer (SCLC), which makes up the remaining percentage. The use of immunohistochemistry has significantly enhanced the accuracy of histological subclassification, facilitating more precise diagnoses and personalized treatment approaches [2].

The most common subtype of NSCLC are adenocarcinoma (40%), followed by squamous cell carcinomas (25%) and large cell carcinoma (10%), among others [3]. Moreover, the application of immunohistochemistry has allowed for greater accuracy in histological subclassification [2].

Lung cancer is associated with smoking, air pollution (which is increasingly prevalent in society), and previous chronic lung diseases. It is important to note that, in the case of all diseases, the risk factor increases with aging due to telomere shortening and disruptions in DNA damage repair [4].

Recent technological advances have enabled rapid detection of lung cancer, leading to earlier treatment and unquestionably improving patient quality of life and survival rates. One of the most significant breakthroughs is molecular profiling using disease- specific gene panels or Next-Generation Sequencing (NGS), which helps predict clinical responses and prognosis, allowing the development of personalized treatment plans. The most relevant genomic mutations susceptible to targeted therapy include *EGFR*, *ROS1*, *ALK*, *MET*, *RET*, *NTRK* and *KRAS* [5, 6]. However, these techniques present certain challenges to achieving optimal results, such as sample quality, sequencing depth, and data interpretation complexities [7].

2 Literature Review

Artificial Intelligence (AI), particularly Deep Learning (DL), has the potential to over- come the limitations of traditional diagnostic techniques. AI enables a faster and more accurate classification of lung cancer subtypes, enhancing diagnostic precision [8, 9]. Studies have demonstrated the effectiveness of DL algorithms in identifying subtle patterns within Hematoxylin and Eosin (H&E) stained images, allowing for the prediction of molecular characteristics without the need for expensive molecular tests like NGS [10, 11]. Despite these significant advances, AI models have not yet reached the level of reliability required for routine clinical practice.

For these reasons, it is crucial to invest in improving the accuracy of new AI models to support pathologists by reducing diagnostic time and improving molecular profiling predictions. This advancement would enable earlier treatment decisions, ultimately leading to better patient outcomes and improved survival rates.

3 Methods and Materials

3.1 Data

The Cancer Genome Atlas (TCGA[1]) public repository was utilized, specifically selecting the Adenocarcinoma (LUAD) and Squamous Cell Carcinoma (LUSC) projects (All projects). Data were downloaded using R (4.4.2 version) with the TCGAbiolinks package (2.32.0). The retrieved datasets included clinical data, imaging slides and mutation profiling.

3.2 Model

The Accurate, Reliable and Active Convolutional Neural Network (ARA-CNN) employed in this study is a Bayesian convolutional neural network designed for histopathological image classification, integrating variational dropout for uncertainty estimation. The architecture comprises sequential convolutional layers with ReLU activation, batch normalization, max-pooling, and variational dropout, followed by fully connected layers leading to a softmax output. The model was trained using the Adam optimizer with an initial learning rate of 0.001, a batch size of 64, and categorical cross-entropy loss over 50 epochs, with early stopping applied.

3.3 Image Normalization

To standardize size and color variations caused by different tissue types and imaging instruments, Python (3.12.8) was used along with the following packages: staintools (2.1.2), openslide-python (1.4.1), numpy (2.2.2), matplotlib (3.10.0), pillow (11.1.0)

[1] https://www.cancer.gov/ccg/research/genome-sequencing/tcga.

and pandas (2.2.3). The Reinhard model for stain normalization, as implemented in StainGAN, was applied to achieve consistency across samples.

4 Findings

4.1 Model Training

The first attempt involved reproducing the ARA-CNN model [12] on our WSI slides. However, we observed that the classification must be improved (Fig. 1). Upon investigation, we found that the results are due to using the entire WSI, whereas the model was designed to process patches as input. Another factor that contributed to the problem was the disparity between the digitalized tissue samples (needle biopsy, bronchoscopy and surgical biopsy), particularly in terms of size and color (could be observed in Fig. 2). Based on these observations, we concluded that normalizing the size and color before training the model was essential to achieve more accurate results and facilitate a smoother workflow in image analysis.

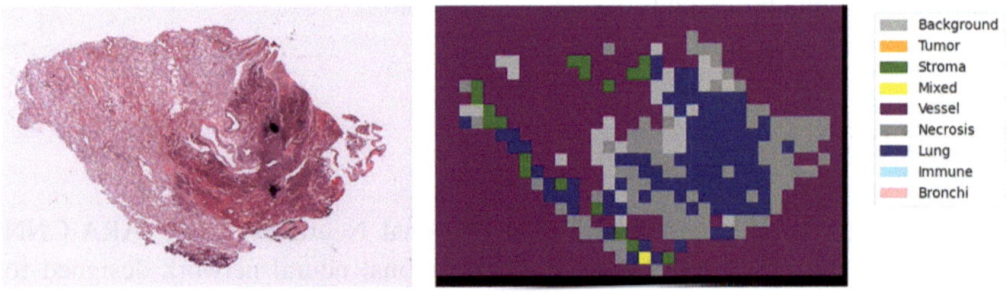

Fig. 1 Reproduced the ARA-CNN model across our slides. *Source* Authors

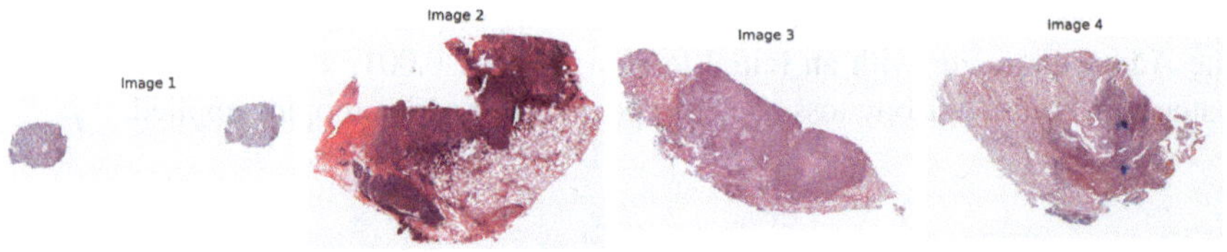

Fig. 2 Original slides. *Source* Authors

4.2 Resize and Color Normalization

Image normalization is a crucial preprocessing step before the model due to variations in the tools used to obtain the images, differences in staining, and variations in tissue sectioning.

It was modify the original slides from the TCGA repository (Fig. 2) by resizing them to the minimum (Fig. 3) and maximum (Fig. 4) dimensions among all slides. However, as observed, this approach is not suitable due to structural changes in the cells, which are critical features for training the model to classify tissue as healthy or tumoral.

Trying to improve this approach, we detected the background color in the original images (Fig. 2) and filled in the images to achieve the maximum size. This approach addresses the cell structure deformation but results in a more pronounced pink background, which could potentially be classified as a feature of the tissue. Additionally, after resizing and normalizing, it was observed that the original tissue of the smaller slides had a stronger purple hue compared to the others (Images 1 and 3 in Fig. 5), which could affect the accuracy of the classification model.

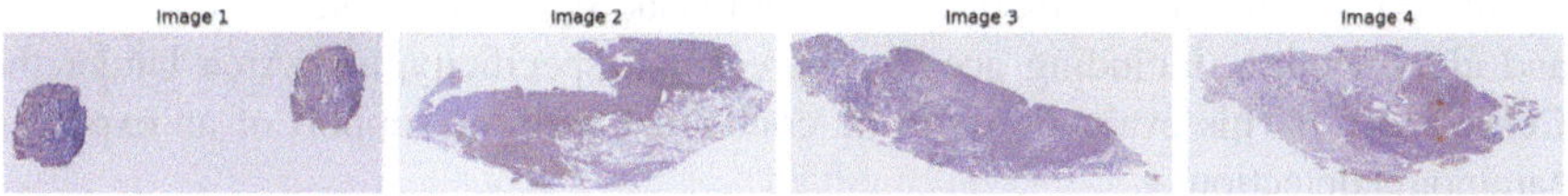

Fig. 3 Resizing all images to the smallest size among the slides to normalize their color. *Source* Authors

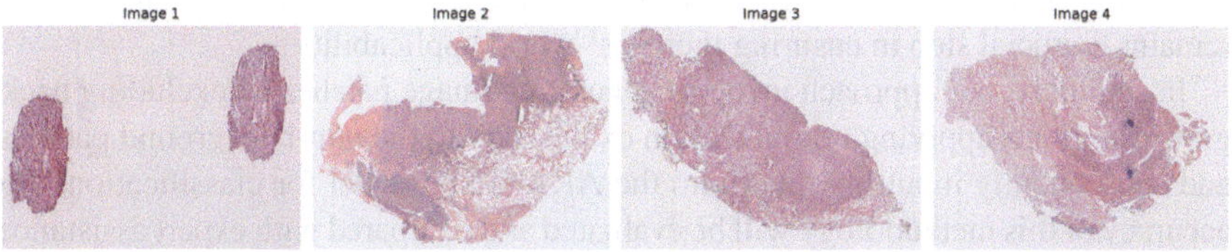

Fig. 4 Resizing all images to the largest size among the slides to normalize their color. *Source* Authors

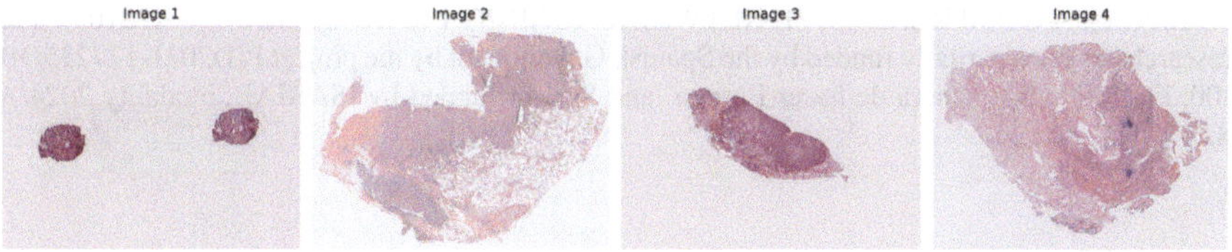

Fig. 5 Resizing with a background color to match the largest size among the slides before color normalization. *Source* Authors

5 Discussion and Conclusions

This study emphasizes the importance of image normalization and patch-based analysis in improving AI performance for NSCLC diagnosis using histopathological slides. Direct use of whole-slide images led to reduced accuracy due to staining variability and structural inconsistencies.

By applying resizing and color normalization techniques, and focusing on image patches instead of full slides, we achieved more consistent inputs and improved model reliability. Techniques like StainGAN and the Ruifrok and Johnston method proved effective in standardizing H&E-stained images.

In summary, our findings support the use of tailored preprocessing and deep learning strategies to enhance diagnostic precision in AI-driven pathology, contributing to more reliable applications in precision medicine.

In further research, the primary focus will be to identify the most effective approach for resizing and normalizing a diverse set of slides. Another essential component is the development of a robust classification model. In upcoming developments, we will report key performance indicators for both normalization—such as Structural Similarity Index (SSIM), Fréchet Inception Distance (FID), Inception Score (IS), Mean Squared Error (MSE), and Peak Signal-to-Noise Ratio (PSNR)—and classification, including accuracy, precision, specificity, and Area Under the Curve (AUC). This evaluation will be conducted with the support of an expert in carcinoma detection.

Another critical area for further investigation is the development of more explainable classification models. Understanding how molecular features contribute to classification outcomes will be essential for gaining trust in predictive models.

Moreover, the validation of classification models using independent datasets remains a crucial step in ensuring their real-world applicability.

In summary, our approach involves generating image patches by excluding background regions, applying normalization exclusively to the non-background patches, and subsequently inputting them into the ARA-CNN model for classification. The accuracy of this methodology will be evaluated and compared with expert assistance.

Addressing these challenges will contribute to the development of more reliable and interpretable AI-driven solutions in molecular biology and precision medicine.

Acknowledgements This article is based upon work from COST Action HARMONISATION (CA20122), A COMPREHENSIVE NETWORK AGAINST BRAIN CANCER (CA22103). This research has been partially funded by the Spanish Government by the project PID2021-127275OB-I00, FEDER 'Una manera de hacer Europa' and Project funded by ISABIAL, modality 2024/A/8.

References

1. Stenzinger, A., Vogel, A., Lehmann, U., Lamarca, A., Hofman, P., Terracciano, L., Normanno, N.: Molecular profiling in cholangiocarcinoma: a practical guide to next-generation sequencing. Cancer Treat. Rev. 102649 (2023)
2. Travis, W.D., Brambilla, E., Noguchi, M., Nicholson, A.G., Geisinger, K., Yatabe, Y., Ishikawa, Y., Wistuba, I., Flieder, D.B., Franklin, W., et al.: Diagnosis of lung cancer in small biopsies and cytology: implications of the 2011 international association for the study of lung cancer/ American thoracic society/European respiratory society classification. Arch. Pathol. Lab. Med. **137**(5), 668–684 (2013)
3. Zappa, C., Mousa, S.A.: Non-small cell lung cancer: current treatment and future advances. Transl. Lung Cancer Res. **5**(3), 288 (2016). https://doi.org/10.21037/tlcr.2016.06.07
4. Thandra, K.C., Barsouk, A., Saginala, K., Aluru, J.S., Barsouk, A.: Epidemiology of lung cancer. Contemp. Oncol./Współ-Czesna Onkol. **25**(1), 45–52 (2021). https://doi.org/10.5114/wo.2021.103829
5. Isla, D., Lozano, M.D., Paz-Ares, L., Salas, C., Castro, J., Conde, E., Felip, E., G´omez-Rom´an, J., Garrido, P., Enguita, A.B.: New update to the guidelines on testing predictive biomarkers in non-small-cell lung cancer: a national consensus of the spanish society of pathology and the spanish society of medical oncology. Clin. Transl. Oncol. **25**(5), 1252–1267 (2023). https://doi.org/10.1007/s12094-022-03046-9
6. Hendriks, L., Kerr, K., Menis, J., Mok, T., Nestle, U., Passaro, A., Peters, S., Planchard, D., Smit, E., Solomon, B., et al.: Oncogene-addicted metastatic non-small-cell lung cancer: ESMO clinical practice guideline for diagnosis, treatment and follow-up. Ann. Oncol. **34**(4), 339–357 (2023)
7. Mayer, C., Ofek, E., Fridrich, D.E., Molchanov, Y., Yacobi, R., Gazy, I., Hayun, I., Zalach, J., Paz-Yaacov, N., Barshack, I.: Direct identification of alk and ros1 fusions in non-small cell lung cancer from hematoxylin and eosin-stained slides using deep learning algorithms. Mod. Pathol. **35**(12), 1882–1887 (2022). https://doi.org/10.1038/s41379-022-01141-4
8. Zhang, Y., Yang, Z., Chen, R., Zhu, Y., Liu, L., Dong, J., Zhang, Z., Sun, X., Ying, J., Lin, D., et al.: Histopathology images-based deep learning prediction of prognosis and therapeutic response in small cell lung cancer. NPJ Digit. Med. **7**(1), 15 (2024). https://doi.org/10.1038/s41746-024-01003-0
9. Ahmed, A.A., Fawi, M., Brychcy, A., Abouzid, M., Witt, M., Kaczmarek, E.: Development and validation of a deep learning model for histopathological slide analysis in lung cancer diagnosis. Cancers **16**(8), 1506 (2024)
10. Coudray, N., Ocampo, P.S., Sakellaropoulos, T., Narula, N., Snuderl, M., Feny¨o, D., Moreira, A.L., Razavian, N., Tsirigos, A.: Classification and mutation prediction from non-small cell lung cancer histopathology images using deep learning. Nat. Med. **24**(10), 1559–1567 (2018). https://doi.org/10.1038/s41591-018-0177-5
11. Park, J.H., Lim, J.H., Kim, S., Kim, C.-H., Choi, J.-S., Lim, J.H., Kim, L., Chang, J.W., Park, D., Lee, M.-W., et al.: Deep learning-based analysis of EGFR mutation prevalence in lung adenocarcinoma H&E whole slide images. J. Pathol. Clin. Res. **10**(6), 70004 (2024)
12. Rączkowska, A., Paśnik, I., Kukiełka, M., Nicoś, M., Budzinska, M.A., Kucharczyk, T., Szumiło, J., Krawczyk, P., Crosetto, N., Szczurek, E.: Deep learning- based tumor microenvironment segmentation is predictive of tumor mutations and patient survival in non-small-cell lung cancer. BMC Cancer **22**(1), 1001 (2022). https://doi.org/10.1186/s12885-022-10081-w

Detection of Gait Alterations Using Mobile Device Sensors and Machine Learning

Emanuel Saldana-Perez, Giovanni Guzmán, and Rolando Quintero

Abstract Sustainability in health services has become a crucial issue in the 21st century, given this sector's environmental, social, and economic impact globally. Sustainable health policies aim to balance providing high-quality medical services with reducing environmental impact, promoting social equity, and achieving economic efficiency. One key aspect in defining sustainable health services policies is providing specialized care to people, especially vulnerable groups and elderly adults, when symptoms may indicate a decline in their health. Traveling to medical centers demands time and effort. Monitoring using smart devices and virtual tools, such as telemedicine, enables medical specialists to provide advice and treatments for managing diseases. One health problem that has increased recently is related to gait or walking disorders. This paper presents a methodology for conducting ten gait tests to classify data collected using four mobile device sensors: acceleration, user acceleration, gyroscope, and magnetometer. Machine learning techniques were applied to a training dataset. Testing on a separate dataset achieved high accuracy in detecting gait alterations. With this approach, gait tests can be performed at patients' homes to detect gait alterations early and monitor the progression of diseases.

Keywords Gait alterations · Monitoring using mobile device applications · Machine learning applications

E. Saldana-Perez · G. Guzmán (✉) · R. Quintero
Centro de Investigac6n en Computaci6n, Instituto Politécnico Nacional, CDMX, Mexico
e-mail: jguzmanl@ipn.mx
URL: http://www.cic.ipn.mx

E. Saldana-Perez
e-mail: jsaldanap2201@alumno.ipn.mx

R. Quintero
e-mail: rquintero@ipn.mx

A. Visvizi et al. (eds.), *Research and Innovation Forum 2025 Volume 1*,
Springer Proceedings in Complexity, https://doi.org/10.1007/978-3-032-09066-9_28

1 Introduction

Worldwide, the health sector is one of the largest consumers of natural resources and generators of waste. According to the World Health Organization (WHO) [13, 15], healthcare systems significantly contribute to greenhouse gas emissions, primarily through energy use, transportation, medical supplies, and waste management. The health sector is estimated to be responsible for approximately 4.4% of global carbon emissions. In addition, health services face several challenges, including increased demand due to an aging population, the prevalence of chronic diseases, and the impact of climate change on public health. Sustainable health service policies are essential to ensure a future in which health systems are resilient, equitable, and environmentally responsible [4, 8]. Although they face challenges, their implementation has a significant benefit for the environment, public health, and economic efficiency. Collaboration among governments, healthcare institutions, professionals, and the community is crucial for a sustainable healthcare model [10]. Implementing sustainable policies reduces the environmental impact and improves the resilience of health systems, ensuring their ability to respond to future crises. There are several key areas for implementing sustainable policies, including the following: managing medical waste [14], energy efficiency, sustainable resource use [6], education and awareness, transportation and mobility, design and construction of sustainable infrastructure, equity, and universal access. Although there are benefits, implementing sustainable healthcare policies faces several challenges, including initial costs, a lack of knowledge and training, resistance to change, inconsistent regulations and standards, and a tendency to prioritize immediate care over long-term benefits.

As a result, various proposals have emerged aimed at defining good practices in health services, such as green hospitals (for instance, the *University Hospital of Skåne* in Sweden or *Kaiser Permanente* in the United States), recycling and reuse programs, telemedicine, sustainable procurement initiatives, and carbon emissions reduction. Telemedicine, particularly during the COVID-19 pandemic, proved to be an effective tool for reducing travel and improving access to healthcare [9]. The developments achieved during the virus lockdown can be utilized for remote monitoring, particularly of chronic degenerative diseases (such as neurodegenerative diseases associated with gait alterations) and conditions that can be monitored through non-invasive processes using sensors or mobile devices.

Gait is a complex movement that requires the coordination of multiple muscle systems. The main muscles involved belong to the pelvic girdle and the lower limbs, although the muscles of the shoulder girdle, trunk, and upper limbs also contribute. Walking autonomously and stably is essential for maintaining a good quality of life, particularly in at-risk groups. However, the appearance of Gait Disturbances (GD) is not limited to older adults; it is also often associated with various pathologies that, if not detected and treated in a timely manner, can lead to mobility problems and reduce the independence of affected individuals [12]. Neurological disorders can alter normal gait by weakening muscles or affecting their coordination, resulting in characteristic gait patterns specific to each disease. Therefore, studying the mechanisms of gait and its alterations is valuable for diagnosing neurological diseases. Gait disorders can alter the speed, regularity, symmetry, or synchrony of body movements and can be caused by injuries or pathologies in the spinal cord, brain, legs, or feet.

Gait abnormalities are one of the leading indicators of mobility problems in at-risk groups, such as older patients, as well as being indicators of various neurodegenerative diseases. In Mexico, the population of older adults was estimated to be 16,421,678 in 2024 [5]. By 2030, more than 10% of the Mexican population will be older adults. However, gait disturbances not only affect older adults but are also related to an increased risk of falls and, consequently, injuries that can lead to hospitalization and loss of autonomy, which is a more significant challenge given that in Mexico, only 2.9 million inhabitants have some form of social security for medical services [11]. This situation increases the burden on health systems, both public and private services, which is reflected in limitations of resources and personnel, especially in rural areas and regions with a high percentage of older adults, people in need of monitoring of neurodegenerative diseases, or after having suffered some accident that requires monitoring and analysis during their rehabilitation. A lack of early assessment and analysis can lead to several problems for individuals who require an early diagnosis, affecting the effectiveness of specialist interventions in treating them. Unfortunately, early intervention by these specialists can sometimes be costly for these patients.

Due to the importance of classifying, analyzing, and monitoring people who suffer from gait disturbances, there is a need to develop affordable and practical technological tools for these risk groups. These tools not only facilitate access to data analysis from these patients but also offer personalized monitoring for each one of them. The research question is whether a mobile application can detect the presence or absence of gait alterations by collecting quantitative data during a walking test, utilizing sensors embedded in the device. Moreover, in this paper, a methodology for the detection of gait alterations is proposed through the use of a mobile application that allows the user to choose from a set of ten gait tests for the quantification of variations in walking through the use of four sensors: acceleration sensor, user acceleration sensor, gyroscope sensor, and magnetometer sensor. Machine learning techniques determine if a patient has any gait disturbances. The aim is to reduce the number of visits to medical centers or hospitals if the patient's condition does not require a detailed evaluation by a health specialist [7]. Collected information provides medical doctors with more detailed insights to monitor the progress of a specific medical condition.

Creating a mobile application, especially in the healthcare sector, involves several stages and requires compliance with additional requirements for both Android and iOS, due to the handling of confidential information and personal data that may be collected. As an initial step, this research focused on developing a mobile prototype to collect anonymous monitoring information during gait tests conducted on individuals with and without a diagnosed gait disorder (Personal data are not processed).

2 Literature Review

Gait analysis is essential in biometric systems and health applications. In the literature, there are several works oriented to the investigation of human gait and its alterations, such as the study [18] which focuses on modelling human gait to identify alterations or affectations in people, through the study of walking style using an IMU sensor placed on the ankle and a mobile application developed with MATLAB to collect data. In a study with 30 volunteers, two databases were created: one included only data from the IMU sensor, while the other included both height and gender data for each participant. Finally, four machine learning algorithms were trained and compared to evaluate the accuracy of gait recognition. The results showed that the configuration used was effective for classification, with both databases offering comparable and satisfactory levels of accuracy in identifying participants.

On the other hand, how the use of mobile phones has made it possible to study aging patients with gait disorders is described in [2]. This study analyzed research published between 2004 and 2024 on the use of smartphones to quantify parameters and enhance diagnosis and rehabilitation in vulnerable individuals. Smartphones have become essential tools for analyzing the gait of older adults, enabling the evaluation of spatiotemporal parameters such as speed and stride length. The results showed that the smartphone-based system achieved accurate spatiotemporal tracking and a reasonable estimation of the trajectories on the x-y axes, as well as of the relative angles of the hip, knee, and ankle joints (θ_1, θ_2, and θ_3, respectively). Compared with the OptiTrack system, the relative hip and knee angles (θ_1 and θ_2) presented an average fit of 88.93% and a normalized root mean square error of 2.71%. However, the precision in the estimation of the relative angle of the ankle theta3 was lower (average fit of 71.04% and error of 4.26%) due to the low capture rate of the retroreflective markers in the mobile system and the higher relative speed in the lower extremities, which generated noise in the calculations of the $x - y$ trajectories [3].

Smartphones stand out as non-invasive, accessible, and low-cost tools for gait assessment, allowing the detection of alterations that can predict health risks in older adults. More accurate detection can be achieved using smartphones, as described in [1], using inertial sensors in smartphones. This approach continues to present various obstacles due to the variability in human gait. In this work, the researchers describe smartSTEP, a new method based on machine learning that enables the processing of accelerometer and gyroscope signals without relying on hand movement classifiers or threshold calibrations. Trained with 9, 000 steps from 12 participants and tested on approximately 2,200 steps from people not included in the training, smartSTEP achieved a recall of 99% and an accuracy of 98.9%. Even in complex scenarios such as asymmetrical gait, uneven surfaces, and stairs. In addition, it achieved a root mean square error of 0.097 s in stride length, a competitive performance compared to current algorithms for this measurement.

Mobile devices offer a portable and low-cost method that can be placed in various locations or spaces. They are not limited to a clinical environment and are not under the supervision of a medical specialist. A study published in 2017 [17] showed that placing a smartphone on the ankle and using an algorithm developed by the study team were viable methods for performing gait analysis, which involves calculating step time, support time, swing time, and cadence. Furthermore, [6] describes how to use the mobile phone non-invasively. This project proposes the development of a smartphone application that allows monitoring participants' gait using sensors integrated into mobile devices, such as the accelerometer, gyroscope, and magnetometer.

Through exercises recommended by specialists, the application collects data to identify potential changes in gait, providing a low-cost and accessible monitoring method for patients, caregivers, and healthcare professionals. The collected data was analyzed using machine learning algorithms to detect possible gait disturbances that require medical attention. Finally, [16] describes the design, implementation, and set of tests for the application, as well as the analysis of data using machine learning algorithms. This evaluation assesses accuracy and performance through volunteers' participation under medical specialists' guidance.

3 Methodology

The proposed methodology is composed of three main modules: mobile application, analysis of gait alterations and model integration, as it is depicted in the Fig. 1.

3.1 Module 1. Mobile Application

The first module deals with the design and implementation of the mobile application. This module requires three stages:

- **Requirements analysis**. The requirements for developing the mobile application are analyzed and defined in this stage. With the support of doctors specialized in the area, the information that must be compiled for the registration of patients is defined, and the set of gait tests must be carried out. The main objective is for the application to facilitate the carrying out of tests so that medical specialists can have quantitative information about the presence or absence of alterations in a patient's gait.
- **Design of the mobile application**. The application's technical structure and main components are established, ensuring that the design is scalable and allows for easy integration of the gait alteration analysis module, taking into account the requirements obtained from the requirements analysis stage. The gait analysis study consists of 10 tests, formally known as Functional Gait Assessment (FGA): gait level surface, change in gait speed, gait with horizontal head turns, gait with vertical head turns, gait and pivot turn, step over obstacle, gait with narrow base of support, tandem, gait with eyes closed and ambulating backwards.

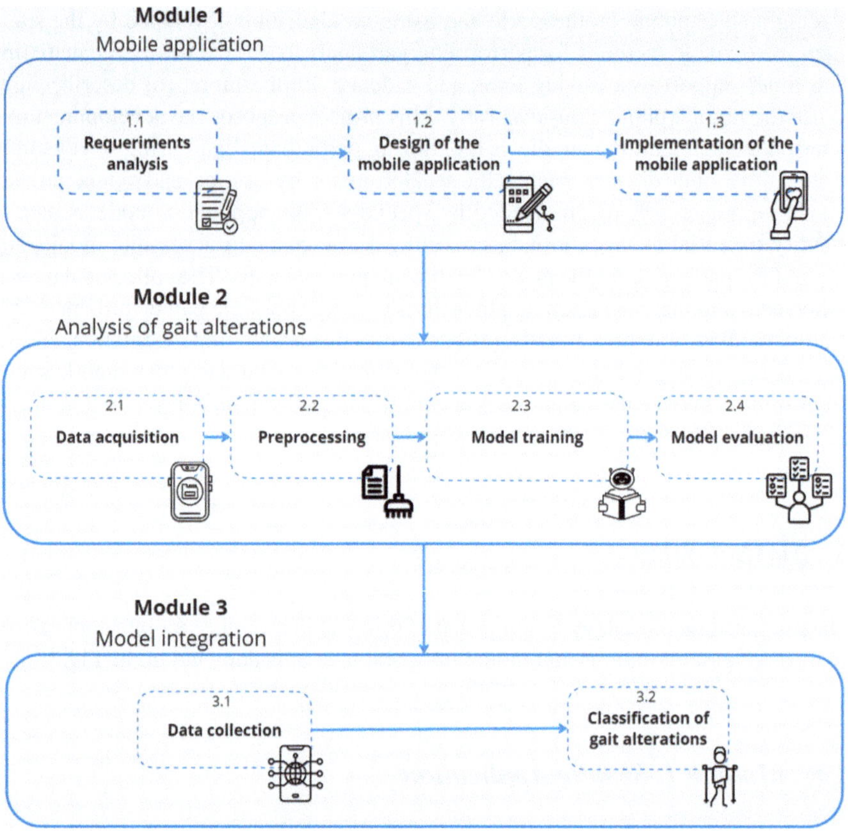

Fig. 1 Proposed methodology (*Source* authors)

- **Implementation of the mobile application**. The implementation uses a multi-platform programming language that allows for periodic adjustments to the application and can generate prototypes for mobile devices running Android and iOS operating systems.

3.2 Module 2. Analysis of Gait Alterations

The study is carried out in this module using machine learning techniques, using the dataset obtained through the mobile application. It consists of four stages:

- **Data acquisition**. A walking test is selected, and data are collected from the mobile device's sensors, including the accelerometer, gyroscope, and magnetometer. Since the data originates from these sensors, the measurements are obtained in meters per square second (m/s^2) on the x, y, and z axes, and a timestamp is added.

Additionally, the magnitude of the vector is obtained by combining the three components.

- **Preprocessing**. Once the data acquisition process is performed using the mobile application, the obtained data are preprocessed, as they may contain noise due to involuntary movements during the walking test. This preprocessing involves applying filters and smoothing techniques to remove noise and enhance data quality, ensuring the information is accurate and reliable for subsequent analysis. Likewise, the measurements in each axis are normalized to facilitate the interpretation of the results and optimize the model's performance.
- **Model training**. In this stage, the model is trained to detect gait alterations. To this end, widely used models in the literature are selected and evaluated, ensuring that the chosen model is appropriate for the type of data and the objective of detecting gait disturbances. This process involves configuring hyperparameters and adjusting the model to maximize accuracy and effectiveness, resulting in reliable results that align with the project objectives.
- **Model evaluation**. An evaluation of the results obtained in training stage is carried out using appropriate evaluation metrics, depending on the nature of the dataset generated during the preprocessing stage. These metrics allow us to measure the accuracy, sensitivity, specificity, and other key aspects of the model's performance, ensuring that the model classifies effectively and meets the research objectives.

3.3 Module 3. Model Integration

The gait disorder detection model is integrated with the mobile application in this module. This integration allows the application to use the trained model in real-time, facilitating efficient and accurate data analysis within the end-use environment, and consists of two stages:

- **Data collection**. This involves generating and storing new data (in a NoSQL database) obtained from walking tests with other patients, increasing the dataset size, and enabling the model to provide feedback.
- **Classification of gait alterations**. Once the gait disturbance analysis module has been integrated, the next stage is for the mobile application to use this model to determine whether gait disturbances exist.

4 Test and Results

This section describes the results obtained with the developed prototype. After defining some of the requirements for the mobile prototype in collaboration with health specialists, the prototype was implemented using the Dart language and the Flutter framework version 3.16. As a first step, users must register for the mobile application,

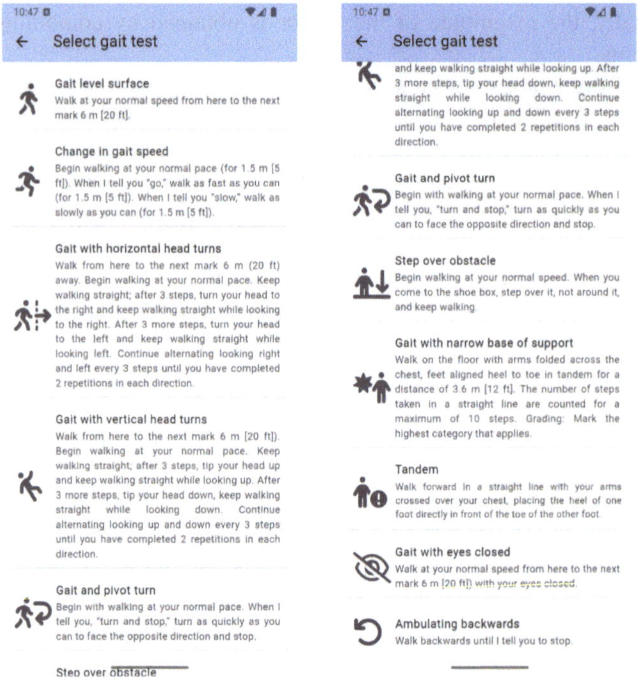

(a) List of gait tests (Part 1). (b) List of gait tests (Part 2).

Fig. 2 List of gait tests (*Source* authors)

provide an email address, and create a password to set up an account. After the user completes the login process, each test on the prototype's main screen is listed, along with a brief description, to help users become familiar with it. See Fig. 2.

After selecting the gait test to be performed, content is displayed on the screen as described in Fig. 3a. First, the *'Start Sampling'* button must be pressed to begin data collection from each sensor. Various tests were carried out with values in the range of 20–200, with sampling with an interval of 20 milliseconds providing the best results. Figure 3b shows a snapshot of the sampling process. When the test is finished, the *'Stop Sampling'* button is pressed. The following two mobile devices with Android operating systems were used: a Xiaomi Redmi Note 11S with Android 13 and a Samsung A31 with Android 12. Verbal consent was obtained from the volunteers participating in the tests, accompanied by a medical specialist in physical rehabilitation.

The information related to the recopilation of the samples for dataset creation is:

1. Demographic details:
   - Total samples: 843 samples.
   - Age: 55 to 75 years old.

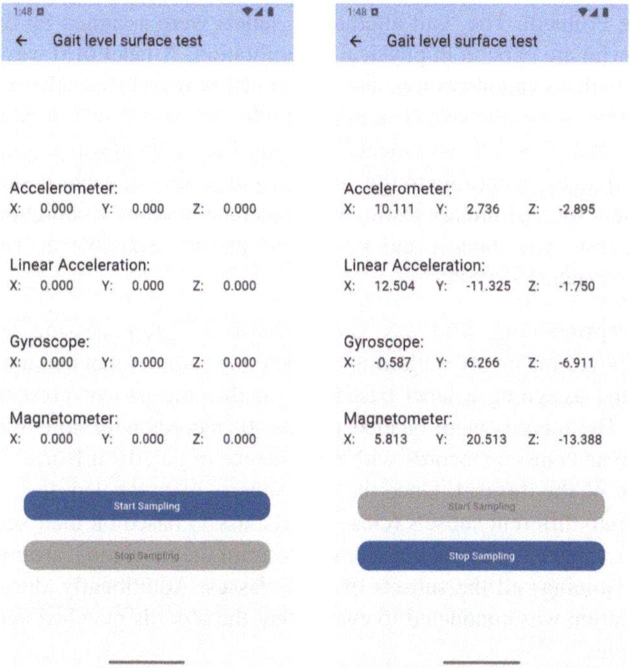

(a) Initial screen. (b) Gait test on execution.

Fig. 3 Gait level surface test (*Source* authors)

- Gender: 320 men, 523 women.

2. Recruitment and inclusion. Participants were recruited through voluntary participation in collaboration with a local hospital, by evaluating patients undergoing treatment at the clinic, or by selecting patients with reduced walking ability.
3. Ethical Considerations. As this study is a preliminary investigation to determine the feasibility of detecting gait alterations using mobile device sensors (accelerometer, user-accelerometer, magnetometer, and gyroscope), formal IRB approval has not yet been sought. The promising results from this study will form the basis for a subsequent application to the Ethics Committee of a federal general hospital. Verbal consent was obtained from all participants, as the study was non-invasive and conducted in accordance with local ethical guidelines and regulations. Participants were informed of the study's purpose, procedures, and anonymous data collection in the presence of a medical specialist in physical rehabilitation, with verbal consent documented by the research team. To ensure data privacy, all data were anonymized by assigning unique participant IDs, with no personally identifiable information collected. Sensor data processing was performed on-device to minimize exposure, and data were stored on a secure server with access restricted to the research team.

4. Labeling Protocol. The 'gait alterations' labels were assigned by three clinical doctors who are experts in physical rehabilitation. A label of 0 was assigned to patients with no gait alteration, and a label of 1 was assigned otherwise.
5. Justification of sample size. The final sample size of 843 was obtained by combining the data from all participants' sessions. Each patient was recorded between 10 and 20 times, according to the duration and quality of each session. This size is sufficient for preliminary testing with machine learning techniques.
6. Data Access. The dataset and python scripts are available at Zenodo DOI: 10.5281/zenodo.15539965.

Data preprocessing. The next step consists of preprocessing the data. As described above, this involves filtering and deleting samples that contain data inconsistencies and assigning a label based on whether the patient presents with gait alterations. The records were labeled in class 0 (records with no evidence of gait disturbance) and class 1 (records with no evidence of gait disturbance).

Creation of the dataset. Once the data subsets were labeled, the records were distributed into different subsets (class 0 and class 1) based on their source or specific characteristics. A dataset was created to train the machine learning models by integrating (joining) all the subsets of both classes. Additionally, during this process, verification was conducted to ensure that the records matched and to prevent inconsistencies.

Figure 4a, b show the results for the gait with horizontal head turns test using the gyroscope sensor, for a patient with no gait alterations and another patient with gait alterations, respectively. The significant difference obtained by this sensor is readily observable. In the case of the patient with a normal gait, the accelerations in each axis are limited to a lower range than those obtained from the patient diagnosed with gait disturbance. Additionally, in a clinically healthy patient, there is no overlap of the magnitudes obtained, as is the case with a patient having a gait disorder.

Data was collected from the mobile application and then preprocessed to implement the machine learning models. As a result, a set of 440 samples was obtained corresponding to class 0, which is the negative class (records with no evidence of gait alterations), and 394 samples for class 1, which is the class of interest or positive class (contains records with proof of gait alteration). A set of 834 samples was obtained 10 samples were deleted in the preprocessing stage). As part of implementing the models and reporting the results, this dataset's BI (imbalance index) must be analyzed, resulting in a value of 1.1. This result indicates that the dataset is balanced and was used as input for the following machine learning techniques: Support Vector Machine (SVM), Logistic Regression (LR), Random Forest (RF), and Long Short-Term Memory (LSTM). For each model used, the hyperparameters needed to be tuned. To achieve this, several tests were conducted using random search and cross-validation, enabling us to determine the configurations that yielded the best results for each model. The hyperparameter values are presented in Table 1.

We use the stratified hold-out method for model evaluation, assigning 70% of the patterns in each class to the training set (308 samples for class 0 and 132 samples for class 1) and the remaining 30% to testing (132 samples for class 0 and 118 for class 1). However, we also implement stratified k-fold cross-validation to ensure a more

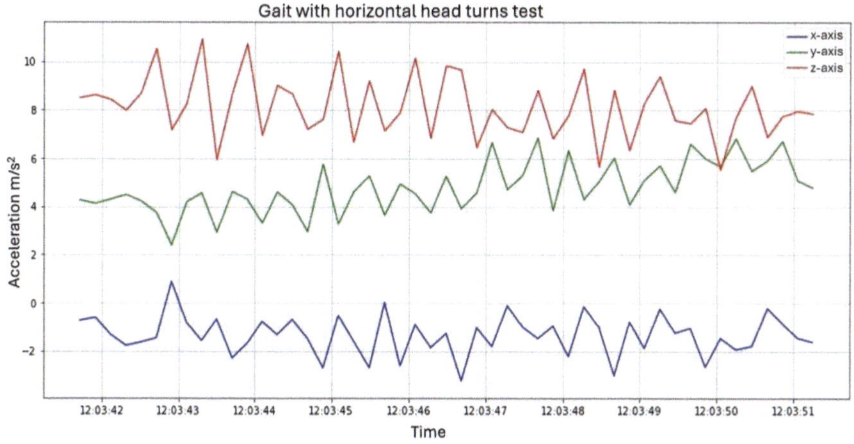

(a) Sensor data from a patient with no gait alterations.

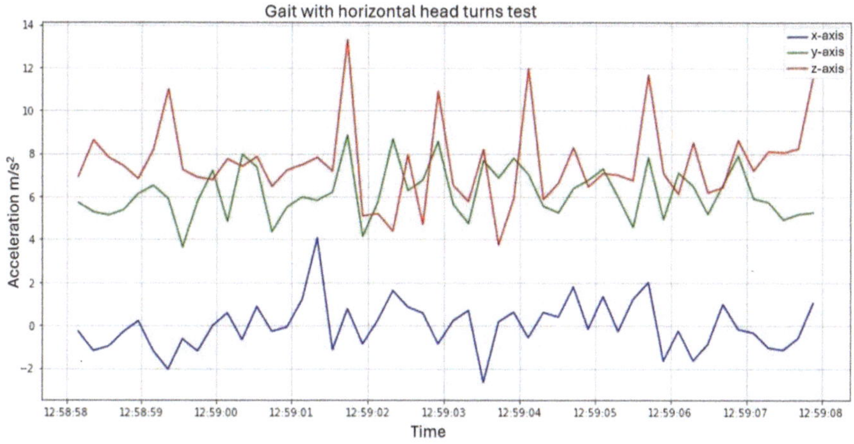

(b) Sensor data from a patient with gait alterations.

Fig. 4 Data obtained by the gyroscope sensor in the gait with horizontal head turns test (*Source* authors)

Table 1 Values used as hyperparameters for machine learning techniques

Algorithm	Hyperparameters
SVM	kernel: sigmoid degree: 3
RF	n_estimators: 100 criterion: gini
LR	maximum number of iterations: 1000 number of splits (folds) in the k-fold cross-validation: 5
LSTM	first layer: 16 units, activation function: relu second layer: 1 unit, activation function: sigmoid
MLP	hidden layers: 16 units activation function: relu maximum iterations: 500

Table 2 Numerical results obtained with each machine learning technique

Model	Accuracy (%)	Precision (%)	Sensitivity (%)	F1 measure (%)
SVM	52	52	49	45
RF	87	80	88	86
LR	87	87	87	87
LSTM	90	90	90	90
MLP	89	89	85	80

robust estimation and avoid bias (using $k = 5$). Additionally, subject-independent protocols (such as leave-one-subject-out) will be explored in future work, especially for scenarios where generalization across participants is critical.

To compare and determine the machine learning most adequate technique for gait alterations detection, the confusion matrices, precision, and recall calculations for each of the techniques implemented were calculated. Table 2 reveals the numerical results obtained.

On the other hand, a confusion matrix is a tool used in artificial intelligence to evaluate the performance of a classification model. It summarizes the model's predictions by comparing the actual labels of the dataset with the predicted labels, displaying the results in a tabular format. The matrix shows the number of true positives, true negatives, false positives, and false negatives, enabling experts to evaluate the model's accuracy, precision, recall, and additional metrics, the confusion matrix helps analyze how effectively the model differentiates between classes, identify spe-

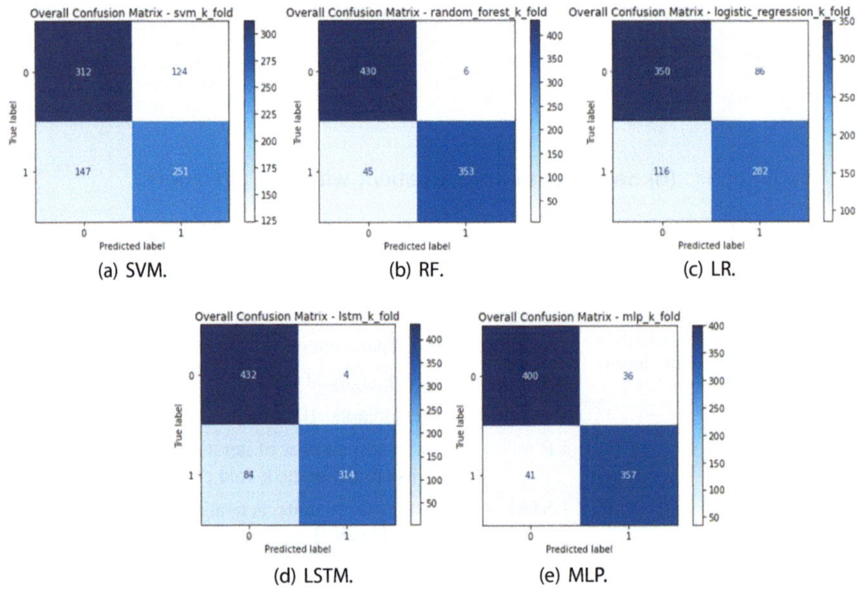

Fig. 5 Confusion matrices (*Source* authors)

cific areas of error, and support model refinement and decision-making. In Fig. 5 the confusion matrices obtained are depicted.

5 Discussion and Conclusions

The results were favorable for detecting gait alterations using a preliminary mobile device application and machine learning techniques. The LSTM model presented the best results, with an accuracy of 90%, a precision of 90%, and a sensitivity of 90%. This indicates that this machine learning technique can adequately classify samples without gait alteration and those that describe some alteration, providing a balance for studying the dataset of both classes. On the other hand, the MLP and RF models also perform competitively. They are an alternative from a computational perspective, as these models require less processing time to obtain the classification results. The data collected by the mobile application provides excellent support for creating a dataset describing gait alterations in Mexico's vulnerable populations. This can help government authorities define care policies based on identifying priority regions. Additionally, a mobile application can provide detailed information when a healthcare specialist needs to analyze a patient's clinical progress.

Our LSTM's 90% sensitivity suggests promise for detecting gait alterations, comparable to clinically accepted tools like the Functional Gait Assessment (FGA) and Timed Up and Go (TUG) test. As this study did not include concurrent testing with FGA or TUG, future work should conduct direct comparisons to establish the clinical utility of these assessments.

Additionally, future work will be focused on developing a protocol for approval by a federal public health institution, integrating confidential data in compliance with current standards and regulations. The data expected to be obtained includes age, existing chronic degenerative diseases, family history, and medications. Additionally, the research will focus on integrating information from other sensors in mobile devices, such as GPS or trilateration, using WiFi data. Since the nature of the data collected limits the classification results, it is also necessary to explore other artificial intelligence techniques, such as transformers or deep neural networks.

Acknowledgements This work was partially sponsored by the Instituto Politécnico Nacional (IPN), the Secretaría de Ciencia, Humanidades, Tecnología e Innovación (SECIHTI) under grant 1275526, and the Secretaría de Investigación y Posgrado (SIP) under grants 20251126, 20250285, and 20251101. Additionally, we are grateful to the reviewers for their invaluable and constructive feedback, which helped improve the quality of the paper.

References

1. Abiad, N.A., Kone, Y., Renaudin, V., Robert, T.: SMARTphone inertial sensors based STEP detection driven by human gait learning. IEEE, pp. 1–8 (2021). https://doi.org/10.1109/ipin51156.2021.9662513

2. Brognara, L.: Gait assessment using smartphone applications in older adults: a scoping review. Geriatrics **9**(4), 95 (2024). https://doi.org/10.3390/geriatrics9040095
3. Castillo, B., Riascos, C., Franco, J.M., Marulanda, J., Thomson, P.: Assessing spatiotemporal behavior of human gait: a comparative study between low-cost smartphone-based Mocap and OptiTrack systems. Exp. Tech. (2024). https://doi.org/10.1007/s40799-024-00716-x
4. Degavre, F., Kieffer, S., Bol, D., Dekimpe, R., Desterbecq, C., Pirson, T., Sandu, G., Tubeuf, S.: Searching for sustainability in health systems: toward a multidisciplinary evaluation of mobile health innovations. Sustainability **14**(9), 5286 (2022)
5. Diario Oficial de la Federación: https://www.dof.gob.mx/nota_detalle_popup.php?codigo=5616097, [Accessed 02-06-2025]
6. Duindam, D.: Transitioning to sustainable healthcare: decarbonising healthcare clinics, a literature review. Challenges **13**(2), 68 (2022)
7. Fan, Z., Yan, Z., Wen, S.: Deep learning and artificial intelligence in sustainability: a review of sdgs, renewable energy, and environmental health. Sustainability **15**(18), 13493 (2023)
8. Gagliardi, A.R., Albergo, F.: Reducing medical errors via data mining techniques: a structured literature review and future research agenda. In: The International Research & Innovation Forum, pp. 723–736. Springer (2023)
9. Holmner, Å., Ebi, K.L., Lazuardi, L., Nilsson, M.: Carbon footprint of telemedicine solutions-unexplored opportunity for reducing carbon emissions in the health sector. PLoS ONE **9**(9), e105040 (2014)
10. Hussain, M., Ahmad, S.Z., Visvizi, A.: Government regulation and organizational effectiveness in the health-care supply chain. Transform. Gov. People Process Policy **16**(4), 649–662 (2022)
11. INEGI: Tabulados Interactivos Genéricos (2025). https://www.inegi.org.mx/app/tabulados/interactivos/?pxq=Derechohabiencia_Derechohabiencia_02_822ebcc5-ef41-40c1-9901-22e397025c64, [Accessed 02-06-2025]
12. Nogueras, A.M., Arenillas, J.L.C., Rodríguez, J.O., Iglesias, F.J.B., Sánchez, C.S.: Fases de la marcha humanal revista iberoamericana de fisioterapia y kinesiología (1999)
13. Organización Mundial de la Salud (OMS): Cambio climático y salud (2024). https://www.who.int/es/news-room/fact-sheets/detail/climate-change-and-health, [Accessed 07-03-2025]
14. Organización Mundial de la Salud (OMS): Desechos de la atención de salud (2024). https://www.who.int/es/news-room/fact-sheets/detail/health-care-waste, [Accessed 07-03-2025]
15. Organización Mundial de la Salud (OMS): Salud por el Clima: Llamado a la acción para lograr sistemas de salud resilientes y bajos en carbono (2024). https://www.who.int/initiatives/cop26-health-programme, [Accessed 07-03-2025]
16. Saldaña, J.: Técnicas de inteligencia artificial aplicada al estudio de alteraciones en la marcha. Tesis de maestría, Centro de Investigación en Computación - Instituto Politécnico Nacional (2025)
17. Shahar, R.T., Agmon, M.: Gait analysis using accelerometry data from a single smartphone: agreement and consistency between a smartphone application and gold-standard gait analysis system. Sensors **21**(22), 7497 (2021). https://doi.org/10.3390/s21227497
18. Talha, M., Soomro, H.A., Naeem, N., Ali, E., Kyrarini, M.: Human identification using a smartphone motion sensor and gait analysis. Digital Library, pp. 197–202 (2022). https://doi.org/10.1145/3529190.3529199

Author Index

A. Visvizi et al. (eds.), *Research and Innovation Forum 2025 Volume 1*,
Springer Proceedings in Complexity, https://doi.org/10.1007/978-3-032-09066-9

MIX
Papier aus verantwortungsvollen Quellen
Paper from responsible sources
FSC® C105338

If you have any concerns about our products,
you can contact us on
ProductSafety@springernature.com

In case Publisher is established outside the EU,
the EU authorized representative is:
Springer Nature Customer Service Center GmbH
Europaplatz 3, 69115 Heidelberg, Germany

Printed by Libri Plureos GmbH
in Hamburg, Germany